FOOD
FOR
LIFE

FOOD FOR LIFE

F. E. DEATHERAGE
Department of Biochemistry
Ohio State University

PLENUM PRESS · NEW YORK AND LONDON

Library of Congress Cataloging in Publication Data

Deatherage, Fred E 1913-
  Food for life.

  Includes bibliographical references and index.
  1. Food. 2. Food supply. I. Title.
TX353.D38 1975   641.3   75-15502
ISBN 0-306-30816-9

© 1975 Plenum Press, New York
A Division of Plenum Publishing Corporation
227 West 17th Street, New York, N.Y. 10011

United Kingdom edition published by Plenum Press, London
A Division of Plenum Publishing Company, Ltd.
Davis House (4th Floor), 8 Scrubs Lane, Harlesden, London, NW10 6SE, England

Printed in the United States of America

To my wife NELLIE LOU

PREFACE

This book is addressed to the university student who is not a science major and to the general reader. An attempt is made to present an integrated view of some of the basic concepts of physical, biological, and social sciences relevant to the problem of providing people with food. The application of these disciplines has led to our present technologies of medicine, agriculture, and food science on which modern civilization rests. Technical information concerning foods has increased enormously in the less than a century that the basic concepts of the science of nutrition have been recognized. Scientific agriculture to provide food for an ever-growing population is scarcely a century and a half old. Feeding oneself is a very personal matter, and at the same time feeding large groups is the concern of society as a whole. Therefore, it is understandable that, in one way or another, the problems of food production and distribution underlie the actions of politicians, bureaucrats, the leaders of government, and business managers. These situations of our modern life make rational and sound solutions to food problems difficult and often contribute to alarmism founded on partial scientific "truth" taken out of context.

The trend toward more "consumerism" is unmistakable. But to serve the individual best, such movements must be based on sound judgments and reasoned scientific principles rather than on the often emotional compromises of opportunistic politicians, lawyers, and businessmen. That man requires wholesome, nutritious food is indisputable. A popular tactic of the scare monger is to recite the chemicals used in the growth, processing, and distribution of food. But all foods are chemical, as are all material things—living and dead. The "good" and "bad" dichotomy applies to chemicals as well as to anything else—there are required chemicals and there are dangerous chemicals. There can also be too much of any good and necessary thing in life or not enough. Similarly, some "bad" things may be good or even necessary at times and insignificant at other times. Under such circumstances, I can only attempt in this book to relate some of the key concepts that are important in providing people with food. I hope that the reader will in some measure become more confident in his own ability to properly evaluate the constant barrage of piecemeal information that has been the result of the communications explosion.

Many college students are required to enroll in science courses even though they are not interested in scientific careers. These beginning courses are discipline oriented and are often taught as though every student were going to be a professional in the discipline. Many students are repelled by the depth of detail necessary in such courses, with the result that for these students the relevance of science to our culture is lost. Some even become antiscience in their attitudes to problem solving. Since everyone must eat and everyone has personal knowledge of food, we have developed a science course conceptually centered around the fundamental problem of survival—feeding ourselves.

The ability of individuals and nations to feed themselves adequately rests on the interacting principles of the biological, physical, and social sciences. No single discipline can show the way to feed the ever-increasing population of the world. As the course has developed, we have limited ourselves to the concepts that can be found in beginning college texts of the various disciplines and have related them to food problems. Although the course was originally intended for the general or nonscience students, surprisingly a significant number of science majors have enrolled, indicating that they wish to understand better how their specialties fit into the total food picture. Our experience in this teaching venture has indicated that students need a text and a ready reference book, and so this book was written. Some of my colleagues would have suggested that this book be useful as an introductory text for students in food science and technology. In addition to the university student, this book is intended for the general reader who is interested in food problems at whatever level—personal, national, or global.

More detailed treatments of the ideas expressed here may be found in textbooks and scientific journals, which will not be referred to in detail. Quite naturally, the text reflects the experiences of the author. At times, my scientific colleagues may condemn my oversimplification. Nonscientific readers may equally cringe at my simplistic view of other areas. I accept this as the harvest one reaps in attempting a conceptualization of such a basic and complex subject as food. But I do not write to make my readers agree with me but rather to inspire critical inquisitiveness from those who care to reflect on the nature of the food problems confronting both the people of highly developed countries and those of the less developed areas.

I want to thank many students, friends, and colleagues who have contributed in diverse ways to this work. Many authors and publishers have kindly granted permission to use some of their materials, and they are gratefully acknowledged as noted. I am particularly grateful to those who have critically read parts of the manuscript and made many pertinent suggestions for improvement. Among these are George Banwart, Jack Cline, Robert Feeney, Masao Fujmaki, Wilbur Gould, Euripedes Malavolta, Harold Olcott, Ivan Rutledge, Bernard S. Schweigert, John Sitterley, Junius Snell, John R. Whitaker, Eugen Wierbicki and Eva Wilson. I hope that readers of these pages will share their comments, criticisms, and suggestions with me, for it is only in this way that improvements can be made.

F. E. Deatherage

Columbus, Ohio

CONTENTS

FOOD
FOR
LIFE

Eat all kind nature doth bestow;
It will amalgamate below
If the mind says "It will be so,"
But, if once you begin to doubt,
Your gastric juice will find it out!

British Medical Journal 1:438 (1937)

INTRODUCTION

Man is a biological system and survives only by consuming other biological systems in whole or in part. His evolution through the ages has led to great variations in his cultural development in relation to the organisms he uses for food. Food habits and attitudes toward eating certain foods may become quite strong and even fixed. Some civilizations have died out, others have survived; through intellectual achievement, some societies of man have become highly developed culturally, economically, and politically, and others have not. When and wherever these developments have happened in history, needed basic food supplies have undergirded them. It is so today.

As yet, there seems to be no perfect way of life—no utopia—but the quest for such seems in some cultures to be a driving force. We in the western world seem to have accepted rapid change as a way of life—such change being based on the advent of modern science and the so-called scientific method of studying the whys and wherefores of every facet of our lives. Other cultures, some highly sophisticated and rich in heritage, have been governed by different attitudes, with the result that, from the American viewpoint, change has been slow. But whatever the culture or subculture—African, Asian, European, American, Spanish, Chinese, Eskimo, Polynesian, Kazak, or Ethiopian—man, the basic unit of society, is still a biological system that must eat other biological systems to survive and to reproduce.

It is sometimes said with good reason that man (like all other organisms) has two basic instincts, survival and propagation of the species. These translate into two driving hungers—for food and for sex. Both food and sex are pleasurable and may be enjoyed without understanding. Each culture has developed such characteristic attitudes and traditions about both of these basic concerns of man that anthropologists have specialized in studying them. In our own culture, the literature on sex is prestigious and burgeoning, with laymen eagerly awaiting bits and pieces of research reports of all kinds. Some critics seem to feel that our so-called intellectual freedom has led a large proportion of our population to be preoccupied with sex. At the same time, however, we are also attaining some measure of understanding of its role in our lives. Similarly, a characteristic of affluent societies is the consumption of delicious food more for social reasons, convenience, mere pleasure, and even gluttony than simply for the maintenance of life processes. Yet as populations grow and cultures become even more complex, it is necessary that all concerned people have some understanding of the fundamental role of food in our lives. In this way, the bits and pieces of food information which each of us receives by way of advertising, labels, radio, TV, newspapers, etc., will have more meaning. Also, we can better sift the seed from the chaff, the food from its packaging, the relevant fact from fad and fancy, and by so doing we can eat in confidence for pleasure as well as for survival.

Interdependence of individuals is a fundamental characteristic of highly developed societies whether composed of ants or humans. Consequently, for modern man, who now depends more than ever on his intellect for survival, understanding of food problems is essential. The continuous flow of cookbooks of all types is evidence of the intricate role that food plays in society. It is important that food be pleasing and enjoyable and at times festive. But these aspects of food represent the end of the long and complex food chain necessary to support life. It is to an understanding of the nature of this vital support system that the following pages will be addressed. But through them all we cannot lose sight of the fact that the survival of all of mankind, as it has been through all generations past and will be in the foreseeable future, is dependent on the consumption each day of other biological systems in whole or in part.

SOME CHARACTERISTICS OF BIOLOGICAL ORGANISMS

Since man and the organisms he consumes are all biological in nature, let us examine some of the simple but basic characteristics of living things. The technologies of agriculture which provide food and the technologies of processing, preservation, and distribution which bring these foods to us depend on these same basic characteristics. Brief reflection on all the different plants, animals, and microorganisms (bacteria, yeasts, and fungi) which are the source of our daily food may be almost overwhelming when we consider the detailed biological complexities of each. (Each group of organisms forms a major branch of the biological sciences.) Also, comprehension of the logistics of getting this varied array of organisms to us each day in usable form may actually inspire a sense of awe. Nevertheless, there are some common denominators almost anyone can understand. Through this volume, these principles may be perceived when we get glimpses of some of the problems involved in the development of ancient civilizations and the modern-day problems of providing food for the world's starving millions or of providing convenience foods in the supermarket. The reader is urged to try to do just that, for, in doing so, one will also gain a profound measure of understanding of the nature of food itself and how food can be most effectively and efficiently used by all.

DYNAMIC NATURE OF ORGANISMS

All living things are dynamic. In other words, they are always changing and go through their own life cycles. Although the life cycle of each organism may have certain particular characteristics, all essentially have these features in chronological order: genesis, growth, maturation, reproduction, and death. For man, genesis is conception, the beginning of the life cycle. For growth, there is the 9 months *in utero* plus the 10–16 years of growth to reach puberty, the onset of the reproductive period. For males, this may span 70 or more years; for females, the span is about 35–40 years. Finally death ends the cycle.

The proverbial human life cycle is three score and ten or 70 years. To be sure, some people live longer than others. But in the history of man the human life cycle has not changed much nor has it been extended to any degree. History notes many people who have lived 70, 80, 90, or even 100 years, the same as now. With all of our modern science and technology of medicine and health care, the human life cycle has not been lengthened. Only the percentage of the total population reaching the mature years has increased as a result of better food and the elimination of pestilence and disease. To have extended the life cycle of man would be to have found the mythical fountain of youth. Yet that quest is being revived as researchers are just beginning to propose theories of the aging process.

In contrast to man, some microorganisms may go through their complete life cycle in hours, or even minutes. Many plants which man uses for food may complete their life cycle in a relatively few days or months; yet some plants such as the giant sequoias may live many centuries. For animals, life cycles can be measured in days, weeks, months, or years, but not many animals exceed the human life span.

It is immediately apparent that if man must eat every day he cannot do so by consuming organisms with long life cycles. Man has survived by consuming only organisms of relatively short life cycles or by using only a portion of the offspring of organisms of longer life span (such as cattle for meat or the fruit of a tree) and by harvesting a part of the nourishment produced for the growing young (milk and honey being perhaps the most important examples of this category). We dare not consume all the wheat, rice, pigs, or whatever, for it is essential that they be maintained generation after generation to support ourselves and our posterity.

Man has learned through the millennia of history that seeds of certain plants can be food sources. After he learned to husband animals for food, man soon learned to cultivate plants—particularly cereals such as wheat and rice—for their seeds. In general, seeds of plants and the spores of microorganisms represent a unique adaptation in biology, a form of dormant or latent genesis, so to speak. As these mature, water is lost and the organism enters a state of "suspended animation." Life processes cannot resume their normal rate until water is made available and germination begins the cycle over again. (Note the role of water, essential for all life; see Chapter 4.) Whereas viable seeds and spores may be kept and stored for a long time, even hundreds of years in some well-authenticated cases, the same type of adaptation is not found in the animals which contribute to our food supply. So animals must be continuously grown and maintained. As we shall see later, this is both an advantage and a disadvantage in providing food.

The dynamic nature of living things requires a continuing input of energy so that an organism can progress through its life cycle. Where does this energy come from? The world as a whole is a dynamic spaceship that travels around the sun. Where does the energy come from to heat the earth, to move the waters of the sea, to cause the rain that permits life to exist on earth? All of this energy comes from the sun, which radiates vast amounts of energy from its continuous nuclear fusion reactions proceeding at temperatures in the millions of degrees. The radiation that strikes the earth provides a continuing supply of energy that supports all life.

At this point, it is necessary to divide all living things into two groups—those that can use sunlight to provide the energy for their growth, maturation, and reproduction and those that must get their energy by consuming organisms which do trap the sun's energy.

Primary to the food chain supporting any species of living things are the green plants, which can make their protoplasm from water, carbon dioxide, nitrogen, and dissolved minerals by the process of photosynthesis. This complex but highly organized series of chemical reactions requires energy in the form of light. Some of the energy is stored in the substance of the plant and its offspring (seed), while some serves to drive the sequential chemical processes of the plant's normal life cycle. As a by-product of photosynthesis, oxygen gas is produced and released into the atmosphere. Air is about one-fifth oxygen.

Many foods we consume come directly from green plants—cereals (wheat, rice, corn, barley, etc.); pulses (peas, beans, peanuts); tubers and roots (potatoes, cassava or mandioca, carrots); fruits (apples, oranges, berries, mangoes); nuts (pecans, cashews, walnuts); and leafy vegetables (broccoli, lettuce, spinach).

The other major group of organisms are those which get their energy for all their dynamic processes from the green plants, directly or indirectly. The animals make up a large part of this group, but many microorganisms, particularly most bacteria, yeasts, and fungi, must also be included. The biologist calls all of these organisms *heterotrophs* (feeders on other organisms) since they depend on other organisms for their sustenance, in contrast to the *autotrophs* (self-feeders), which get their energy from the sun. Animals known as *herbivores* or plant eaters (rabbits, cattle, sheep, elephants, many species of birds, fishes, insects, etc.) are direct users of plants. The *carnivores* or animal eaters must consume animals of one type or another to survive. There are many such preying insects (praying mantis); birds (hawks, owls); fish (bass, marlin); and mammals (lions, mink). Then there are the *omnivores* that live by consuming both plants and animals. Man is in this group, as are the dog and the pig.

How do the heterotrophs get energy from the autotrophs? They get it generally by oxidation, whereby the organism consumed is essentially burned, using oxygen from the air, to yield energy, carbon dioxide (put back into the atmosphere), and water. This burning which takes place in biological systems is a very orderly process that efficiently uses the energy released. Yet overall the amount of heat released is the same as if the burning took place in a hot flame. When grass is burned, it is largely converted to carbon dioxide and water to give the same amount of energy that a cow might get from using it for feed. If we will but look at ourselves, do we need to doubt that we consume food and produce heat and other forms of energy from it as we go about our daily lives? Just to add a touch of perspective, an adult man requires the energy equivalent to $\frac{1}{3}$ ounce (9 grams) of gasoline, $\frac{3}{7}$ ounce (12 grams) of fat, or $\frac{4}{5}$ ounce (23 grams) of sugar per hour. In contrast, a 2-ton automobile operating at its most efficient speed requires about 500 ounces (14,000 grams) per hour. On an equivalent weight basis, the car would use about 20 ounces (560 grams) or 60 times more than man. To be sure, these are not strictly equivalent analogies. However, when we compare the complex feats that man can accomplish in relation to the essentially single task performed by the car, man is indeed a formidable machine.

We might represent the life cycle of any biological system by a line:

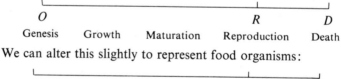

$$O \qquad\qquad\qquad R \qquad D$$

Genesis   Growth   Maturation   Reproduction   Death

We can alter this slightly to represent food organisms:

$$O \qquad\qquad\qquad H \qquad C$$

Genesis   Growth   Maturation   Harvest   Consumption

where up to harvest (OH) we are concerned with food production and from harvest to consumption (HC) we are concerned with food preservation, storage, distribution, and preparation for eating.

We can illustrate the energy source for the autotrophs (green plants, etc.) as follows:

Water + carbon dioxide + nitrogen + minerals + sunlight $\longrightarrow$
$$\text{autotrophs (green plants)} + \text{oxygen}$$

and similarly for the heterotrophs:

Green plants + oxygen $\longrightarrow$ water + carbon dioxide
$$+ \text{heterotrophs (animals and certain other organisms)}$$

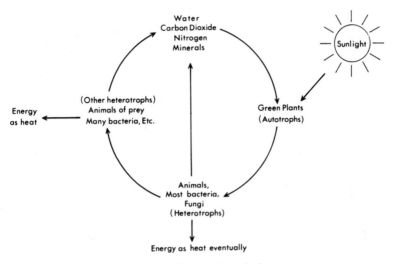

Fig. 2.1. Energy flow in biological systems.

In the latter case, it must be remembered that for many organisms, such as meat-eating animals, other heterotrophs may replace the green plants. In such a case the food chain may be lengthened, but the origin of the energy is the green plants and ultimately the sun. If the two expressions above are combined, we can schematically indicate the dynamic cyclic phenomenon which makes all life possible on the earth (Fig. 2.1). Light energy from the sun drives the whole complicated system and leaves as heat. From the standpoint of physical concepts of energy, the whole cycle is thermodynamically possible because light is at a higher level of energy than heat.

CELLS, THE FUNDAMENTAL BIOLOGICAL UNITS

Cells are the basic biological units. They vary in size from about 0.1 micron (10^{-7} meters) to the large eggs of birds. Though small in terms of volume, nerve cells may be over a meter in length. Even the smallest cell is a highly organized complex structure and functions as a chemical machine. Cells were first recognized by Robert Hooke, who, in his *Microscopia* in 1665, described the cellular organization of cork. Almost two centuries elapsed before the cell theory was accepted as a foundation of modern biology. Today biologists usually mark the beginning of modern biological sciences in the middle of the last century. Theodor Schwann stated in 1839 that "all organisms are composed of essentially like parts, namely, of cells." Rudolph L. K. Virchow, recognized by many as the founder of the science of pathology (diseased tissue), wrote in 1858: "Where a cell exists, there must have been a pre-existing cell, just as an animal arises only from an animal and the plant only from a plant." These concepts plus the theory of natural selection, often called *evolution*, so effectively presented by Charles Darwin in 1859, had a profound effect on biological thought.

With the advent of modern chemistry in the late eighteenth and early nineteenth centuries, biological organisms were recognized as chemical systems. Antoine Lavoisier, pioneering scientist and great intellect who was guillotined in the French Revolution in 1794, showed that both a candle and an animal produce heat in almost equal amounts in accordance with the oxygen consumed and the carbon dioxide produced. Joseph Priestley, the eminent English scholar, clergyman, and scientist who discovered oxygen and who emigrated to the United States in 1794 in his late years, showed that green plants produce oxygen and consume carbon dioxide. Thus from these beginnings and through the life work of thousands of scientists, we are able to explain and interpret biological phenomena in chemical terms.

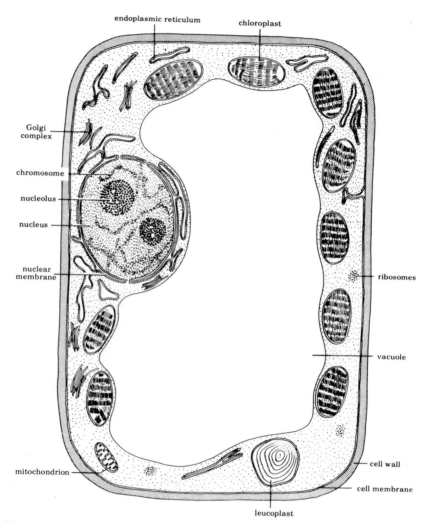

Fig. 3.1. Diagram of the structure and various organelles of a plant cell. (Reprinted from *Biological Science*, First Edition, by William T. Keeton. Illustrated by Paula DiSanto Bensadoun. By permission of W. W. Norton & Company, Inc. Copyright © 1967 by W. W. Norton & Company, Inc.)

It should be recalled that there are two major classes of organisms—the photosynthetic autotrophs (or self-feeders), which use light energy, and the heterotrophs, which get their energy from consuming the autotrophs or other heterotrophs in whole or in part. Let us briefly look at a typical cell of each. We, as Schwann did, can recognize unmistakable similarities in their organization. However, the task is easier for us, as we are the beneficiaries of scientific

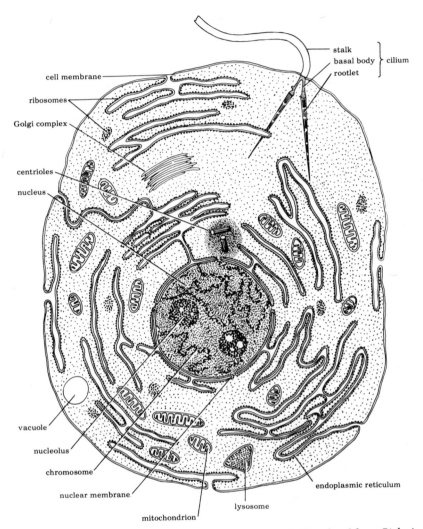

Fig. 3.2. Diagram of the various organelles of an animal cell. (Reprinted from *Biological Science*, First Edition, by William T. Keeton. Illustrated by Paula DiSanto Bensadoun. By permission of W. W. Norton & Company, Inc. Copyright © 1967 by W. W. Norton & Company, Inc.)

tools such as the electron microscope undreamed of in the nineteenth century. Figure 3.1 is a simplified sketch of a generalized green plant cell, and Fig. 3.2 similarly represents a generalized animal cell. The various commonly recognized organelles (structural components within cells whose functions are analogous to those of organs) are labeled in these figures.

Both types of cell have a nucleus with nucleolus, nuclear membrane, and chromosomes. The last are principally composed of *deoxyribonucleic acid* (DNA). The nucleic acids are extremely large linear molecules which carry all the information needed by the cell or the entire organism, if it is multicellular, to perform all of its functions throughout its entire life cycle. These most unique of all molecules are made of four different nucleotide subunits and these in turn are built from only six smaller molecules or building blocks— phosphoric acid, a simple sugar (deoxyribose), and four nitrogen bases (see Chapter 6, pp. 150–155). These four nucleotides represent a four-letter "alphabet." All the information a cell or organism needs is coded by sequential arrangement of the nucleotides (letters). To the modern computer buff, the nucleic acids are much like the tape on a supercomplicated computer controlling an organized series of chemical processes necessary to build a manufacturing plant and control the functioning of all its complicated machinery and workers during its entire period of usefulness. But there is a difference; the cell, since it has a life cycle, must have a mechanism for reproducing this vital informational nucleic acid and transferring it to the next generation to start the life cycle over again.

Let us look at these remarkable nucleic acid molecules another way. They carry information using only an alphabet of four letters arranged usually in three-letter words. They do this much more precisely and carry more information than I am able to in writing this book using a 26-letter alphabet with upper- and lowercase letters arranged in words of varying lengths, punctuation marks, assorted special symbols for numbers, etc., plus illustrations.[1]

Both types of cells contain an outer membrane which allows nutrients and other substances to pass in and out. The membrane is not a passive sieve but actually performs what the physiologist calls *active transport*, wherein energy is expended to carry certain substances in and out of the cell. These cells have a *Golgi apparatus* and *endoplasmic reticulum*, which seem to be tubular or membranous in character, and *ribosomes*. These organellas perform many chemical tasks. The ribosomes themselves are largely composed of nucleic acids which differ slightly from those of the nucleus and are called *ribonucleic acids* (RNA). RNA contains the sugar ribose instead of deoxyribose and one of the nitrogen bases is different (see pp. 148–150). The RNAs encoded by the nuclear DNA (genetic material of the chromosomes) contain the information for the intricate structure of the many different vital proteins necessary for carrying on life processes.

[1] As remarkable as the chromosomes, nucleus, and attendant structures are, they are relatively unimportant as a source of food. This is because quantitatively they account for a very small part of the total mass of a food and the nucleic acids are not a source of any required nutrient except phosphoric acid, which occurs in many other forms in all living things. You became able to synthesize your own nucleic acid requirements at conception, when you received from your parents the genes (nucleic acids) necessary to start your life cycle.

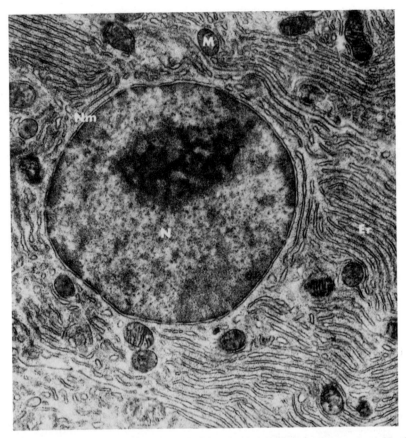

Fig. 3.3. Electron micrograph of a portion of a liver cell showing the nucleus, N; the nuclear membrane, N_m; mitochondria, M; and the endoplasmic reticulum, Er. × 11,000. (Courtesy of D. W. Fawcett.)

Proteins occur in many forms, including enzymes, the extraordinary catalysts which specifically permit certain chemical reactions to proceed so that all the needed chemicals are available as required by the organism; contractile tissue, or muscle, to do work; protective tissue such as skin and hair; photosensitive proteins associated with the visual process. The biological functions of various proteins seem endless, but as far as human food is concerned they are the source of the most critical nutrients required to feed the world's population. Proteins will be discussed in detail later.

Both types of cells have *mitochondria*.[2] These organelles appear to have a central role as far as most cellular chemistry is concerned. Some people have

2 Plural; singular is *mitochondrion*.

rightfully dubbed the mitochondria as the powerhouses or dynamos which keep vital processes going. The primary fuel for many vital processes is the simple sugar glucose. In Figs. 3.1, 3.2, and 3.3, mitochondria appear to be solid with internal mazelike partitions. These organelles are capable of step-wise oxidation of glucose to carbon dioxide and water, and in so doing they trap a proportion of the energy in the form of *a*denosine *tri*phosphate, ATP, which is the energy-packed molecule cells use to make the chemicals they need and to perform many other functions such as muscular contraction.

AMP is *a*denosine *mono*phosphate, one of the nucleotides (or letters) in the RNA, with two additional phosphoric acid groups making a total of three. The phosphate groups at the right in the formula below have considerable chemical energy located in the oxygen linkages marked with $\sim$.

$$\text{Adenine—ribose}\underbrace{\qquad}_{\text{adenosine}}\text{O}-\overset{\overset{\text{O}}{\|}}{\underset{\underset{\text{OH}}{|}}{\text{P}}}\sim\text{O}-\overset{\overset{\text{O}}{\|}}{\underset{\underset{\text{OH}}{|}}{\text{P}}}\sim\text{O}-\overset{\overset{\text{O}}{\|}}{\underset{\underset{\text{OH}}{|}}{\text{P}}}-\text{OH}$$

Triphosphoric acid itself can be made by heating (adding energy to) ordinary phosphoric acid thus:

$$3\,\underset{\text{phosphoric acid}}{\text{HO}-\overset{\overset{\text{O}}{\|}}{\underset{\underset{\text{OH}}{|}}{\text{P}}}-\text{OH}} + \text{heat} \longrightarrow \underset{\text{triphosphoric acid}}{\text{HO}-\overset{\overset{\text{O}}{\|}}{\underset{\underset{\text{OH}}{|}}{\text{P}}}-\text{O}-\overset{\overset{\text{O}}{\|}}{\underset{\underset{\text{OH}}{|}}{\text{P}}}-\text{O}-\overset{\overset{\text{O}}{\|}}{\underset{\underset{\text{OH}}{|}}{\text{P}}}-\text{OH}} + \underset{\text{water}}{2\text{H}_2\text{O}}$$

When the reverse reaction takes place, the triphosphoric acid hydrolyzes (water is a reactant) back to simple phosphoric acid, and heat or energy is released. In an analogous manner, when ATP loses or transfers the end phosphoric acid (or the second from the end) energy is lost or transferred to another molecule.

Many compounds made by cells require energy. Often these substances are formed by the removal of the constituents of water from the reactants (other chemical compounds) which themselves are in water solution or aqueous medium. Stated another way, water must be removed in a chemical process which must take place in water. This sounds like double talk. It is in a sense, and chemically can be accomplished in a circuitous manner. For example, glucose is made into starch by removing water:

$$n\,\underset{\text{glucose}}{\text{C}_6\text{H}_{12}\text{O}_6} \longrightarrow \underset{\text{starch}}{(\text{C}_6\text{H}_{10}\text{O}_5)_n} + \underset{\text{water}}{n\,\text{H}_2\text{O}}$$

This cannot be done with simple glucose in water solution but rather is done by making glucose phosphate from glucose and ATP and combining the glucose phosphate molecules to give starch. Thus

$$n \text{ Glucose} + n \text{ ATP} \longrightarrow n \text{ glucose phosphate} + n \text{ ADP}$$

$$C_6H_{12}O_6 \qquad\qquad C_6H_{11}O_5-O-\overset{\overset{\displaystyle O}{\|}}{\underset{\underset{\displaystyle OH}{|}}{P}}-OH$$

$$n \text{ Glucose phosphate} \longrightarrow \text{ starch } + n \text{ phosphoric acid}$$

$$C_6H_{11}O_5-O-\overset{\overset{\displaystyle O}{\|}}{\underset{\underset{\displaystyle OH}{|}}{P}}-OH \qquad (C_6H_{10}O_5)_n \qquad H_3PO_4$$

Although in many ways this scheme is an oversimplification, it does show how energy is trapped and used in cells through the use of phosphoric acid as a part of ATP. Many other energy-requiring reactions in cells are possible through the intermediary ATP; hence calling the mitochondrion the power-house of the cell is an apt analogy. Whereas the mitochondria may burn or oxidize a molecule of glucose to carbon dioxide and water to produce ATP, the same ATP has sufficient energy to *polymerize* (make large molecules from small ones) other molecules of glucose to form starch. Starch is a major food ingredient for man and is the major energy storage compound in the seeds of such cereals as corn, wheat, and rice. The energy-giving compounds in seeds must be sufficient to initiate growth of the next generation until the new plant can sustain itself.

The location of mitochondria in cells coincides generally with the location where energy is needed for work to be done. Such work may be mechanical, chemical, or even electrical. The number of mitochondria differs in cells in accordance with their work requirements. For example, a liver cell may contain a thousand or more, accounting for one-fifth of its volume. Mitochondria are often adjacent to fuel sources such as fat and carbohydrate. These organelles play such an important role in many of the fundamental chemical processes of cells that a simplified schematic presentation is warranted. In Fig. 3.4, many details important to the biochemist and physiologist have been omitted in order to emphasize certain aspects of importance in food and nutrition, and to anchor the scheme on more commonly known substances. In order to understand the scheme, it must be kept in mind that generally the reactions are reversible in accordance with biological requirements at any particular time.

The outer mitochondrial membrane will admit selectively many small molecules that can be oxidized to produce energy in the form of ATP, which

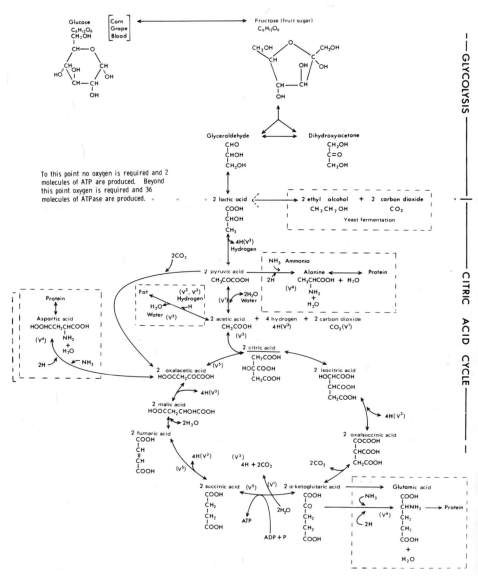

Fig. 3.4. Glycolysis and the citric acid cycle.

leaves the mitochondria to go where it is needed. Among these are the simple sugar glucose, lactic acid, and acetic acid. These and other substances are fundamental intermediates in the biochemical transformations of biological systems. We know them more commonly when they accumulate or are isolated. Respectively, we associate glucose with sugar of grapes, corn, and blood; lactic acid with sour milk, sauerkraut, and pickles; and acetic acid with vinegar. These and other simple compounds such as citric acid (citrus fruit), malic acid (apples), and even ethyl alcohol (wine, beer) fit into the pattern.

Whereas this scheme shown in Fig. 3.4 (known as glycolysis in tandem with the citric acid cycle) may be the source of chemical energy by oxidizing (burning) the many kinds of molecules entering into it, the chemical processes also serve as the source of many molecules needed to build protoplasmic material. One of the key compounds is acetic acid, a molecule of two carbon atoms. Acetic acid is the primary compound from which fats, waxes, sterols, rubber, and many other vital substances are made biologically. The cycle is also the bridge between proteins and carbohydrates. Proteins themselves are highly complex molecules made by joining smaller ones together. These small ones are of only 20 different kinds—but sometimes proteins contain 200–300 up to 100,000 or more of these 20 kinds of building blocks (see Chapter 6). Several of the most common blocks are directly related to the intermediates shown in Fig. 3.4, and these have been noted. Alanine comes from lactic or pyruvic acid, glutamic acid (the very flavorful substance widely used as its sodium salt, monosodium glutamate) comes from α-ketoglutaric acid, and aspartic acid comes from oxalacetic acid or malic acid. Recall again that most biological reactions are reversible, so these protein building blocks, amino acids, can enter the cycle and serve as energy sources in the same way that fats can serve as energy sources via acetic acid or that even alcohol can be used to give energy. (In the human, 95–98% of alcohol consumed is burned preferentially to carbon dioxide and water.) Study the scheme to see if you can observe many of the points just mentioned. For those familiar with chemistry, formulas have been inserted to add meaning, but it is not necessary to know the chemical details to get an overall view of these remarkable cellular processes.

All reactions in the cycle are made possible by enzymes which are specific proteins. Some of these enzymes must have attached vitamin-containing groups (called *coenzymes*). Where these are important, a V has been inserted. V^1 represents vitamin B_1 or thiamine; V^2, vitamin B_2 or riboflavin; V^3, the pellagra-preventing vitamin, nicotinamide; V^5, vitamin B_5 or pantothenic acid; V^6, vitamin B_6 or pyridoxine.

These reactions are not all that goes on in mitochondria, but if you follow the cyclic scheme of reactions you will observe that one molecule of glucose containing six carbon atoms is converted to two molecules of lactic acid, which in turn give a molecule of oxalacetic acid with four carbon atoms and a molecule of two carbon atoms which is acetic acid. The acetic acid may be used to build large fat molecules (and others, too) or may be oxidized and burned to carbon dioxide and water, as we shall see presently.

In the overall scheme, one molecule of glucose, $C_6H_{12}O_6$, plus six of water, H_2O, yields six molecules of carbon dioxide, CO_2, and 12 pairs or 24 hydrogen atoms:[3]

$$C_6H_{12}O_6 + 6H_2O \longrightarrow 6CO_2 + 12H_2$$

$$\text{glucose} \quad \text{water} \qquad \text{carbon dioxide} \quad \text{hydrogen}$$

We know, of course, that most biological systems do not produce hydrogen gas. (Some bacteria do, however, as when they live in oxygen-free atmospheres.) We also know that when hydrogen is burned in oxygen to produce water much heat or energy is liberated. For every gram (0.04 ounce) of hydrogen, 34,000 cal (calories) or 34 Cal (kilogram calories or Food Calories)[4] is produced. For instance, when 1 g (gram) of fat, carbohydrate, or protein from your food is oxidized to H_2O and CO_2, the energy produced is 9, 4, or 4 Cal, respectively.

Within the mitochondria, there is coupled with the foregoing chemical system shown in Fig. 3.4 another to oxidize, stepwise, the hydrogen as it is produced. This oxidation, eventually to water, requires oxygen from the air, and the oxidation process is carried on in a very orderly manner so that as much energy as possible is trapped by the synthesis of ATP which the cell can use to perform its vital functions.

Some of the energy resulting from the stepwise oxidation of the hydrogen originating from the oxidation of the original molecule of glucose is trapped in 36 high-energy phosphates. These plus the two from the conversion of glucose to lactic acid give 38 high-energy phosphate groups from the complete burning of a single glucose molecule. These 38 are trapped in ATP.

Considering the muscle as a heat engine, it is capable of trapping almost 40% of available energy of glucose or sugar combustion. This efficiency exceeds that of many man-made machines such as automobiles and locomotives.

It should be stated here that although ATP and a few related organic phosphates are the energy-giving intermediates for many vital chemical reactions requiring energy (uphill reactions in contrast to downhill reactions, which will be referred to later) the quantity of ATP present at any one time is not great; rather it is made (synthesized) and used over and over again. To illustrate, ordinary muscle (meat) in the living animal carries its ready reserve energy as glycogen (animal starch, a large molecule made up of hundreds of glucose units). In terms of weight, there is about 1 g of glycogen in 100 g of

---

[3] In biological systems, oxygen is often put into a molecule by insertion of water and removal of hydrogen atoms.

[4] A calorie is the heat necessary to raise the temperature of 1 gram of water 1° Centigrade (1.8° Fahrenheit). The energy equivalent of food is usually given in kilogram calories or Calories, which is the heat required to raise the temperature of 1000 grams or 1 kilogram (2.2 pounds) of water 1°C, or to raise 10 grams (0.35 ounce) of water from its melting point to its boiling point.

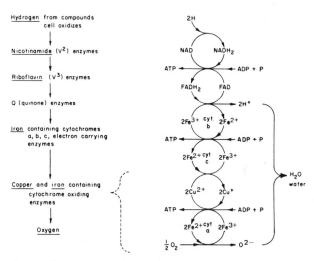

Fig. 3.5. Hydrogen-electron transport system. P indicates *p*hosphoric acid; ADP, *a*denosine *di*phosphate; ATP, *a*denosine *tri*phosphate; NAD, *n*icotinamide *a*denine *di*nucleotide; FAD, ribo*f*lavin *a*denine *di*nucleotide; H, *h*ydrogen atom; H^+, *h*ydrogen ion or proton, positively charged hydrogen atom; Fe^{3+}, Fe^{2+}, Cu^{2+}, and Cu^+, iron and copper ions, respectively. Note the term "nucleotide" and recall that both DNA, *d*eoxyribo*n*ucleic *a*cid, and RNA *ri*bo*n*ucleic *a*cid, are polymers or long chains of nucleotides. AMP, *a*denosine *m*ono*p*hosphate, is one of the nucleotides of RNA. The *di*nucleotide terms "NAD" and "FAD" indicate that the enzymes or protein catalysts have attached to them a coenzyme of two nucleotides, one of *a*denine, ribose, and phosphoric acid and the other of *n*icotinamide, ribose, and phosphoric acid (NAD) or ribo*f*lavin and phosphoric acid (FAD).

muscle. If each of these glucose units potentially represents 38 units (molecules) of ATP, this would mean that the same 100 g of muscle should have approximately 107 g of ATP.[5] This for a number of reasons is impossible. So the muscle cell has a system for oxidizing its stored glycogen through glucose to form the ATP as it is needed. At any one time only a small amount, about 1.5%, of the available energy of muscle is present as ATP.

The stepwise oxidation of the hydrogen atoms is illustrated in Fig. 3.5. The hydrogen atoms coming from lactic acid or other intermediates of the citric acid cycle usually are transferred directly to a hydrogen-carrying molecule, an enzyme (protein catalyst) containing nicotinamide (niacin), one of the most important vitamins. The hydrogens are then passed by the nicotinamide-containing enzyme to an enzyme containing another vitamin, riboflavin or B_2. The riboflavin passes its hydrogen on to other enzymes (called Q enzymes) which pass only the single electrons of the hydrogen atoms to the first of several cytochromes, electron-transferring enzymes containing an atom

[5] 1 g glucose ÷ 180 (molecular weight of glucose) = 107 g ATP ÷ [38 × 507 (molecular weight of ATP)].

of iron or copper in ionic, or electrically charged, form. In doing this, the hydrogen is made into a hydrogen ion (H^+) and electrons of negative charge, e^-, and the ferric ions (Fe^{3+}) of the cytochromes are changed to ferrous ions (Fe^{2+}). After several of these cytochromes operating at successively lower energy levels, the Fe^{2+} ions deposit their electrons to the oxygen atoms, $O + 2e^- \rightarrow O^{2-}$, which then react with the hydrogen ions of the water in the cell to form more water ($2H^+ + O^{2-} \rightarrow H_2O$). At several of these reaction sites down the chain, adenosine triphosphate (ATP) is synthesized from *a*denosine *di*phosphate (ADP) and inorganic phosphate (P). All of this is represented in Fig. 3.5.

A note of interest here is the biological accommodation that hydrogen atoms and hydrogen ions are small and can be moved or transferred fairly easily. Yet the electrons, about 1/1850 of the weight of a hydrogen atom or ion, can move much more rapidly than an oxygen atom, whose weight is 16 times that of the hydrogen atom or 30,000 times that of the electron. Within the cell, the electron from the hydrogen is carried to the oxygen (which reacts with the hydrogen ions which are always present in water where cells must live) rather than the relatively heavy oxygen being moved to the hydrogen.

The fundamental role of the oxidation of carbohydrate is to provide the energy for most of the vital processes of both autotrophic and heterotrophic cells. At the same time, many of the small molecules, intermediate in the carbohydrate oxidation (Fig. 3.4), can serve also as small molecular building blocks for many large and complex molecules associated with the intricate structure and function of cells.

Biological organisms often have more than one system for accomplishing the same thing. In so doing different substances may be made which are important to the organism. And so it is with the oxidation of glucose, a sugar with six carbons, to a smaller sugar of five carbons, ribulose, and carbon dioxide:

$$6C_6H_{12}O_6 + 6H_2O \longrightarrow 6C_5H_{10}O_5 + 6CO_2 + 12H_2$$

glucose   water   ribulose   carbon dioxide   hydrogen

As before, (page 20) effectively one molecule of glucose has been oxidized to yield six of carbon dioxide and 12 pairs of hydrogen atoms. The organism can convert the ribulose to ribose, already mentioned with respect to nucleic acids, to fructose (fruit sugar) as in Fig. 3.4, to some other sugars, and to glyceraldehyde.

When we consider the orderly and efficient way that mitochondria can carry on many chemical transformations as they trap energy for use by the various organelles of the cell and at the same time the way that they make a number of other important intermediates for the cell itself, it is easy to understand why mitochondria have received a great deal of attention from biological scientists. It is also easy to understand why mitochondria play equally important roles in both autotrophic and heterotrophic cells. In the latter, which cannot use light as their energy source, the energy-giving compounds must have been produced outside the cell by other heterotrophs or autotrophs. In the autotrophs, the mitochondria work in tandem with another system which is capable of trapping light energy and converting it to chemical energy in the form of glucose.

This brings us to some of the differences between plants and animals and between autotrophic and heterotrophic cells.

ORGANELLES OF PLANT CELLS

Cell Wall

As may be seen by comparing Figs. 3.1 and 3.2, plant cells uniquely have a rather heavy cell wall outside the cell membrane itself. This is usually composed of carbohydrate material which is essentially insoluble in water and of high molecular weight. The cell wall gives rigidity and strength to the plant. Interestingly, the small molecule which the cell uses to make the large molecules of its wall is glucose—the same molecule that the mitochondria can use for energy production if necessary.

Chemically, the most prominent compound of the cell wall is cellulose. In terms of tonnage, cellulose is perhaps the most important of all chemical compounds produced in the entire biological cosmos of this earth. Cellulose itself is a long, linear molecule and wood is about two-thirds cellulose; cotton is almost pure cellulose. Man through the eons of his development has learned to use cellulose in diverse ways. What could modern civilization do without paper, for instance? It is largely cellulose. A number of great industries are based on cellulose. Later in this book, cellulose will be mentioned again, for it is the major energy source for many herbivorous animals—particularly the ruminants such as cattle, sheep, goats, and deer which man since prehistoric times has used as food, even before he learned to cultivate plants for food.

Vacuole

Plant cells usually contain a *vacuole*, which often gets larger as the cell ages. The vacuole seems to be a repository for many compounds the plant cell produces. Some investigators feel that the vacuole contains the residual compounds which the cell has made that have been transferred there after their usefulness has ended, since most higher plants have only a limited secretory capacity. In this they are not like most animal cells, which have very small vacuoles.

Chloroplasts

We have already mentioned several times that green plants are the primary source of all biological activity. What makes them green are the *chloroplasts*—the green organelles that are capable of taking water, carbon dioxide from the air, and sunlight and producing glucose, which is the key compound in all biological processes, and oxygen, which supports all animal life. Figure 3.6 is a photograph of the highly structured chloroplast taken through an electron microscope. From the time man first learned to use fire—burning the wood made by green plants in air containing oxygen also produced by green plants—he has been the user of the sun's energy stored in plants by the chloroplasts. Indeed, all fossil fuels—coal, petroleum, and natural gas—are of biological origin and represent stored energy originally trapped from the sun by chloroplasts.

Before proceeding, it should be pointed out that a number of organisms besides green plants have the ability to trap light energy in order to go through their life cycles. These are the blue-green algae, brown algae, purple photosynthetic bacteria, etc. In other words, the green plants are not the only autotrophs. But although there are some other photosynthetic systems, that associated with green plants is the most common. All these organisms contain colored pigments capable of absorbing light.[6] Some organisms contain more than one pigment—even several such compounds, which are sometimes called *photosynthetic pigments*. Whatever the system an organism uses to trap light as its primary source of energy, there are many common details. Discussion of most of these is beyond our scope. Furthermore, the scientific community has yet to learn many of the secrets of this fundamental process on which all biology rests. Nevertheless, sufficient information exists to give some meaningful insights.

The similarity between the process to get energy from use of oxygen of the air to burn glucose to produce water and carbon dioxide and the converse process, to produce glucose and oxygen by means of light, water, and carbon dioxide, seems quite striking to most of us. On the other hand, the sophisticated scientific investigator working in this area is aware of much that is

---

[6] They appear colored in white light because white light contains all colors and color is apparent in any object which reflects some of the colors and absorbs others. For example, when an object appears to be red when seen using white light, the object itself absorbs green light and reflects red. If viewed in pure green light, the object would appear black because no light would be reflected.

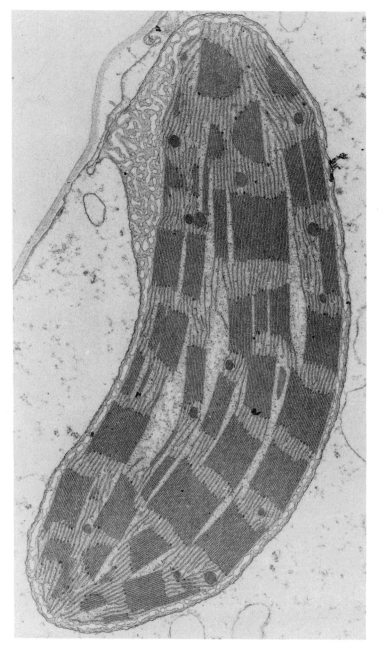

Fig. 3.6. Electron micrograph of a chloroplast from the leaf of a corn plant. The dense-layered lamellas are the grana where chlorophyll is located. × 10,700. (Courtesy of L. K. Shumway.)

different between the two processes. Let us look at some of the essential features of the photosynthesis of glucose in green plants in relation to the oxidation of glucose to furnish energy for many biological activities. As we do, we will see that some important nutrients needed by man are related basically to the process of photosynthesis which takes place in the chloroplasts. Yet since man is not a photosynthetic organism these molecules serve man differently than they serve the plants.

The green color of chloroplasts is due to chlorophyll, which occurs in a few very closely related but slightly different molecular forms. Usually associated with chlorophyll are some carotenoid (yellow, orange, and/or red) pigments, including the carotenes (color of carrots) which are the source of vitamin A for man. The human liver and the livers of most animals including fish and birds can chemically convert carotenes into vitamin A. β-Carotene, along with other carotenes, gives butter its natural yellow color, and pure β-carotene is sometimes used to color butter and other foods brilliant yellow to orange, depending on the amount used. The molecular structures of chlorophyll and β-carotene (Fig. 3.7) are of interest, for they are usually found together in such things as spinach. Popeye the sailor promoted the nutritional value of spinach and spinach leaves are often used by researchers studying photosynthesis. Both chlorophyll and β-carotene contain many double bonds (two atoms joined together by two linkages $=$ rather than only one —). When there are a large number of double bonds, molecules are colored and absorb light energy. These two compounds absorb the energy of the sun in such a way that the molecules become excited and lose an electron, which is passed through hydrogen-electron transfer compounds somewhat similar to those noted in Fig. 3.5 for the transfer of electrons for the conversion of hydrogen and oxygen to water with formation of ATP. However, in the case of photosynthesis, light absorbed by the photosensitive molecules imparts sufficient energy to an electron so that it can effectively change water to hydrogen ions, oxygen gas, and electrons:

$$2H_2O \longrightarrow 4H^+ + O_2 + 4e^-$$

The electron-transferring molecules are capable of forming ATP from ADP as before. The electrons reach molecules similar to the Q enzymes (Fig. 3.5), where the H^+ and electrons are passed to the hydrogen-carrying enzymes containing nicotinamide and riboflavin as before.

Interestingly, at the Q stage (Fig. 3.5) we often find, among other quinones, vitamin K, the vitamin discovered to be important in blood clotting.

The hydrogen-carrying enzymes with the ATP produced photosynthetically are able to convert carbon dioxide and water to glyceraldehyde phosphate and eventually to glucose. Notice in Fig. 3.4 that glyceraldehyde is a primary intermediate of carbohydrate (glucose) oxidation in the citric acid scheme.

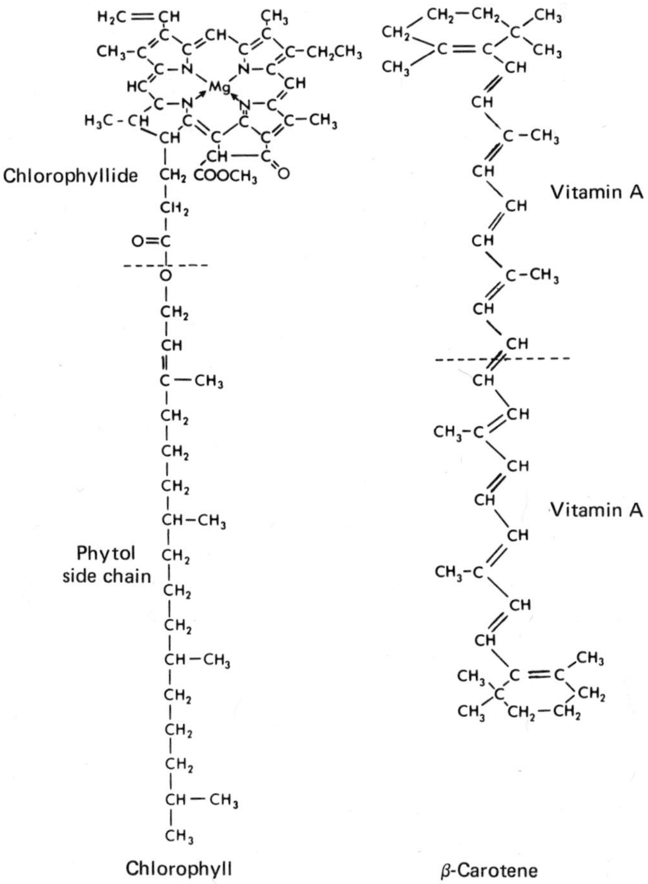

Fig. 3.7. The molecular structure of chlorophyll and β-carotene. C indicates a carbon atom; H, hydrogen atom; O, oxygen atom; N, nitrogen atom; Mg, magnesium atom.

Another key reaction in the fixation of carbon dioxide in photosynthesis is strikingly similar to the reverse reaction shown on p. 22. Carbon dioxide adds to the five-carbon sugar, ribulose, and forms two molecules (each containing three carbons) of glyceraldehyde. These, in turn, can be converted to fructose (fruit sugar) and glucose by suitable enzymes (see Fig. 3.4). Thus it is appropriate here to note again that most biochemical reactions are reversible, their direction dependent on whether energy is being consumed or liberated.

The photosynthetic process consumes the energy of the sun to produce carbohydrate (glucose) and oxygen from water and carbon dioxide:

$$6CO_2 + 6H_2O + sunlight \longrightarrow C_6H_{12}O_6 + 6O_2$$
$$glucose$$

and cells can use that carbohydrate to give them the energy to live:

$$C_6H_{12}O_6 + 6O_2 \longrightarrow 6H_2O + 6CO_2 + energy$$

The focal role that the sugar glucose plays in all biology is obvious. This sugar is produced by the chloroplasts of green plants (and other photosynthetic organisms) and is consumed by animals and other nonphotosynthesizing organisms to provide the necessary energy for life. Although we have surveyed only a few details of these processes, it is also obvious that there is a good deal of similarity in the chemical processes and the compounds produced in all biological systems.

ORGANELLES OF ANIMAL CELLS

Two characteristics that laymen often associate with animal cells are movement and the replacement of old or damaged cells with new ones. Every child runs, falls, skins his knee, and sees his knee heal. These two characteristics are attributes of two of the organelles in the simplified animal cell in Fig. 3.2, the *cilium*[7] and the *lysosome*.

Cilia

Cilia are hairlike contractile fibers. Many unicellular organisms use the cilia to propel themselves through their aqueous medium. The cilia of some fixed cells can beat in unison, causing liquid to pass by the cell surface. In many higher animals, ciliated cells serve all kinds of functions where fluid movement is essential. Of course, in higher animals there are cells organized into tissues that are specifically and primarily designed for movement of the animal itself or of substances within the animal, such as the muscles of the arm or leg and those of the heart, stomach, and intestine. Whether movement of cilia, entire cells, or groups of muscle cells is involved, that ubiquitous compound ATP is the energy-giving molecule.

Lysosomes

Lysosomes are packets of many kinds of enzymes encapsulated and kept in inactive form so that they do not normally harm the cell and its function. However, if for any reason the lysosomal membrane breaks, due to injury, etc.,

[7] Singular; plural is *cilia*.

enzymes or catalysts are released which are capable of digesting or hydrolyzing the cellular constituents of the injured cell to give the simple molecular building blocks from which they were made. In a sense, the cell has a mechanism for destroying itself (as well as reproducing itself) in such a way that necessary molecular building blocks are conserved for use by other cells. For example, proteins are digested (hydrolyzed) to amino acids; nucleic acids to ribose, deoxyribose, phosphoric acid, and the constituent nitrogen bases; fats to glycerol and fatty acids; complex carbohydrates to simple sugars. These processes can be respectively illustrated as follows:

$$\text{Protein} + H_2O \xrightarrow{\text{hydrolyzing enzymes}} \text{amino acids}$$

$$\text{Nucleic acids} + H_2O \longrightarrow \text{ribose or deoxyribose} + \text{phosphoric acid} + \text{adenine} + \text{guanine} + \text{cytosine} + \text{uracil or thymine}$$

$$\text{Carbohydrates} + H_2O \longrightarrow \text{simple sugars (glucose, fructose, ribose, etc.)}$$

$$\text{Fats} + H_2O \longrightarrow \text{fatty acids} + \text{glycerol (glycerin)}$$

These processes promoted by the enzymes of the lysosomes of the cell are similar in function to the digestive processes man uses to obtain the same building blocks from food in his gastrointestinal tract. (This subject will be covered in more detail in Chapter 7.) Such hydrolytic processes, cleaving or splitting of large molecules into smaller ones by reaction with water, generally do not require energy but rather liberate energy. The same hydrolytic reactions are often associated with the degradative processes that occur upon the death of an organism. Even though this association exists, it must be kept in mind that many of these reactions do occur as needed in an orderly fashion in a living organism.

CONCLUSIONS

There are other organelles in cells whose functions are less well known. Indeed, the functions of those we have discussed have been simplistically and incompletely described. For example, the vacuoles in animal cells are relatively minor in size, and organelles such as leucoplasts seem to be associated with the production of certain cell constituents. Nevertheless, the present state of knowledge permits some degree of understanding of the nature of cells as the fundamental units of biological activity. In unicellular organisms—bacteria, protozoa, plankton of the sea, etc.—the cell must perform all necessary functions of the life cycle—genesis, growth, maturation, reproduction, and death. In multicellular organisms, cells, though originating from a single cell at genesis, differentiate to perform more specialized functions so that the more complex organism as a whole may complete its life cycle. We need only consider the different organs and tissues of our own bodies to comprehend the degree of cellular specialization. Notwithstanding these complexities, all biological processes derive from the chemical processes in individual cells. Many of the ideas developed later in this book are easier to understand when the requirements and functions of cells themselves are appreciated.

4

ENVIRONMENTAL CONDITIONS FOR LIFE

Living things, as we know them, do not exist on the sun or the moon, and scientists have not found proof that there are living things on the other planets of the solar system. Many places on the earth itself are inhospitable. What then are the environmental conditions which permit an organism to grow, reproduce, and start another generation?

There are five basic conditions, and they are chemically related and definable because all life processes are chemical in nature or are manifest through chemical reactions. It is not necessary to know all the details to appreciate these environmental factors, but as we discuss each in turn two things should be kept in mind. First, the overall life process of any living thing is a magnificently ordered series of reactions. Second, each reaction is quantitative and is required in well-defined amounts. Some reactions may be relatively simple, such as that of oxygen with hemoglobin of the blood, which we can easily see in the change of dark blue-red color of venous blood to the brilliant crimson of arterial blood. (A similar type of reaction gives the bright red color on the cut surface of fresh uncooked meat.) Others may be extremely complicated, such as the synthesis or manufacture of hemoglobin itself and the packaging of it in the red cell so that it can do its job without interfering with other life processes. The number of ordered or sequential reactions in any living thing is in the hundreds or thousands. Not only is it essential that a living organism be able to carry on these reactions, but also the reactions must occur in the proper sequence and cause the correct amount of chemical compounds to be produced or consumed. With such a system, if for any reason any single vital chemical reaction is prevented from occurring the organism dies. This situation may also result if some vital reactions get out of control. Cancer may be considered an example of this.

As one begins to comprehend the details of the chemistry of living things (biochemistry), two facts become clear: (1) Some organisms have built-in systems of redundancy; that is, some systems of chemical processes may be circumvented or replaced by another if something goes wrong. Thus when a person becomes diabetic due to faulty combustion of carbohydrate to provide energy, fat combustion may supply the energy for a time; if the missing chemical (insulin) is properly supplied the diabetic person may expect to live a relatively normal life. Corollary situations exist for those microorganisms and some higher organisms which seemingly can adapt to other food sources if the usual ones become unavailable. We have already noted in the previous chapter two interlocked systems for glucose metabolism. (2) Organisms have chemical defense mechanisms at their disposal. Indeed, the control of one organism by another by chemical means is common. If a vital chemical reaction of an organism is arrested, destroyed, or put out of control by a chemical produced by an adversary organism, the one dies and the other lives. This is a most fascinating field of biology. The insect and snake venoms, the bacterial toxins, etc., operate in devious chemical ways. Man himself produces certain proteins (for example, immune bodies) as protection against invading bacteria. Molds exude antibiotics (penicillin, etc.) to prevent attack by many bacteria, and many bacteria have similar capabilities. The chicken synthesizes a special protein antibiotic (avidin) in the egg white to defend the young embryo from bacteria which might get into the egg. After all, eggs are sometimes laid in untidy places, and the eggs must be porous to permit the oxygen–carbon dioxide exchange necessary for the young chick embryo. Insects attract mates by releasing specific chemicals into the air. The leaf-cutting ants produce a specific chemical to kill certain fungi and permit others to grow in their well-tended fungal gardens. Even the bombardier beetle can, without burning itself, actually kill its enemies by a scalding spray which it produces chemically. Examples of chemical control of one species by another are endless, and from this vast reservoir of chemical biology many useful tools have been developed for use in medicine to guard our health—antibiotics, vaccines, immune globulins, etc.—and in agriculture to protect the organisms man needs for his food—antibiotics again, insecticides, plant growth regulators, etc. Here also may be found some avenues to protect and preserve our foods after they are produced. Some varieties of cheese produced by the activity of certain bacteria on milk keep better than others because the bacteria produce a chemical antibiotic which inhibits the growth of food spoilage organisms.

Now as we return to the major environmental factors controlling biological systems, let us remember that biological organisms are highly organized, intricately structured chemical systems which use chemical processes to perform all vital functions. Let us also remember that while we must nurture certain living things to provide us with food, the essence of food preservation and processing is to retard or stop the chemical processes in the food organism and to preserve the substance of the organism which contains the essential nutrients needed by man.

WATER

There is no life without water. All biological systems can complete their life cycles only in the presence of liquid water. There are all types of marine and aquatic organisms, fish, insects, crustaceans, mammals, and micro-organisms, that can only live in water. Terrestrial plants have elaborate systems of roots, etc., to transport rainwater from the soil to all of their working cells. Terrestrial animals must carry all necessary water with them.

To be sure, many organisms have remarkable adaptations to the lack of water. Many microorganisms and plants can go into a state of suspended animation when sufficient water is absent. Bacterial and fungal spores and seeds of plants are examples. Also, many plants stop a number of their vital processes in the absence of sufficient water and begin again when water is available. For evidence of this, we need only observe the accommodations of indigenous plants to the rainy and dry seasons of the tropical and semitropical regions of the world. This has much in common physiologically with the accommodation to temperate and arctic regions, where many plants simply seem to stop growing when their vital water is frozen in winter and resume when the spring and summer make this water liquid and available again for life processes. Many insects have accommodations of this sort, also. Some, indeed, are capable of synthesizing glycerol from sugar (see Chapter 3: glyceraldehyde $+ 2H \rightarrow$ glycerol). Glycerol is an effective antifreeze and has remarkable biological properties. For example, animal and human sperm may be placed in glycerol solutions, frozen, and kept for years without loss of viability. This fact has had tremendous effect in providing high-quality animal protein for food. The semen of bulls whose daughters have proven to be healthy, efficient producers of milk is preserved in this way so that it can be diluted and used to inseminate cows far removed from the bull. In this way, some bulls have sired many hundreds of high-producing offspring even after their death.

Terrestrial plants and animals also possess some remarkable accom-modations to conserve water for maintaining vital functions. Cacti and desert mammals have evolved in such a way that their physiology and behavior conserve water, which is available only infrequently. Some animals get their water only from that stored in plants they eat and the water they produce by oxidizing their food to carbon dioxide and water. Such water appears to be the only water for insects which thrive in dry places. Some species of the microscopic mites (the common chigger is a large mite), the ubiquitous creatures living on and with animals including man, have a truly unique system of water acquisition. These mites, whose diet seems to be sloughed and dead skin tissue and which live, thrive, reproduce, and die in such diverse places as the human body, pillows, mattresses, carpets, and furniture, apparently have the ability to convert water vapor from the air to liquid water for their own use. This can be done at relative humidities as low as 60%. Chemically and thermodynamically, this is a notable biological accommodation to the fundamental problem of providing liquid water to support life processes.

Our own thirst for water and the obvious necessity of rain for growth of the plants we depend on emphasize to us in a personal way the importance of water. Civilizations developed only where there was water. Nomadic peoples must continuously seek water. Water is a central theme of history, and the well or fountain was the social center of biblical times, as it is still in some parts of the world. Wars have been fought over water supplies, a major source of international tension. With the urbanization resulting from our modern industrial economy, the true value of abundant clean water is being more generally, and even emotionally, recognized by everyone. Water pollution is a perennial political issue. We must be concerned with our precious water lest we destroy its life-giving potential. Let us now proceed to see what is required for water to support life.

Since all cells live and function in water and since life processes are essentially chemical, it is possible to define environmental factors necessary for life in terms applicable to the aqueous solutions in which cells or entire organisms live.

TEMPERATURE

The fact that living cells function only by carrying on vital processes in liquid water restricts the range of temperatures at which life is possible. Normally, water freezes (changes to solid) at $0°C$ ($32°F$) and boils (changes to gas) at $100°C$ ($212°F$) at normal atmospheric pressure. If substances are dissolved in the water, the freezing point may be lowered somewhat and the boiling point elevated. But temperature is much more important in biology than whether water is liquid or not. This is because all organisms are consumers of energy in one form or another and the laws of thermodynamics (the science of energy movement) apply to biological systems as well as to power plants with respect to all the chemical reactions that go on in either of them. We do not intend to treat thermodynamics in detail, but many references have already been made to heat and other forms of energy in biological systems. All forms of energy are interchangeable and eventually end up as heat.

Temperature itself is a fundamental thermodynamic concept and relates directly to chemical processes. It is a measurement of the average kinetic energy (energy of movement) of all molecules in any particular system. All molecules move except at absolute zero, $-273°C$ ($-491.4°F$). The higher the temperature, the faster they move and the more kinetic energy they have. (The kinetic energy of any particle equals the mass of the particle times its velocity squared, divided by 2.) Everyone knows that a fast-moving auto has more energy than a slow-moving one. The faster one requires more braking to stop and also causes more damage on crashing.

Temperature is related to chemical processes in two basic ways. First, the energy consumed or liberated by a reaction is related to the temperature at which the reaction takes place. This factor is perhaps less important in biological systems because of the limited range of temperatures which permit life. This range is called the *biokinetic zone* and is generally considered to be from $0°C$ ($32°F$) to $60°C$ ($140°F$) for most organisms, even though a very few simple ones can live at slightly below $0°C$, because dissolved matter lowers the freezing point, and a few can live at several degrees above $60°C$. From the biological and food point of view, the second effect that temperature has on chemical reactions is much greater. This is the role of temperature on rates of reactions, or how fast a chemical process may proceed.

We so commonly observe that temperature controls rates of biological activity that we take it for granted. One example is the use of the household refrigerator to slow the deterioration of fresh foods and the rates of growth of bacteria and molds which compete with us for our food. Many other examples can be given. Many microorganisms, plants, and so-called cold-blooded animals (reptiles, fish, insects) can tolerate a relatively wide range of temperatures, and it is possible to observe the effects of temperature on their activity. In the cold, the housefly or frog is less active, plants may grow slower, and yeast may ferment grape juice or bread slower, and so on. Many higher animals, including man, are warm-blooded. This means that their body temperature is usually above the ambient temperature and that they have a system to maintain their body temperature within some very narrow range. We may sweat, a dog may pant, a bird may fluff its feathers and spread its wings when body temperature begins to rise. Conversely, we may put on a coat, the dog may curl up, and the bird may tightly fold its wings and crouch on its nest if body temperature begins to drop below the desired level. Warm-blooded animals usually have body temperatures in the range $36.4–41.5°C$ ($97.5–107°F$). Table 4.1 shows usual body temperatures for a number of

Table 4.1. Body Temperature of Selected
Animals

	Average (°C)	Normal range (°C)
Man	36.9	36.2–37.6
Dairy cow	39.2	38.2–41.0
Elephant	36.4	36.2–36.7
Cat	38.0	37.2–39.0
Sparrow	41.5	37.3–43.5
Cardinal	41.5	41.0–42.5
Chicken	41.4	41.0–42.5
Polar bear	37.4	37.0–38.0
Humpback whale	38.2	38.0–38.4

animals. There are some small fluctuations in daytime and nighttime temperatures. Also, there are slight variations in individuals, but even so each species has a very narrow temperature range. Birds tend to have somewhat higher temperatures than mammals. Some warm-blooded animals hibernate and have a built-in device for lowering body temperatures in winter. As we shall soon see, this is a means of conserving available energy. When this happens, normal activity is greatly slowed and the animal lives on its stored energy—its fat. This phenomenon of lowering body temperature to conserve energy is simply a reflection of a basic thermodynamic principle that rates of chemical reaction decrease with temperature. Since all biological systems live by chemical processes, the temperature of the cell or complex organism determines the rate at which vital processes take place.

Just by observing yourself, you can see and feel the profound and dramatic effect of temperature on body processes. One characteristic of many diseases is elevation of body temperature. So when you become ill you are concerned about your temperature. If you go to see a physician, he will take your temperature and note your pulse rate (rate of heart action) and increased frequency of breathing. But why do these changes occur? They are simply a reflection of the fact that your energy requirements are increasing, and so your heart must pump more blood and you must consume more oxygen per unit of time. As your body tries to accommodate to increased energy requirements, you will feel cold and may shiver to generate more heat. When your energy requirements drop, you will feel hot and begin to sweat. The amount of increase in heat produced and in oxygen requirement indicates just how much effect temperature has on these vital chemical processes. For every degree Centigrade your body temperature goes up, you will require 13% more oxygen and produce 13% more heat energy. (On the Fahrenheit scale, the

Table 4.2 Percentage Increase of Body Energy
Requirement with Temperature

Temperature °C (°F)		Percent of normal
36.9 (98.4)		100
37.9 (100.2)	$100 \times 0.13 + 100$	113
38.9 (102.0)	$113 \times 0.13 + 113$	128
39.9 (103.8)	$128 \times 0.13 + 128$	145
40.9 (105.6)	$145 \times 0.13 + 145$	164
41.9 (107.4)	$164 \times 0.13 + 164$	185

figure is 7.2% per degree.) However, the increase is compounded, like interest rates are, so the real increase is somewhat more rapid than a simple 13% rate. If normal energy requirements at normal body temperature are taken as 100%, the increase in energy requirements with temperature can be represented as shown in Table 4.2. Thus at the modestly elevated body temperature of 102°F, you produce 28% more energy and consume as much more oxygen; at 104°F, 50% more; and at 106°F, 67% more than normal. In view of the relatively large excess of heat, it is understandable that survival is difficult because the body has no effective mechanism for dissipating such excessive heat.[1] Temperature, heat production, and rates of vital processes tend to continue upward until some vital process collapses and death occurs.

When we compare changes with temperature of heat production in our body and a simple physical phenomenon such as gas expansion or pressure, we can readily see that there is a fundamental difference. The data in Table 4.2 indicate that for a 5°C increase in temperature the body produces 85% more heat. If we take a tank (closed volume) of air and represent its pressure as 100 (1.00 atmosphere) and raise its temperature 5°C (from 36.9 to 41.9°C) as above, we will find that the pressure has increased only to 101.6 (1.016 atmospheres). In other words, for the same increase in temperature body heat production goes up 85% while the gas pressure goes up only 1.6%. Generally speaking, the rates of chemical reactions increase 2 to 3 times for every 10°C increase in temperature, whereas physical reactions such as gas expansion change only about 3% in the same interval. Thus it is clear that since biological organisms live by chemical processes, temperature control within certain limits is essential. For an organism as complicated as man, rigid temperature control is obligatory in order to maintain all the integrated complex chemical reactions in proper balance, because each particular reaction responds some-what differently to temperature changes.

[1] The body can get rid of heat only through its surface—skin, lungs, mouth.

There are sound scientific bases for relating temperature to rates of chemical reaction. Detailed treatment of this is beyond our scope. Nevertheless, biologists, chemists, and others use a simplified expression for the temperature coefficient of various processes. The symbol for the temperature coefficient is Q_{10}, and it is defined as follows:

$$Q_{10} = \frac{\text{rate of reaction at temperature } T + 10°C}{\text{rate of reaction at temperature } T°C}$$

When we work this out for individual chemical processes, we find Q_{10} to vary from 2.0 to 3.0, with many values around 2.3. However, there are some important exceptions.

You will recall that we pointed out that many enzymes, the catalysts for most life processes, are proteins. We noted also that proteins are large molecules which themselves are specifically and intricately structured. Most proteins are sensitive to elevated temperatures. It seems that when a giant protein molecule moves too rapidly and has too much kinetic energy, due to high temperature, it can change in a great many ways—any one of which may destroy its original architecture so that its vital catalytic powers are lost. Because proteins are so sensitive to increases in temperature, they have exceedingly high temperature coefficients for what the biochemist calls *heat denaturation*. We see the results of heat denaturation in our ordinary life—the coagulation of egg or meat on cooking and heating or cooking foods to kill organisms, etc. When we determine the Q_{10} for heat denaturation and/or coagulation of protein, we find that in the range of 50–70°C (122–158°F) the Q_{10} may be not 2–3, but rather 10, 20, 50, 200, or more. These extremely high values reflect the great sensitivity of proteins to these temperatures. The practical use of these temperatures, slightly above the biokinetic zone, to kill harmful or pathogenic bacteria was discovered by Louis Pasteur, which is why mild heating (below the boiling point) to kill many organisms competitive with man is called *pasteurization*. Such processes are very important in supplying wholesome milk, meat, beer, and other foods to the public. As we shall see later, heat denaturation of proteins is absolutely essential in many food processes, whether for long-time preservation or for ordinary cooking in the home.

In Table 4.3, a few data are given showing the effects of temperature on a number of chemical and biological processes. These effects are expressed as Q_{10}. Most biological processes appear to be thermochemical reactions from their Q_{10} values. This is true even for the incorporation of molecules into cells. This is called *active transport* across cell membranes (see p. 14) because the Q_{10} values are higher than would be predicted from simple diffusion of salt and sugar through a water solution. Note that the heat killing of organisms appears to parallel heat denaturation and coagulation of protein, and

Table 4.3. Increase in Rate of Reaction with a 10°C Increase in Temperature, Q_{10}

Types of reaction	Reactions in inanimate system	Q_{10}	Temperature range (°C)	Reactions in organisms	Q_{10}	Temperature range (°C)
Thermochemical	Most reactions	2–3				
	Digestion of starch by enzyme (malt amylase)	2.2	10–20	Photosynthesis, high light	1.6	4–30
				Growth of bacterium E. coli	2.3	20–37
				Contraction of muscle (small intestine)	2.4	28–38
	Digestion of casein by enzyme (trypsin)	2.2	20–30	Respiration of sugar beets	3.3	15–25
				Respiration of oranges	2.3	10–20
				Respiration of pea seedlings	2.4	10–20
				Penetration of selected substances into cells	2.4–4.5	10–25
	Protein coagulation:			Heat killing:		
	Egg albumin	625	69–76	Spores	2–10	40–140
	Hemoglobin	14	60–70	Bacteria	12–136	48–59
				Protozoa	890–1000	36–43
Photochemical	Photographic film exposure	1.05	−85–30	Killing of bacteria by ultraviolet light	1.06	5–36
				Photosynthesis, limited light	1.06	15–25
				Bleaching of light-sensitive pigment of the eye	1.0	5–36

Adapted in part from A. C. Giese, Cell Physiology, W. B. Saunders Co., 1957.

remember that the loss of a single vital protein by heat denaturation can kill a functioning organism.

There are some photochemical reactions in biological systems. These are reactions due to the direct absorption of light by a molecule. In this class of reactions, we are familiar with the effect of light on photographic film. The Q_{10} of this reaction is quite small and is analogous to the light bleaching of the isolated photosensitive pigment from the retina of the eye. Note also that the Q_{10} for photosynthesis in plants at very low light intensity indicates that when light is limiting plant growth the rate of photosynthesis does not change much with temperature. When there is plenty of light, the effect of temperature is typical of thermochemical reactions. This means that the sequential reactions of ATP and sugar production of the overall photosynthesis process are thermochemical even though the initial step involving light is photochemical. Even the amateur photographer can observe an analogous situation. Film can be exposed in hot or cold weather, but the chemical processes of picture development depend on temperature. Just look at the directions for taking "instant" *Polaroid* pictures.

Organisms can live and complete their life cycles only in a narrow range of temperatures. The temperatures at which organisms live profoundly affect most biological processes. Consequently, it follows that temperature manipulation is one of the most important parameters in providing and conserving food. We are all familiar with cooking, canning, refrigeration, freezing, pasteurization, etc. These are uses of temperature to control biological systems for our own consumption.

HYDROGEN ION CONCENTRATION, ACIDITY, pH

Now we will proceed to discuss three more basic environmental factors—the fundamental characteristics of water solutions which control biological activity. The first of these is the hydrogen ion, H^+, concentration. The second is the total number of particles—molecules, ions, or groups of molecules, ions, etc.—which occur in solution. The third is the kinds of molecules, ions, etc., in the water which are required to support the organism. Another very important way of looking at this third characteristic is that the substances dissolved or suspended in the water are the nutrients which the organism must have to live and to complete its life cycle. Study of these substances is the science of nutrition, and each organism, whether man or microorganism, has its own particular set of nutrients. In the next chapter, we will discuss in some detail the nutrients required by man.

In Chapter 2 we noted that water always contains hydrogen ions, H^+ or protons, and that these are important reactants in the formation of water with oxygen of the air to give energy. Hydrogen ions are also important in many other biological situations in that their presence or absence has a profound effect on the architecture and shape of proteins, particularly, and other large molecules which make up functioning organisms. Since hydrogen ions affect proteins and enzymes (which are proteins), the concentration of protons, H^+, has a great effect on rates of biochemical processes. Now let us proceed to see what are the effects of the concentration of these important ions always present in water. First, we need to discuss a few simple principles of chemistry that will help us understand better not only the role of hydrogen ions in biological systems but also many other aspects of man's requirements in order to feed himself adequately.

Let us begin with the nature of acids and their opposites, bases, which control the number of hydrogen ions in water. If we dissolve a molecule in water and protons or H^+ ions are produced, we call this molecule an *acid*. Acids generally taste sour. A strong acid is one which gives a high percentage of its ionizable hydrogen as H^+. A weak acid is one which gives a low percentage of its ionizable hydrogen as H^+. For example, when the gas hydrogen chloride, HCl, dissolves in water it ionizes thus:

$$HCl \xrightleftharpoons{H_2O} H^+ + Cl^- \qquad (1)$$

| hydrogen chloride | | proton, or hydrogen ion | | chloride ion | |

Because most of the hydrogen of HCl when dissolved in water occurs as protons, H^+ (and also most of the chlorine occurs as Cl^-), the water solution of HCl, known commonly as hydrochloric acid, is a strong acid.

Acetic acid, the acid of vinegar, like many of the acids of metabolic importance already mentioned in Chapter 3, is a weak acid because only a small number of available hydrogens in the molecule occur as protons, H^+, in water solutions. This can be represented by

$$CH_3COOH \rightleftharpoons CH_3COO^- + H^+ \qquad (2)$$

acetic acid   acetate ion   proton

Certain substances can react with water to form acids or bases. An important example is carbon dioxide, which forms with water a weak acid, carbonic acid:

$$CO_2 + H_2O \rightleftharpoons H_2CO_3 \rightleftharpoons H^+ + HCO_3^-$$

carbon dioxide   water   carbonic acid   proton   bicarbonate ion

The bicarbonate ion itself can yield another proton as follows:

$$HCO_3^- \rightleftharpoons H^+ + CO_3^{2-}$$

bicarbonate ion   proton   carbonate ion

We are familiar with carbonic acid in carbonated water of soft drinks. The salts (see below) of carbonic acid are seen as sodium bicarbonate, $NaHCO_3$, baking soda; calcium bicarbonate a mineral in so-called hard water; sodium carbonate, sold in stores as washing soda; and calcium carbonate, which occurs as marble, limestone, chalk, and as the shells of eggs, oysters, etc. Thus when a weak acid dissolves in water, relatively few protons are formed.

Now let us take a similar look at bases. Being the opposite of acids, they reduce the concentration of hydrogen ions in water or produce something which reacts with protons to remove them. A strong base is sodium hydroxide or caustic soda, $NaOH$. Since water is H_2O, or HOH, $NaOH$ may be thought of as water with a sodium atom in place of a hydrogen. In water, sodium hydroxide gives

$$NaOH \rightleftharpoons Na^+ + OH^- \qquad (3)$$

$$\text{sodium} \qquad \text{sodium ion} \qquad \text{hydroxyl}$$
$$\text{hydroxide} \qquad\qquad\qquad \text{ion}$$

The large amount of hydroxyl ions, OH^-, tends to remove protons. In water, ammonia gas, NH_3, reacts with water to give ammonium hydroxide, NH_4OH, thus:

$$NH_3 + H_2O \rightleftharpoons NH_4OH \qquad (4)$$

$$\text{ammonia} \qquad \text{water} \qquad \text{ammonium hydroxide}$$

Ammonium hydroxide does not ionize as strongly as sodium hydroxide does and so it gives a relatively lower concentration of ions:

$$NH_4OH \rightleftharpoons NH_4^+ + OH^- \qquad (5)$$

$$\text{ammonium} \qquad \text{ammonium} \qquad \text{hydroxyl}$$
$$\text{hydroxide} \qquad \text{ion} \qquad\quad \text{ion}$$

Note that in this instance not only OH^- is produced but also the ammonia, NH_3, effectively ties up protons to form NH_4^+ and so depresses acidity.

When an acid reacts with a base, a salt is produced. Thus if we take solutions of hydrochloric acid and sodium hydroxide and mix them, a rapid reaction results:

$$H^+Cl^- + Na^+OH^- \longrightarrow Na^+Cl^- + H_2O \qquad (6)$$

$$\text{hydrochloric} \qquad \text{sodium} \qquad \text{sodium} \qquad \text{water}$$
$$\text{acid} \qquad \text{hydroxide} \qquad \text{chloride}$$

Sodium chloride is common table salt, without which we cannot live. In addition to giving sodium chloride and water, this reaction releases energy, which is liberated as heat because the water is, for the most part, un-ionized. However, as we shall see shortly, water is always ionized to a slight degree, and this is essential for life.

Now although acetic acid and ammonium hydroxide are a weak acid and a weak base, respectively, because they ionize only weakly in water, they do indeed react with each other to form another salt:

$$NH_4^+OH^- \quad + \quad CH_3COO^-H^+ \longrightarrow CH_3COO^-NH_4^+ + H_2O$$

| ammonium hydroxide (household ammonia) | acetic acid (vinegar) | ammonium acetate (a salt) | water |

$$(7)$$

In this reaction, even though the ammonium hydroxide and acetic acid are themselves poorly ionized, the proton (H^+) and hydroxyl ion (OH^-) are so reactive that as water is formed the NH_4OH and CH_3COOH continue to ionize until the reaction is essentially complete. Here again the salt, ammonium acetate, is largely ionized in water solution, but the water for the most part is not. This fact that water itself is poorly ionized is a driving force for many chemical reactions which form water.

You should note that the equations (6) and (7) resulted from the reaction of hydrogen ions and hydroxyl ions:

$$H_2O \; \xrightleftharpoons{\quad} \; H^+ \quad + \quad OH^- \qquad (8)$$

| water | hydrogen ion | hydroxyl ion |

The equilibrium for this reaction is far to the water side. This does not mean, however, that there are no hydrogen or hydroxyl ions in water. Indeed, the reactions for ammonia and water (equations 7 and 8) clearly show that some hydrogen ions must have been present to react to form the ammonium ions. And if there were hydrogen ions to form the NH_4^+ there had to be hydroxyl ions, OH^-, as well, which must have come from water.

Equation (8) indicates that for pure water, although most of it is not ionized, there are a small number of hydrogen and hydroxyl ions present, and furthermore that in pure water for every hydrogen ion there is a hydroxyl ion. An equilibrium exists between H^+, OH^-, and H_2O. Now if we dissolve anything in water which gives hydrogen ions, such as hydrochloric acid, acetic acid, or carbon dioxide, the hydroxyl ion concentration in water must be depressed in order for equilibrium (8) to be maintained. Conversely, if we add anything to water to depress hydrogen ions, such as hydroxyl ions from sodium hydroxide, ammonia, or sodium bicarbonate (baking soda), we automatically increase hydroxyl ion concentration. Thus the equilibrium concept of water represented in equation (8) is fundamental. Since all biological systems require water to live, they do indeed depend on the concentration of hydrogen and hydroxyl ions. Many organisms have elaborate systems for control of these ions. But before dwelling further on this point, we must introduce a means for quantitizing the amounts of these ions in water.

Fortunately, equation (8) permits a simplification because for any equilibrium such as 8 if we can measure two factors we can compute the remaining one. Because almost all the water is not ionized, if we measure hydrogen or hydroxyl ion concentration we automatically measure the other. As these concepts evolved concerning the nature of water, it became easier to determine hydrogen ion concentration; consequently, hydrogen ion concentration only is used for most applications. Let us now look to see a further simplification, which introduces a new term, *pH*. pH is a scale of hydrogen ion concentration analogous to the degree scale used to define temperature.

It is beyond our scope to delineate all the theory and how the measurements are made to prove equation (8). But in order to make the simple term pH comprehensible, we must begin by considering a few conventions. This will make not only the concept of pH easier to understand but also many other parts of this and the following chapters.

In science, as in other areas, certain conventions are arbitrarily adopted to give order and meaning to communication. In this country, we drive our cars on the right-hand side of the road. We read English from left to right. In other cultures, the reverse conventions can be found. In the science of chemistry, a few conventions of quantitation or measurement have been developed as a starting point toward understanding how substances relate to each other.

Atoms have both size and weight and are the basic units of matter. They are made up of protons, neutrons, and electrons, and most of their weight is in the nucleus, which is made up of protons and neutrons. The electrons are in external orbits around the nucleus. The number of electrons, each of which carries a single negative electric charge, is equal to the number of protons, which have a positive charge, so atoms themselves are electrically neutral. The simplest atom is hydrogen, H. It has a nucleus of a proton, H^+ or hydrogen ion, and one orbital electron. Heavy hydrogen or deuterium has one proton and one neutron in its nucleus and one orbital electron. (Neutrons have no electrical charge, as their name implies.) A proton weighs 1.674×10^{-24} grams and an electron weighs 9.107×10^{-28} grams or 1/1837 of the weight of the proton. A neutron weighs 1.676×10^{-24} grams. As you can see, these are all very small and to use such cumbersome figures is difficult.

Atoms combine in definite proportions to form molecules. This is accomplished by electron giving and receiving and/or sharing in order to have stable electronic configurations. An atom such as sodium has a single electron in its outer orbit that it can give away to form a sodium ion, which, although positively charged, is stable. An atom such as chlorine needs one additional electron to form a more stable structure, although in doing so it may become a strong negative ion:

$$Na \longrightarrow Na^+ + e^- \tag{9}$$

$$Cl + e^- \longrightarrow Cl^- \tag{10}$$

Thus when metallic sodium reacts with chlorine gas, these reactive materials form sodium chloride or common salt. By adding equations (9) and (10), we get

$$Na + Cl \longrightarrow Na^+Cl^- \tag{11}$$

The definite proportion is that one atom of sodium reacts with one atom of chlorine to give one molecule of sodium chloride composed of a positive sodium ion and a negative chloride ion. Thus ions, electrically charged atoms or molecules, are formed when one atom or molecule gives or receives electrons. Such compounds[2] are called *polar*.

But atoms may also share electrons to form molecules. This is very common with hydrogen, carbon, oxygen, nitrogen, phosphorus, etc., which make up biological systems. Carbon needs or gives four electrons for stability in molecules, oxygen needs two or gives six, nitrogen needs three or gives five, etc. Note the theme of eight electrons for stability. This is true for all except hydrogen; its stable electron shell is 0 or 2, reflecting its small size. So it can share electrons as well as form hydrogen ions, H^+, or protons. Such compounds formed from shared electrons are said to be *covalent* or *nonpolar*.

The compound water is, if the dots represent electrons,

$$H \colon \ddot{O} \colon H \qquad H_2O$$

Similarly methane, natural gas or the gas produced by many bacteria, is

$$\begin{matrix} H \\ H \colon \ddot{C} \colon H \\ H \end{matrix} \qquad CH_4$$

Carbon dioxide is

$$\ddot{O} \colon\colon C \colon\colon \ddot{O} \qquad CO_2$$

Ammonia is

$$\begin{matrix} \ddot{} \\ H \colon \ddot{N} \colon H \\ H \end{matrix} \qquad NH_3$$

We can even have carbon joined with itself and with other elements as in acetic acid:

$$\begin{matrix} & :\ddot{O}: \\ H & \colon\colon \\ H \colon \ddot{C} \colon C \colon O \colon H \\ H \end{matrix} \qquad CH_3COOH$$

[2] The term *compound* is used in chemistry to denote a substance composed only of the same molecules. Thus water is a compound because it is made up only of molecules containing two atoms of hydrogen and one of oxygen—H_2O. The term *mixture* denotes a substance composed of unlike molecules. Thus when sugar, a pure compound, is dissolved in water, a pure compound, the resulting solution is a mixture. Flour is a mixture of all the compounds (kinds of molecules) which make up the milled wheat. The term *element* is used to denote a substance composed of only one kind of atom. Thus hydrogen, oxygen, carbon, etc., are elements. Elements refer to atoms, compounds refer to molecules.

Chemists often use — to indicate a pair of electrons shared by two atoms in a molecule. Thus acetic acid is

$$
\begin{array}{ccc}
\text{H} & \text{O} \\
| & || \\
\text{H—C—C—O—H} \\
| \\
\text{H}
\end{array}
$$

or more simply, CH_3COOH. (The latter formula is more concise and therefore easier to use. Such notation introduces no ambiguity for simple compounds. Each carbon atom is written followed by the atoms joined to it.) To the trained chemist, these formulas contain much information. For example, this one shows that the molecule contains two carbon atoms, four hydrogen atoms, and two oxygen atoms. The formula also tells how the atoms are arranged and what the molecule's properties are and to what other molecules or compounds it is related. Further, the formula shows the structure

$$
\begin{array}{c}
\text{O} \\
|| \\
\text{—COOH(—C—O—H)}
\end{array}
$$

characteristic of organic[3] acids, and the chemist knows that the terminal hydrogen may ionize in water solution as shown in equation (2). Going on, since the weights of the carbon, hydrogen and oxygen atoms are known and are 20.04×10^{-24} gram, 1.674×10^{-24} gram, and 26.55×10^{-24} gram, the total weight of the acetic acid molecule is

$$
\begin{array}{l}
2C = 0.00000000000000000000004008 \text{ gram} \\
4H = 0.00000000000000000000000670 \text{ gram} \\
2O = \underline{0.00000000000000000000005310 \text{ gram}} \\
 0.00000000000000000000009988 \text{ gram}
\end{array}
$$

You will agree that such figures are so cumbersome as to be almost impossible to use. So a convention was developed of atomic and molecular weights in which the weight of the most common isotope[4] of carbon was assigned the value 12.00. On this basis, the weight of a hydrogen atom is

---

[3] Note the word *organic*. In the early days of chemistry, it was thought that the chemistry of biological organisms was different from that of minerals or nonliving matter. As the science developed, the same principles were found to apply to the chemistry of all things—nonliving and living. But most compounds of protoplasm contain carbon and the chemistry of carbon compounds is now known as organic chemistry, even though much of organic chemistry is unrelated to biological systems.

[4] Because some atoms of an element may have more neutrons than others, atoms of the same element may have different weights. These atoms are called *isotopes* of the element. Chemical properties of atoms are not determined by weight as much as by the protons in the nucleus and the orbiting electrons.

1.0078 and that of oxygen 16.00. With only small error (less than 1%), the atomic weights of the common elements in biological systems are

Carbon	12	Phosphorus	31
Hydrogen	1	Sulfur	32
Oxygen	16	Sodium (Na)	23
Nitrogen	14	Chlorine	35.5

Using these figures, the molecular weight of acetic acid is

2C	24	2 × atomic weight of carbon
4H	4	4 × atomic weight of hydrogen
2O	32	2 × atomic weight of oxygen
	60	molecular weight of acetic acid

You will agree that using the convention of atomic and molecular weights is much simpler.

In Chapter 2 and elsewhere, we have stressed the quantitative nature of the chemical processes in biological systems. We have noted this in equations, also. Atoms and molecules react in definite proportions. One molecule of hydrochloric acid reacts with one molecule of sodium hydroxide (equation 6) or one molecule of acetic acid with one of ammonium hydroxide (equation 7). There are other examples where one molecule can react with one, two, three, or more of one or another compound. Recall that in Chapter 3 we noted that energy was obtained by oxidation of the sugar glucose where one molecule of glucose was oxidized with six molecules of oxygen to give six molecules of water and six of carbon dioxide (see p. 28). Whatever the chemical reaction, a specific proportion of molecules are involved.

Although atoms and molecules are so small that we cannot see one, we can readily observe chemical reactions. In other words, we characterize atoms and molecules by their behavior. How many atoms or molecules are represented in a pound, a ton, or a gram of a substance? The concept and conventions of molecular weight lead directly to an idea first proposed by the Italian chemist Amedeo Avogadro in 1811 in quite a different way. This idea is that the *number* of molecules (atoms) in an amount of a compound (element) equal to its molecular (atomic) weight expressed in grams, pounds, tons, or other units is the same. It is such an extremely useful idea that the term *mole* is used universally. It is the amount of a pure substance equivalent to its molecular weight expressed in grams. For example, 1 mole of water equals 18 grams or 1 mole of acetic acid equals 60 grams.

We might ask how many molecules or atoms are in a mole. The answer is obtained simply by dividing the atomic weight or molecular weight by the actual weight in grams of the particular atom or molecule. Thus there are 6.02×10^{23} (602,000,000,000,000,000,000,000) atoms or molecules in a mole of anything. This number is known as *Avogadro's number* and is very large, indeed, emphasizing once again the small size of atoms and molecules.

Since the molecule is the reactive unit in chemistry, the mole represents the fixed number of the reactive molecules. By using it, we can say, for example, that 1 mole of hydrochloric acid reacts with 1 mole of sodium hydroxide to give 1 mole of water and 1 mole of salt. Now since 36.5 grams, 40.0 grams, 18 grams, and 58.5 grams are the respective weights of a mole, equation (6) becomes

$$H^+Cl^- + Na^+OH^- \longrightarrow Na^+Cl^- + H_2O \qquad (12)$$
$$36.5\ g \qquad 40\ g \qquad\qquad 58.5\ g \qquad 18\ g$$

Note that all matter is conserved; 76.5 grams of reactants gives 76.5 grams of products. The mole represents an easy way to weigh and measure molecules which interact without the cumbersome arithmetic illustrated on p. 46.

Let us now use this idea to see how much carbon dioxide you will produce in oxidizing 1 pound, 454 grams, of glucose (see Chapter 3):

$$
\begin{array}{cccc}
454\ g & 484.3\ g & 665.9\ g & 272.4\ g \\
C_6H_{12}O_6 \ + & 6O_2 & \longrightarrow \quad 6CO_2 \ + & 6H_2O \qquad (13) \\
180 & 6 \times 32 = 192 & 6 \times 44 = 264 & 6 \times 18 = 108
\end{array}
$$

$$\frac{454\ g\ \text{glucose}}{180} = \frac{g\ \text{oxygen}}{192} = \frac{g\ \text{carbon dioxide}}{264} = \frac{g\ \text{water}}{108} \qquad (14)$$

(mol wt glucose) (6 × mol wt oxygen) (6 × mol wt carbon dioxide) (6 × mol wt water)

Thus in oxidizing the pound of glucose, you will use about 1.06 pounds of oxygen and produce 1.46 pounds of carbon dioxide and 0.6 pound of water. In ordinary terms this is slightly over a pint of water. Since air is only one-fifth oxygen, you will use approximately 500 gallons or 60 cubic feet of air and exhale 100 gallons of carbon dioxide gas—enough to carbonate 40–50 gallons of soft drinks.

Now that we have an idea of what the mole is, and its usefulness, let us return to our consideration of the very important environmental condition of water which must be controlled to permit life to proceed—the hydrogen ion concentration and pH.

We have already noted in equation (8) that there are always some hydrogen and some hydroxyl ions in water and that by measuring one we can compute the other. This is because when one of the ions increases the other decreases and vice versa. How many moles of each ion are present at any one time? A very small number! It can be shown experimentally that in a mole of pure water there is only 0.0000001 or 1×10^{-7} mole of H^+ or OH^-.

As acidic substances are added to water, the hydrogen ion concentration

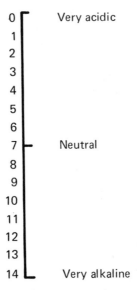

0	Very acidic
1	
2	
3	
4	
5	
6	
7	Neutral
8	
9	
10	
11	
12	
13	
14	Very alkaline

Fig. 4.1. pH scale.

will increase from this value of 1×10^{-7}, and as basic substances are added the hydrogen ion concentration will decrease from 1×10^{-7}. For example, the H^+ concentration in tomato juice may be 0.0002 or 2×10^{-4} and in drinking water it may be 0.00000001 or 1×10^{-8}. These numbers, as in the case of actual weights of molecules, are almost too cumbersome to use. They are not only small but in relation to each other these numbers differ in size by many orders of magnitude. The tomato juice has 20,000 times more H^+ ions than the drinking water. However, if the H^+ concentration is expressed as the negative logarithm or exponent of 10 rather than in moles of H^+ directly, we get an easily usable scale for expressing hydrogen ion concentration. This is the pH, defined as the negative logarithm of the concentration of hydrogen ions in water:

$$pH = -\log[H^+]$$

Using this device, our tomato juice has a pH of 3.7 and the drinking water a pH of 8.0.

Because of its definition as a logarithm, the pH scale is limited to between 0.0 and 14.0, and also there are some limitations on its use near 0.0 and 14.0. This aspect is of little concern to us since most biological phenomena take place between pH 1.5 and 11.5, and most organisms live in even narrower ranges of pH, as shown in Tables 4.4 and 4.5. The pH scale is illustrated in Fig. 4.1 and is an extremely useful method of quantitizing the acidity (or alkalinity) of water solutions.

Table 4.4. pH Values of Environmental Media and Some Organisms
Found Therein

Type of habitat	pH range	Medium	Characteristic organisms
Very acid	2.0–3.0	Organic	The fungus *Merulius lacrymans*
	3.2–4.6	Peat bogs	*Sphagnum* and *Drepanocladus* mosses
	3.4	Vinegar	Vinegar eels (nematodes), vinegar bacteria
	3.2	Soil water	High moor vegetation
Somewhat acid	5.2	Sour milk	Lactic acid bacteria
	4.7–5.8	Distilled water, Fresh rain	(Slightly acidic due to dissolved carbon dioxide from air)
Somewhat acid to alkaline	5.2–9.0	Soil water	Peat pit mosses
	6.2–9.2	Neutroalkaline lakes	Mosses at edge
	4.5–8.5	Soil water in most soils	Various forest and field plants
Nearly neutral	6.8–7.0	Freshly boiled (cooled) distilled water	
	6.5–8.0	Drinking water	
	6.8–8.6	River water	River plants and animals
	6.2–8.2	Springs	Algae around springs
	7.8–8.6	Seawater	Marine plants and animals
Alkaline	7.5–9.1	Soil water	Greasewood, rabbit bush, salt and brome grass growing on alkaline flats
	9.3–11.1	Organic medium	Several species of the common *Penicillium* and *Aspergillus* molds

Adapted in part from A. C. Giese, *Cell Physiology*, W. B. Saunders Co., 1957.

As we relate this scale to biological environments, a reasonable question to ask is how pH is determined. The pH concept and scale were proposed by the Danish biochemist S. P. L. Sorensen, in 1909. Accurate measurement of pH in the early days was tedious and difficult. Suitable, easy to use methods did not become readily available until just before World War II. This is in spite of the fact that the essential principle of pH measurement is quite simple to understand. You are familiar with batteries used in autos, flashlights, radios, etc. All such batteries get their energy from chemical reactions. Now imagine a battery where the voltage of the battery is determined by the H^+ concentration of the solution placed between its two poles or electrodes. By putting a highly sensitive voltmeter across the electrodes of such a battery, we

Table 4.5. pH Values of Some
Biological Fluids

Fluid	pH range
Blood, human	7.30–7.45
Cerebrospinal fluid, human	7.35–7.40
Blood, body fluids, other animals	6.7–7.9
Milk, human	6.4–7.6
Milk, cow	6.4–6.8
Saliva, human	6.4–7.6
Saliva, cow	8.3–8.5
Gastric contents, human	1.5–3.0
Duodenal contents, human	4.5–8.0
Urine, human	4.8–8.0
Apple juice	2.9–3.3
Lemon juice	2.2–2.4
Lime juice	1.8–2.0

can determine pH of a solution simply by measuring the voltage produced when we place our solution between the electrodes.

Now let us proceed to observe the role of pH in biological systems. Table 4.4 gives some data on the limiting pH values of the aqueous medium which permit growth of various organisms. Many microorganisms and other lower forms have limited or wide ranges of pH in which they may live. Higher organisms tend to be more fastidious in their pH requirements. Also, when we observe the pH of cells and specific tissues of higher multicellular organisms, we find usually narrow ranges of tolerance. Normal human blood is rigidly controlled between pH 7.30 and 7.45, and if pH of blood drops as low as 7.0 or goes as high as 7.75 death is not long delayed. Since pHs are logarithms, in essence what we are saying is that man cannot tolerate either the doubling or the halving of the normal limits of hydrogen ion concentration of the blood. One of the fundamental stimuli that control our rate of breathing is the pH of our blood. When we work more or run, we produce more carbon dioxide, which is acidic and tends to depress blood pH. We breath faster and eliminate more CO_2. Deficiency of oxygen is another stimulus for breathing.

Table 4.5 gives an idea of the range in pH values in some fluids of biological origin. Note the very narrow range of pH permitted for cerebrospinal fluid, which bathes vital tissues as does blood. Even the fluid expressed from plant cells has a narrow range within a particular tissue. The fluids secreted by higher animals have a wider range of pH values, and this may reflect the compensatory response to food consumed while maintaining much more rigid control of the pH of tissues for proper functioning.

The few data of Tables 4.4 and 4.5 could be expanded with almost an infinite number of examples of the importance of pH to the growth of organisms. Without going further, we can conclude that pH is a fundamental factor in defining the total environment for any single-celled or multicellular organism. pHs near the neutral point favor the greatest variety of organisms, but even in the more acidic or alkaline ranges some organisms may live. pH adjustment and control are necessary to the growth of any organism, and man has done this consciously, although more often unknowingly, as he has developed cultivation techniques for his food organisms. We shall see later that pH can also be manipulated to kill organisms competitive with man and to preserve those organisms or parts of organisms we need for food.

OSMOTIC PRESSURE, WATER ACTIVITY

The second characteristic of water solutions basic to life is osmotic pressure. The phenomenon of osmotic pressure of solutions was discovered in 1877 by the German botanist Wilhelm Pfeffer, who showed its importance in plants, particularly with reference to water movements, and related it to the concentration of dissolved substances in water. The osmotic pressure of a solution is directly related to its freezing point and boiling point. These three properties of any solution are due, not to what is dissolved, but only to the *total number* of ions and molecules dissolved in the water. Of course, life does not exist in boiling water or ice, even though boiling point and its elevation and freezing point and its depression are necessary in preserving food and are basic to food processing and storage. But osmotic pressure itself, even though important in food preservation and storage, is fundamental in all living things.

You will recall that all cells live in water, and that all cells have an external membrane through which water and other substances, nutrients and metabolites,[5] must pass. Many such membranes will permit water—a small molecule, H_2O, with molecular weight 18—to pass in both directions, in and out of the cell. Other substances in the cell are usually larger molecules and cannot readily pass through the membrane by diffusion. Such being the case, the cell membrane acts as a semipermeable membrane or one-way street for many substances, but not for water, which can diffuse in and out as needed. A crude analogy might be a bird in a cage, where the bird is confined but air and small things such as insects can move in and out of the cage. Another analogy with which laymen are more familiar is that a room or theater

---

[5] *Metabolite* is a term which denotes a chemical compound or element (molecule or atom) involved in the vital processes of a living organism. These processes are referred to as *metabolism*.

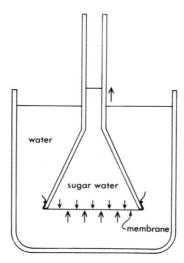

Fig. 4.2. Effect of osmotic pressure.

assumes certain characteristics such as echo, ease of movement, etc., according to whether they are crowded or contain only a few people irrespective of whether all the occupants are lawyers, football players, men, or young girls or a grand mixture of all types of people.

The osmotic pressure of a solution can be illustrated by placing a sugar solution in a vessel that has one wall which will allow water to diffuse in both directions but will not allow substances dissolved in the water to pass. Some types of cellophane or other films or sausage casings could serve this function as a semipermeable membrane. Figure 4.2 shows a container of water into which is inserted an inverted funnel containing sugar solution. A semipermeable membrane is sealed across the face of the funnel. Water will begin to diffuse into the funnel, raising the level of sugar solution in the neck of the funnel, until the weight of the rising column of sugar solution increases the pressure inside the funnel to the point that the same number of water molecules are going out of the funnel through the pores of the membrane as come in by diffusion. You may visualize the membrane as a sieve. On one side are tiny rubber balls covering some of the holes, while on the other side of the holes there is only air. Through each hole that is uncovered air can move in and out, but each covered hole acts like a valve on a tire—air can go in but not out. Osmotic pressure is analogous to that increase in pressure necessary to drive the same amount of air in the same amount of time through the available holes on the partially covered side of the sieve as can be driven through all the uncovered holes on the other side. Figure 4.3 shows a cross-section of a sieve with 12 holes, three of which are covered with rubber balls. Air may be blown

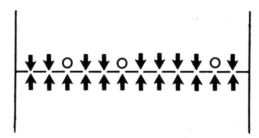

Fig. 4.3. Sieve model of osmotic pressure.

upward through all 12 holes, even though three are covered, because in this direction the air can blow the balls off the holes. But the air can be blown down through only nine of the 12 holes because three are blocked by rubber balls. Thus to blow a fixed amount of air through in a given length of time, more pressure would be required for the downward direction than for the upward. This increase in pressure is analogous to osmotic pressure.

We are all familiar with the phenomenon of osmosis. We like the lettuce and celery in our salads to be crisp, and if they are limp or semiwilted we can restore the crispness by placing them in water. Since most of the cell contents cannot leave the cell but water can diffuse in, the cells become engorged and the lettuce or celery becomes turgid or crisp. If to our crisp salad we add a salad dressing containing vinegar, salt, and sugar, the salad will soon become limp and sloppy. This is because the osmotic pressure of the dressing is higher than that of the cells of lettuce and celery, which causes water to leave the cells and dilute the dressing. Figure 4.4 shows schematically a turgid plant cell and a shriveled one. Soaking in water for an extended time may eventually damage the cells, but remember that we noted in discussing Fig. 3.1 of Chapter 3 that plant cells have a tough wall which gives strength to withstand greater ranges of osmotic pressure. This is not true of animal cells, which generally are much more sensitive to differences in osmotic pressure. Figure 4.5 shows, in side view, a normal red blood cell, a swollen one ready to burst

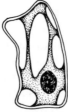

Turgid   Shriveled

Fig. 4.4. Turgid and shriveled plant cells.

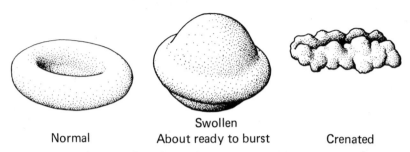

Normal   Swollen / About ready to burst   Crenated

Fig. 4.5. Effect of osmotic pressure on red blood cells.

because osmotic pressure on the outside is too low, and a shriveled or crenated one caused by too high osmotic pressure. When we wash a fresh wound with water, it hurts, but if we wash it with a salt solution of the same osmotic pressure as blood (such a solution is called physiological saline) it will not hurt as much. It is imperative that fluids injected into our bloodstream or muscles for medical reasons be at the same osmotic pressure as blood.

How much pressure are we talking about? We have already said that osmotic pressure along with freezing point lowering and boiling point elevation are due solely to the number of ions or molecules in solution. When 1 mole of substance (defined in the previous section) is dissolved in 22.4 liters of water at 0°C, the osmotic pressure of the solution is 1 atmosphere or 15 pounds per square inch. If the 1 mole is dissolved in 1 liter (slightly more than 1 quart), the solution would have an osmotic pressure of 22.4 atmospheres or 336 pounds per square inch, which is equal to a column of water (see Fig. 4.2) 672 feet high. What is the osmotic pressure of your blood? Your blood *must* be kept at a concentration of 0.289 ± 0.008 mole of dissolved material per 1000 grams of water. So the osmotic pressure of your blood is 0.289 × 22.4 or 6.47 atmospheres or approximately 95 pounds per square inch. At body temperature, it is somewhat higher—about 105 pounds per square inch. If you do not maintain proper osmotic pressure, you cannot maintain the proper fluid balance in your vital tissues. Edema or swelling is one symptom of such a physiological difficulty. Your sense of thirst helps control your water and osmotic pressure balance.

The osmotic activity varies more in plant tissues. There are daily fluctuations in foliage and differences in foliage and root tissue. In microorganisms, the osmotic pressure is a very well-defined condition for growth. If other nutrients are present, many microorganisms can thrive on media of various osmotic pressures, but for each species there is a limit beyond which life is impossible. As we shall see in Table 4.7, bacteria tend to grow where the osmotic pressure is less, generally yeasts can tolerate higher osmotic pressures, and some fungi are quite formidable in growing in media of extremely high osmotic pressure.

Let us return for the moment to the general consideration that osmotic pressure, freezing point, and boiling point of a solution are all basically related to a single parameter—the number, not the kinds, of molecules and ions dissolved in the water. Let us now see if we can find a simple quantitative way of expressing this relationship. After all, we have only two factors to consider—number of water molecules or moles of water and number of dissolved molecules or moles of dissolved matter.

First let us say that pure water has an *activity* of 1.000—something like a perfect batting average of 1000 in baseball. The activity of water with any dissolved matter would be something less than 1.000. For example, let us take 1000 grams of water, a convenient amount, and in this dissolve 1 mole of material. The activity of the water, denoted as a_w, is calculated as follows: 1000 grams of water of molecular weight 18 (mol wt of $H_2O = 2 \times 1 + 1 \times 16 = 18$) is 55.5 moles, and 55.5 moles of water plus 1 mole of dissolved substance gives a total of 56.5 moles. The *water activity*, defined as the relative number of water molecules to the total number of molecules present, is 55.5/56.5 or 0.982. This value is the water activity for any solution where 1000 grams of water dissolves 1 mole of any material such as glucose, citric acid, or salt. However, since salt ionizes in water into Na^+ and Cl^-, giving in a sense two particles from one molecule, $\frac{1}{2}$ mole of salt has the same osmotic effect as 1 mole of sugar, which does not ionize. Another way at looking at water activity, a_w, is that if we multiply a_w by 1000 we will have the number of water molecules in every thousand molecules of the solution.

In Table 4.6, the activities of water are assembled in decreasing order. With each activity, the next column gives the ideal or theoretical number of moles of dissolved material corresponding to the activity; next the actual number of moles of salt, sodium chloride (molecular weight 58.5), or sugar, sucrose or cane sugar (molecular weight 342), to produce the activity; next the corresponding weight percentage of salt or sugar generally used in foods, followed by the osmotic pressure, the boiling point, and the freezing point of the solution. You will note some discrepancy between ideal and actual values. This is because water can, under certain circumstances, attach so strongly to certain ions or molecules that it does not act to dissolve other ions or molecules.

Particular attention is called to the a_w of human blood, which, like that of the blood of most other higher animals, must be controlled quite closely and may not fluctuate very much from this value of 0.995. This is essentially the water activity in all vital tissues. Remember that a water activity of 0.995 actually means that 995 of every 1000 molecules are water. In other words, the remaining 5 molecules of every 1000 represent all manner of substances

Table 4.6. Water Activity, Osmotic Pressure, Boiling Point, and Freezing Point of Various Concentrations of Salt and Sugar

Water activity (a_w)	Moles of dissolved material in 1000 g (55.5 moles) of water			Weight percent		Osmotic pressure[a] (atmospheres)	Boiling point[b] (°C)	Freezing point (°C)
	Ideal	Salt	Sugar	Salt	Sugar			
1.000	0	0	0	0	0	0	100.00	0.00
0.995c	0.281	0.147	0.272	0.857	8.50	6.29	100.16	−0.52
0.990	0.566	0.304	0.534	1.75	15.45	12.7	100.29	−1.05
0.982	1.00	0.550	0.91	3.11	23.75	22.4	100.52	−1.86
0.980	1.13	0.618	1.00	3.50	26.0	25.3	100.59	−2.10
0.965	2.00	1.10	1.69	6.05	36.6	44.8	101.04	−3.72
0.960	2.31	1.27	1.92	6.92	39.6	51.8	101.20	−4.30
0.940	3.54	1.91	2.72	10.0	48.2	79.3	101.84	−6.58
0.920	4.83	2.56	3.48	13.0	54.4	108.2	102.51	−8.98
0.900	6.17	3.16	4.11	15.6	58.4	138.3	103.21	−11.47
0.850	9.80	4.63	5.98	21.3	67.7	219.5	105.09d	−18.23
0.800	13.9	—e	—e	—e	—e	311.0	107.22	−25.8
0.750	18.5	—e	—e	—e	—e	414.0	109.62	−34.4
0.700	23.8	—e	—e	—e	—e	533.0	112.38f	−44.3
0.650	30.0	—e	—e	—e	—e	672.0	115.60	−55.8

a At 0°C. Multiply atmospheres by 15 to get pressure in pounds per square inch.

b At 1 atmosphere or 760 mm of mercury.

c Water activity of human blood is 0.99483 ± 0.00014.

d Approximate boiling point for jellies and jams, 220°F.

e No longer soluble at room temperature. But these lower a_w are possible by using more soluble substances such as calcium chloride and glycerin (glycerol) or by heating to increase solubility of sugar.

f Approximate boiling point for cooking home made fudge or fondant, 234–238°F.

necessary for life. Even casual observation of the amount of water in muscle (lean meat) will indicate that about 80% of the weight of muscle is water and 20% is other material, mostly protein. Why the discrepancy that 0.5% of all molecules are not water but 20% of the weight is not? The answer is that water is a very small molecule compared to most of the other molecules. Whereas a water molecule weighs 18, a protein molecule may weigh 50,000, 100,000, 500,000, or more. So it is understandable that all the molecules present in tissue, even allowing only 5 out of 1000 to be something other than water, permit infinite biological variations. Returning to muscle, 100 grams of muscle contains 80 grams of water, or 80/18 moles of water, or 80/18 × $6.02 \times 10^{23} = 2.68 \times 10^{24}$ water molecules. If this is 995 out of every 1000 molecules, then there are 1.34×10^{22} molecules of other kinds. You will agree that 13,400,000,000,000,000,000,000 is a formidable number of molecules at the disposal of the muscle to perform its function. Indeed, it is these molecules more than the water that are studied in search of answers to many biological (including medical) problems. In the following chapters, much of our attention will be focused on those nonwater molecules of biological systems, for they are literally the food on which we depend.

When, for medical reasons, we wish to bypass all of those organs which serve to maintain an internal cellular environment (gastrointestinal tract and liver, particularly) and inject fluids directly into the bloodstream or tissues, it is essential that the material injected have an a_w of 0.995. This means that if fluid is needed, physiological saline, 0.85% salt, is used; if calories are needed, 5.00% glucose is used, etc. Pharmaceutical preparations are usually similarly adjusted with respect to water activity.

Among the microorganisms in the world are many helpful ones and many competitive ones to us and our food. Water activity of the material on which they live is often critical. Usually there is a limit of a_w below which organisms cannot grow, mature, and reproduce. In Table 4.7 are listed several kinds of microorganisms important to us and the wholesomeness of our food supplies. While some organisms are competitive in that they may grow on or even kill the organisms we need for our food or may grow on food to make its flavor or color unpalatable, there are other organisms that can make our food toxic or even lethal. *Clostridium botulinum* is one of the most dangerous. It is a rather ubiquitous type of soil bacterium which grows in the absence of oxygen and forms spores. If this bacterium or its spores are not killed in the process of canning foods on which it is likely to occur, it can produce, in the absence of oxygen within the can, one of the most lethal toxins known. *Clostridium botulinum* does not appear to tolerate an a_w below 0.950. From Table 4.6, we can determine that this organism cannot live in salt concentrations greater than 8% (too salty for most foods) or sugar concentrations of 45%. Table 4.7 also shows that many yeasts can thrive at a_w values lower than most bacteria

Table 4.7. Minimum Water Activity (a_w) for
Growth of Some Important Food
Microorganisms

	Minimum a_w
Groups of organisms	
Most spoilage bacteria	0.91
Most spoilage yeasts	0.88
Most spoilage molds	0.80
Salt-tolerant (halophilic) bacteria	0.75
Dry-tolerant (xerophilic) molds	0.65
Dry-tolerant (osmophilic) yeasts	0.60
Specific organisms	
Pseudomonas	0.97
Achromobacter	0.96
Escherichia coli	0.96
Bacillus subtilis	0.95
Clostridium botulinum (spores)	0.95
Bacillus cereus	0.94
Salmonella	0.93
Staphylococcus aureus (anaerobic)	0.90
Staphylococcus aureus (aerobic)	0.86
Aspergillus niger	0.85
Saccharomyces rouxii	0.62

can tolerate. Many molds can tolerate very low activities. For example, the common black mold *Aspergillus niger* can grow on media equivalent in water activity to 65% sugar or 18% salt.

So, like pH, there is a simple and meaningful scale to use to define the concentrations of dissolved material which will permit organisms to live. Conversely, we have a scale which can help us understand the food-preserving action of salt, sugar, and alcohol, techniques which have come to us from antiquity.

NUTRIENTS

The third essential factor concerning water solutions is the nature of the substances in water which permit organisms to grow. We call these substances *nutrients*, and the science of nutrition is concerned with the identification and function of those things essential for the growth, development, maturation, and reproduction of a species. Once the nutrients needed by man are identified, it is the responsibility of agriculture to produce the organisms which contain these nutrients and it is the responsibility of the food industry to preserve, process, store, and distribute the food containing the required nutrients.

Fundamentally, for each species of organism there is a particular group of substances classified as nutrients. The nutrient requirements of some organisms are relatively simple; for man, a greater number are needed. Even within species individual strains or genetic groups may have unique requirements. Nutrient requirements sometimes seem to reflect the evolution of the biological species. For example, most animals do not need ascorbic acid or vitamin C because they synthesize their own. However, neither man nor most of the primates nor the guinea pig is capable of doing so. This inability to make ascorbic acid reflects the nature of the food organisms man and other primates and the guinea pig use for food. Fruits and vegetables contain vitamin C.

Detailed treatment of nutrient requirements for just a few organisms would be tedious, although comparative study of the nutrients required by various species of organisms is quite interesting and important. In the next chapter, the nutritional requirements of man will be discussed in some detail. However, there are a few generalizations concerning nutrients which should be kept in mind.

Water and some of its simpler characteristics have been emphasized so far. Some would not classify water as a nutrient, but we will because it is indeed essential for life in any form. Beyond water, there is increasing divergence in the types of nutrients.

There must be a source of energy. Already we have devoted many pages to discussion of energy production and consumption in organisms. The source of energy must be in the form required by the particular organism. For the autotrophs, green plants and other photosynthesizing organisms, the fundamental energy source is light and the source of light which keeps the biological cosmos of this earth operating is the sun. For all other organisms, energy must come directly or indirectly from the oxidation of protoplasmic materials produced by the autotrophs. In other words, the energy source for the autotrophs is light which penetrates the transparent water of the tissues and impinges on certain molecules capable of absorbing or trapping light energy and converting it to chemical energy, which in turn is used to synthesize the nonwater components of the organism (see Chapter 3). The heterotrophs, of which man is one, are those organisms which get their energy from consuming in whole or in part autotrophs or other heterotrophs.

All organisms must have a source of the element carbon, the basic element of most of the compounds of biological importance. For the autotrophs, the carbon source generally is carbon dioxide; for the heterotrophs, it is the carbon-containing compounds of the protoplasmic material of food organisms. Some organisms, and even specific tissues, are more fastidious than others as to the kinds of carbon compounds they can use for energy. For instance, brain cells seem to be able to use only glucose, lactic acid, or pyruvic acid for their energy or carbon source, whereas certain microorganisms use a wide variety of substances—even such hydrocarbons as fuel oil and lubricating oil, or carbon monoxide or cyanide, both of which are quite poisonous for man.

We noted in Chapter 3 that a fundamental process in getting energy from glucose or other substance is the formation of water from the hydrogen (produced by metabolic oxidative processes) and oxygen of the air. Further, a key to this process is that oxygen serve as an electron acceptor. For many heterotrophs, such as man, oxygen is a required nutrient. Not all heterotrophs require oxygen for this purpose, however. The microbiologist classifies organisms as aerobic, which require oxygen; facultative, which may live with or without oxygen; and anaerobic, which can live only in the absence of oxygen. When organisms live without oxygen, they produce compounds such as methane, CH_4; hydrogen, H_2; hydrogen sulfide, H_2S; ammonia, NH_3 (all gases); or ethyl alcohol or other higher alcohols, hydrocarbons, sulfur and nitrogen compounds; or even elemental sulfur (solid) or nitrogen (gas). In such compounds, carbon, nitrogen, sulfur, and even hydrogen itself act as electron acceptors. It should be said, however, that such end products still contain much potential chemical energy in that they can be further oxidized in the presence of air, which supports more complete combustion. From a chemical point of view, then, heterotrophs require one or more nutrients which act as electron acceptors. For man, this is oxygen. Conversely, autotrophs require nutrients which act as electron donors—water and carbon dioxide.

All organisms must have a source of nitrogen because all proteins, nucleic acids, and many other essential compounds in cells contain nitrogen. Remember that enzymes, the catalysts which make the chemical processes of life possible, are proteins. Organisms vary quite widely in the types of nitrogen compounds which are nutrients. Man and many higher animals require nitrogen in the form of protein or amino acids. Some are called *essential amino acids* because the animal dies if any one of the eight or ten is missing. In lower forms of life, we find greater variations in the nitrogen compounds required as nutrients. Some microorganisms need only nitrogen gas, N_2, from the air, while others may use nitrogen in relatively simple forms such as ammonia (NH_3) or ammonium (NH_4^+) salts, nitrate (NO_3^-) salts, or urea (NH_2CONH_2), and still others may use only more specific and complex sources of nitrogen such as amino acids. Autotrophs including higher plants require nitrogen usually in the form of ammonia or ammonium salts or as nitrate salts. Many plants, notably leguminous plants, live symbiotically with microorganisms which can produce nitrogen in a form usable by the plants from the nitrogen of the air. Such microorganisms live among and on the roots of the plant. In general, as we define the nitrogen source for any organism, we must do so in rather detailed chemical terms.

All organisms require phosphorus—usually as some type of salts of phosphoric acid (H_3PO_4). Free phosphoric acid is too acidic and must be at least partially neutralized as the sodium, potassium, calcium, or ammonium salt. The universal biological function of phosphoric acid was emphasized in Chapter 3.

Sulfur in some form is required by all organisms. Two of the 18 amino acid building blocks of proteins contain sulfur. This sulfur is usually sufficient for most heterotrophs, although many microorganisms and green plants may require sulfur in a simpler form such as sulfate. Sulfuric acid, like phosphoric acid, is too strong and so the salts of sulfuric acid are the usual form of this nutrient.

The mineral nutrient requirements, other than phosphate, vary greatly with organisms. In animals, the skeletal parts are often composed of large amounts of calcium. Also, animal tissue must contain the chlorides of sodium and potassium. Interestingly enough, the sodium requirement of animals generally exceeds the potassium requirement, but plants require larger amounts of potassium. The skeletal portions of plants are generally organic, usually cellulose with varying amounts of lignin, and so the mineral requirements of plants and animals differ. All organisms, large and small, require small amounts of minerals for specific purposes. Generally, such minerals are attached to enzymes and thus are intimately related to the chemical processes within living cells. One exception to this in higher animals, including man, is fluorine or the fluoride ion, F^-. (Fluorine, like chlorine, sodium, potassium, etc., is too reactive to occur in nature as the free element.) This ion is obligatory for sound bones and tooth enamel. When we look at other elements required by this or that organism, we get a formidable list: magnesium, Mg; iron, Fe; zinc, Zn; copper, Cu; boron, B; cobalt, Co; chromium, Cr; arsenic, As; molybdenum, Mo; tin, Sn; vanadium, V; selenium, Se; and still others. Boron, B, is required by plants, but this has not been demonstrated for man. Some of these elements are required by certain organisms in such small amounts that their necessity is difficult to prove. But proving the essentiality may be easier than learning the function of the nutrient in the organism.

Organic compounds other than carbohydrates, amino acids, etc., may be required by a particular organism. This is especially true for heterotrophs. The autotrophs, and particularly the green plants, require few if any organic nutrients—even amino acids. They seem to be able to make whatever organic compound needed. Not so for the heterotrophs. Most of them require some organic nutrients; if not specific amino acids, usually other compounds. With respect to man and other higher animals, within this group of nutrients are found the substances we commonly call *vitamins*. The rather universal metabolic role of some of these substances was made clear in Fig. 3.4 of Chapter 3. From this point of view, it is not difficult to understand that in the process of evolution a heterotroph might indeed lose the ability to make some essential compound if that compound was always present in the organisms which made up its food supply.

Many microorganisms (bacteria, yeasts, and fungi) are able to use very simple sources of carbohydrate for energy and nitrogen for protein synthesis, yet require one or more vitamins or related compounds. The understanding of the diversity of nutrient requirements among microorganisms has provided the means of studying the detailed biochemical processes involved in such fields as genetics, physiology, and pathology as well as in nutrition itself. Almost any organism has its unique requirements, although there is a commonality of nutrients among all living things. Chapter 5 will deal with the nutritional requirements of man.

SUMMARY

Water, temperature of the water, pH of the water, activity of the water, and nutrients in the water are five basic environmental factors that permit any biological organism to complete its life cycle—genesis, growth, maturation, reproduction, and death. Since all organisms, whether unicellular or multicellular, are composed of cells, the fundamental environmental parameters for life can be defined with respect to the cell. Furthermore, as shown in Fig. 4.6, these five environmental conditions can be defined in quantitative terms, which reflects the fact that biological processes are chemical processes.

We have defined the overall conditions that permit life, even though there are quantitative variations in the five basic environmental conditions for each organism. We can use these conditions as guidelines to an understanding of the means available to us to culture, use, preserve, process, and store our food and to inhibit the growth of organisms competitive with us. We are always in collusion with our biological allies and at war with our biological enemies which want our food organisms for their own benefit. These competitive organisms may be the ubiquitous microorganisms in the soil or transported by air, insects, noxious plants, etc. Since we are interested in the problems of feeding ourselves, let us conclude by listing common ways in which the five basic environmental parameters are manipulated to provide for more efficient use of our food organisms:

1. Water removal or drying.

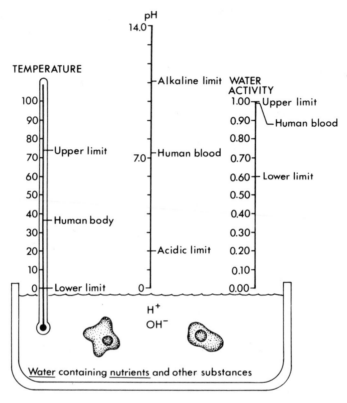

Fig. 4.6. Fundamental environmental parameters of water solutions which control all cells and permit life.

  a. Dry seeds such as wheat, rice, beans, and their products such as flour, chocolate, coffee, peanut butter.
  b. Salad oil, shortening, sugar, gelatin, purified soy protein, etc.
  c. Dried milk, dried potatoes, corn flakes, potato chips, etc.
2. Temperature manipulation.
  a. Use of household refrigerator or refrigerated storage of fruits, vegetables, meat, and dairy products.
  b. Cooking.
  c. Canning—heat sterilization in sealed containers, which prevents contamination of heated and cooled food.
  d. Freezing. May also be considered under (1) above and (4) below.
3. pH control—preservation by acid or alkali.
  a. Fermented products such as sauerkraut, pickles, vinegar.
  b. Soft drinks, usually by phosphoric acid.
  c. Salad dressings and condiments.

    d. Water purification such as in the lime softening process used by certain municipalities; water may be essentially sterile because of high pH during purification.

4. Decrease of water activity.

    a. Salt curing of meat and fish.

    b. Jellies and jams.

    c. Candies, some cake frostings, cookies, cake, etc.

    d. Wines, liqueurs, and alcoholic beverages of high alcohol content.

5. Addition or removal of nutrients.

    a. Fertilizers in production of agricultural crops.

    b. Mineral, vitamin, and protein supplementation of food or animal rations.

    c. Certain insecticides and rodenticides.

    d. Removal of oxygen to prevent growth of aerobic spoilage bacteria.

    e. Addition of sodium and potassium nitrates to meat, which serve as an oxygen source and prevent the growth of anaerobic bacteria within ham, corned beef, or cured sausage.

    f. The protein avidin, occurring naturally in egg whites, which creates a deficiency of biotin in an otherwise excellent medium for many spoilage bacteria which require biotin, a vitamin required also by man.

5

NUTRITION

To the student of the history of science, nutrition offers many fascinating stories. References to what is now considered to be in the realm of nutrition can be found in the earliest recorded history. This is not surprising since food is so fundamental to man. His daily dependence on food quite naturally was associated with his well being. Famine, night blindness, the discomforts of overindulgence, the scourges of seafarers and armies—all problems of nutrition—are frequently mentioned.

Hippocrates, 460–359 B.C., observed that "growing bodies have the most innate heat; they require the most food, for otherwise their bodies are wasted. In old people, the heat is feeble and they require little fuel" (see Table 5.3). His observation that "persons who are naturally very fat are apt to die earlier than those who are slender" is often heard today, as an admonishment against overeating. Leonardo da Vinci (1452–1519), the genius of the Rennaissance, observed that there is continual need of adequate nourishment and that "where flame cannot live, no animal can sustain its existence." The real meaning of such observations had to await the development of the science of chemistry.

We have already mentioned the work of Lavoisier and Priestley in Chapter 3 which helped to explain da Vinci's observation. These men benefited, of course, from the contributions of others, but in our urge to focus on specific concepts we often identify individuals with the clarification of certain ideas. The Scottish naval surgeon James Lind published in 1752 a *Treatise on Scurvy* which clearly showed that scurvy, the disease which plagued sailors, could be cured by eating oranges and lemons. In spite of this knowledge, Robert Scott's Antarctic expedition of 1912 failed in large measure because of scurvy. In the early 1800s, Francois Magendie showed that dogs could not survive on a diet of carbohydrate (sugar) and fat (olive oil, butter) and that some sources of nitrogen seemed to be required. Justus von Liebig later showed that proteins, "the noblest of the food elements," were needed since carbohydrate and fat appeared to supply mostly energy.

Before the scientific era, disease and pestilence were associated in the minds of people with demons or evil spirits. This idea still persists with regard to mental diseases, even though many appear to be metabolic disorders (biochemical malfunctions). When Pasteur, the French chemist considered the founder of modern microbiology, and Robert Koch, a German physician who was attracted by the work of Pasteur, showed that many diseases were actually caused by microorganisms, the demon theory of disease seemed to be well on its way out. But it was found that many diseases could not be explained on the basis of an invasion of microorganisms. One such malady was beri-beri, which was common among peoples whose diet was polished rice and little else. The newly created Japanese navy of the 1880s was plagued with this often fatal disease. K. Takaki, a Japanese naval surgeon, discovered that beri-beri responded to a change of diet. He may or may not have been aware of the report of Magendie in 1826 that a dog fed on white bread and water would die in 50 days and that one fed the coarse bread of the military "keeps his health." Nevertheless, the time was ripe for the science of nutrition as we know it to come into its own. It was found that diet could cure some diseases.

The concept that certain molecular entities were needed for normal growth and health appeared to have special merit in view of the great advances of the science of chemistry, which elucidated the chemical nature of many substances of biological and nonbiological origin. The closing years of the nineteenth century saw the beginning of the long and complicated search leading to the unequivocal identification of the essential nutrients—those molecules necessary for man (or other organisms) to complete his life cycle— genesis, growth, maturation, and reproduction.

Christiaan Eijkman, a Dutch physician, was sent to Indonesia in 1886 as an army surgeon. Beri-beri was quite prevalent there, particularly among prisoners who were fed only polished or milled rice. Some chickens as well as humans were confined in the prison, and both had the same diet. The chickens came down with a polyneuritis somewhat similar to beri-beri. For some reason the prison rations were changed—polished rice for the prisoners, unpolished rice for the chickens. The chickens no longer contracted their disease. When Eijkman observed this, he devised experiments to prove that the disease was a true dietary deficiency disease and not a bacterial disease. (Eijkman himself was an accomplished bacteriologist.) He fed chickens polished rice or unpolished rice. Those receiving the latter did not get sick. The sick chickens on polished rice were cured by the addition of only rice polishings (hulls, seed coats, and germ) to their diet. When men afflicted with beri-beri were given rice polishings along with polished rice, the beri-beri did not appear. This simple and dramatic proof that the deficiency of some unknown dietary component definitely caused a specific widespread human disease was, to many people, a revolutionary scientific idea.

Eijkman's discovery came at a time when a number of workers were trying to determine more specifically the roles of fat, carbohydrate, and protein in the diet. Now besides these three major constituents, Eijkman had shown that some other substances must be required. Indeed, the concept of nutrition is both simple and complex: certain molecules must be in the diet to assure good health. But how many different molecules are critical nutrients for man?

It became abundantly clear at this time that in order to identify any nutrient the first order of research was to find a test animal and a test diet, because experiments on human beings are complicated by a variety of limitations. In our value structure, it is considered wrong to jeopardize the life of a human experimental subject—a principle sometimes unfortunately neglected. This limitation on nutritional experiments with humans pointed up the value of experimental animals in identifying nutrients. But what animals and under what conditions? Animals respond differently than humans.

Just to find an experimental animal and a reproducible procedure permitting meaningful interpretation presented formidable obstacles. Table 5.1, a list of nutrients that man requires, indicates the complexities confronting scientific investigators at the beginning of this century. We cannot begin to relate in any degree of completeness the interesting tales of human frustration, exhilaration, disappointment, rivalry, and romance among the hundreds of scientists who devoted their lives to our current understanding of the science of nutrition.

Modern biochemistry has grown beyond the simple concept of nutrition enunciated at the turn of the century, yet tracking down, identifying, and determining the function of essential nutrients formed a major part of biochemistry until World War II. We have not yet solved all the nutritional and food problems, and much more needs to be done. However, the success of the basic scientific philosophy developed in biochemistry during this period clearly showed the way for the solution of many biological problems in addition to those directly related to nutrition.

The modern technologies of medicine and agriculture are, in great measure, the result of the quantitation and understanding of biological processes from the biochemical point of view. One of the ironies of this part of scientific history is that whereas nutrition and food are primary concerns of man, relatively little money was made available for research on human food and nutritional problems during the first half of the twentieth century. Research was directed to agricultural nutritional problems rather than to human ones. Even today, human nutrition receives very little attention in medical school curricula and is almost completely absent from college and university curricula for general students. It was often said in the early decades of this century, half facetiously but with some truth, that if men were sold like cattle, pigs, and chickens at so much a pound human nutrition research could get support and not be neglected. This is not to say that there is not a great deal of carryover of knowledge from farm animals and plants to man, but rather that many nutritional programs were directed toward feeding animals properly and only indirectly toward the more complex problem of feeding man. We will discuss these unique complexities in more detail later. Let us now return to a few more early developments in the science of nutrition.

While Eijkman was doing his work, others were working on the nutritive properties of proteins, carbohydrates, and fats—the major categories of protoplasmic constituents which make up our food. Rats, mice, and other animals were often used in these studies. E. V. McCollium of the United States and Gowland Hopkins of England found that when animals were fed diets of relatively pure fats, carbohydrates, and proteins they did not thrive but needed some "accessory factors." These were identified as "fat-soluble accessory factor A" and "water-soluble accessory factor B." In the period 1903–1912, several workers trying to cause beri-beri in guinea pigs produced scurvy instead. This disease was cured by feeding the animals fresh plant material. The curative nutritional factor was labeled "C."

Casimir Funk, the eminent Polish biochemist who came to the United States in 1915, coined the term *vitamin*. While still in his twenties, he had recognized that not only beri-beri and scurvy but also ricketts, pellagra, and sprue were nutritional deficiency diseases. In the course of his attempts to isolate the compound that would cure beri-beri, he obtained a quite potent concentrate, a small amount of which cured the deficiency in his test animals. He noted that the concentrate gave strong reactions for the class of organic compounds known as amines. From this he got the word *vitamine* or "amine of life," and this vivid descriptive term for the accessory growth factors became both a scientific and a household word. So strongly did this new term become entrenched that when it was later found that all the so-called accessory factors were not amines, the final *e* was dropped and the word *vitamin* was retained.

This early work of Funk on beri-beri took an interesting twist. He succeeded in preparing from his concentrate a crystalline material—nicotinic acid amide. But, disappointingly, the active principle for preventing beri-beri was lost. The nicotinic acid did not cure his test animals, although improvement was observed in their polyneuritis. He had indeed isolated one of the most important vitamins (see Table 5.1), but it was not recognized as the pellagra-preventing vitamin until almost a quarter of a century later, by Conrad Elvejhem and coworkers at the University of Wisconsin. In Japan, Umetaro Suzuki and his coworkers isolated nicotinic acid in 1912, but they did not realize its importance, either.

The goal of research in nutrition was to identify all the nutrients. If this could be done, it would be possible to experiment with various carefully controlled diets and then begin to elucidate the metabolic role of each nutrient. But isolation of the vitamins was a formidable task. In time, it was found that not only are there vitamins A, D, E, and K, which are fat soluble, but also the water-soluble vitamin B, which was eventually found to consist of a series of vitamins from B_1 to B_{12}. The last is clinically a very important cobalt-containing vitamin.

But, as is clear from Table 5.1, not all the nutrients are vitamins, for research on the nutritional adequacy of proteins soon showed that there are different kinds of proteins. Animal proteins, such as those of eggs, meat, and milk, seemed to be more nutritious than vegetable proteins. Some proteins would not even support life in test animals. Since proteins are made of amino acids and since not all proteins have all 18 common amino acids, it became necessary to identify the nutritionally essential amino acids—a difficult problem because all the amino acids of proteins were not yet known. William C. Rose of the University of Illinois identified the last essential amino acid, threonine, in 1936.

Identifying essential nutrients made for exciting times in biochemistry. Rivalries developed among various groups. These rivalries occasionally took on some aspects of snobbishness—establishment versus nonestablishment, and so on. Scientists, like everyone else, are subject to human frailties, which occasionally affect scientific progress.

Sometimes an essential nutrient was discovered but not called a vitamin until later. When in the later stages of nutritional research it was found easier to use microorganisms as test organisms instead of rats, mice, dogs, etc., scientists discovered that some nutrients for microorganisms, such as para-aminobenzoic acid and inositol, were not required by higher animals. And then there was always the difficult problem of making the final jump from test animal to the human. The importance of vitamin E with respect to reproduction in rats was discovered 50 years ago. It was almost immediately christened the "antisterility factor" by the popular press. Such misdirected publicity actually discouraged work on this nutrient, because the subject of fertility was too close to the then delicate topic of sex. At any rate, it has been difficult to prove that vitamin E is required by man. Only now is its function being elucidated.

The isolation of any proven nutrient was a formidable task. After a nutrient had been isolated and purified, it had to be chemically identified, its molecular structure proved, and a synthetic form prepared. The rigorous standards on which the science of chemistry is founded demanded that all of these steps be followed. The isolated nutrient's efficacy was not accepted as proved until the synthetic molecule was shown to be equal in activity to the natural molecule.

Biochemists of this period, when many of the fundamental concepts were being promulgated, gave much attention to the nature of the enzymes which make the chemical processes of life possible. So remarkable are these catalysts that at one time their power was thought by some to be mystical. At first, it was thought that they could function only in living cells or organisms. Eduard Buchner, one of the early Nobel Prize winners, showed that enzymes could act outside the cell and cause alcoholic fermentation to take place. In 1897 he isolated zymase, thought to be a pure enzyme. But some years later Arthur Harden and W. J. Young showed that zymase requires phosphate in order to function and that zymase is composed of a heat-labile fraction and two water-soluble heat-stable fractions. This was prophetic work, for it was subsequently shown that the heat-labile part was protein and that the alcoholic fermentation of glucose took place by means of phosphoric acid esters. Next, one of the two water-soluble heat-stable fractions turned out to be a mixture of adenosine mono-, di-, and triphosphates—AMP, ADP, and ATP. The other was found to contain nucleotides of nicotinamide, one of the vitamins. We have already seen in Chapter 3 that many vitamins function as integral parts of certain enzymes and that phosphoric acid esters are of fundamental importance in a large proportion of metabolic processes.

One of the most significant of the early workers in enzymes was a very perceptive and independent biochemist, J. B. Sumner, who became convinced that enzymes themselves were proteins and that, consequently, enzymes should show the properties of proteins as pure chemical compounds subject to all the established principles of the science of chemistry. This was no mean task or in many ways a popular one. He spent most of his career at Cornell, where from 1914 to 1929, he was never promoted from his original appointment as assistant professor. Persevering, in 1926 he succeeded in purifying the enzyme urease[1] and crystallizing it. Crystallization of proteins of any kind is a difficult task. To crystallize a protein and show that the crystalline material retained its powerful catalytic properties was an almost unbelievable accomplishment at that time. It proved that enzymes were capable of being studied as any other chemical substance. Sumner had dispelled any mysticism surrounding those potent catalysts of life processes. His work was confirmed and he was promoted to professor in 1929. He subsequently received many honors, including the Nobel Prize in 1946. John H. Northrop crystallized trypsin and pepsin, important digestive enzymes (see Chapter 7) during the period 1930–1933, and shared the Nobel Prize with Sumner and with Wendell M.

---

[1] Urease hydrolyzes urea, the most important nitrogen excretory product in human urine: $NH_2CONH_2 + H_2O \rightarrow CO_2 + 2NH_3$. The ammonia is characteristic of contaminated urine, reflecting the fact that the enzyme is present in many bacteria and plants. Urease is not found in animals.

Stanley, who crystallized the first virus. All three had found that viruses are nucleoproteins—that is, they are composed of nucleic acids and proteins. The viruses, classified according to whether they contain DNA or RNA, are those gigantic molecules which can invade cells and use the cell's metabolic apparatus for their own reproduction to cause disease and perhaps destroy the organism.

The foregoing discussion sketches a few of the more important aspects of the development of the science of nutrition and indicates how nutrients are related to the structure and function of biological systems. It should be clear also that our knowledge of nutrition emphasizes the commonality of the chemical processes of life among all biological systems—the plants, animals, and microorganisms that supply us with our food, the organisms that compete with us, and we ourselves. Let us now turn our attention to these essential chemical substances which we must have to exist.

OUR NUTRIENTS

Table 5.1 is a list of the known nutrients required by each of us. It has been compiled from various sources and reflects the recommended intake of nutrients proposed by the Food and Agricultural Organization (FAO) and the World Health Organization (WHO) of the United Nations, the Food and Nutrition Board of the National Research Council of the United States, and similar agencies of other countries. Not all such groups agree on the quantities of individual nutrients needed for optimal health. Furthermore, the amounts of nutrients required for mere maintenance of life are less. Some of the required nutrients are usually not even listed in recommendations of governmental agencies because they are so common in foods and because, although they are known to be required, it is difficult to establish the minimum amounts necessary for life. Humans are fastidious eaters and what they are willing to consume depends on esthetics, social customs, and personal opinions about palatability. People do not relish synthetic diets for the extended periods of time necessary in some types of nutrition studies.

In some cases where official recommendations are not available, some values have been derived from information in the scientific literature. In the case of trace elements, it is most difficult to develop rations free of them. Analytical techniques have in recent years become so sensitive that multitudes of different molecules and atoms of various elements can be detected in almost all biological systems. Consequently, when it has been found that certain vital biological molecules such as enzymes do contain or depend on a particular ion to perform their function, this element has been included among the nutrients.

Table 5.1. Human Nutritional Requirements

| | Daily requirement | | | |
| | Grams (unless noted) | | Moles | |
Nutrient	Man[a]	Woman[b]	Man[a]	Woman[b]
Oxygen	800	593	25.0	18.5
	560 liters[c]	414 liters[c]		
As air 21% O_2	2930 liters[d]	2173 liters[d]		
Water[e]	2800	2000	155	111
Energy	2700 Calories	2000 Calories		
If all from glucose	675	500	3.75	2.78
If all from fat	300	222	0.338	0.25
Protein	56	46	0.47[f]	0.38[f]
Essential amino acids				
Tryptophan	0.5		0.0025	
Phenylalanine[g]	2.2		0.0133	
Lysine	1.6		0.0110	
Threonine	1.0		0.0085	
Methionine[h]	2.2		0.0150	
Leucine	2.2		0.0168	
Isoleucine	1.4		0.0106	
Valine	1.6		0.0137	
Sodium chloride[i]	3.0		0.051	
Potassium	1.0		0.026	
Phosphorus	0.8		0.026	
Calcium	0.8		0.020	
Magnesium	0.35		0.015	
Essential fatty acids[j]	3.0		0.010	
Iron[k]	0.018		3.2×10^{-4}	
Ascorbic acid, vitamin C	0.045		2.5×10^{-4}	
Zinc	0.015		2.2×10^{-4}	
Niacin, nicotinic acid amide, vitamin B_3	0.018		1.5×10^{-4}	
Fluorine	0.002		1.0×10^{-4}	
Vitamin E[l]	0.015		3.5×10^{-5}	
Copper	0.0015		2.4×10^{-5}	
Pantothenic acid,[n] vitamin B_5	0.005		2.3×10^{-5}	
Manganese[n]	0.001		1.8×10^{-5}	
Vitamin K[l]	0.005		1.1×10^{-5}	
Pyridoxine, vitamin B_6	0.002		1.0×10^{-5}	
Retinol, vitamin A[l]	0.0015		5.2×10^{-6}	
Riboflavin, vitamin B_2	0.0016		4.2×10^{-6}	
Thiamine, vitamin B_1	0.0014		4.2×10^{-6}	
Iodine	0.00012		9.4×10^{-7}	
Folic acid	0.0004		9.0×10^{-7}	
Biotin[n]	0.00015		6.0×10^{-7}	

Table 5.1—*continued*

	Daily requirement			
	Grams (unless noted)		Moles	
Nutrient	Man[a]	Woman[b]	Man[a]	Woman[b]
Vitamin D[l,m]	0.00001		2.6×10^{-8}	
Chromium	0.000001 (1×10^{-6})		2.0×10^{-8}	
Cobalamin, vitamin B_{12}	0.000005 (5×10^{-6})		3.7×10^{-9}	
Cobalt, as vitamin B_{12}	0.0000002 (2×10^{-7})		3.7×10^{-9}	
Molybdenum[o]				
Vanadium[o]				
Selenium[o]				
Tin[o]				

[a] Man: weight 70 kg (154 lb), height 175 cm (5 ft 8 inches), age 22–50 years.

[b] Woman: weight 58 kg (128 lb), height 163 cm (5 ft 5 inches), age 22–50 years.

[c] At 0°C (32°F) and pressure of 1 atmosphere or 760 mm of mercury; 19.8 and 14.6 cubic feet, respectively.

[d] At room conditions of 20°C (68°F) and 740 mm of mercury, 103.5 and 76.8 cubic feet, respectively.

[e] Includes water consumed directly and as food component and water produced from oxidation of food.

[f] Average molecular weight of amino acids is 120.

[g] 0.22 g if tyrosine is available.

[h] 0.35 g if cystine is available.

[i] 2–3 g is minimum but salt requirement as well as water requirement may go up in warm climates and in heavy work due to perspiration losses.

[j] As linoleic, linolenic, and/or arachidonic acid.

[k] Recommended for menstruating women; for men about 0.01 g is recommended.

[l] A number of closely related chemical compounds may serve this vitamin function. 1 IU (International Unit) of A = 0.3 μg retinol or 0.6 μg β-carotene. 1 IU of E = 1.0 mg DL-α-tocopherol acetate or 0.81 mg D-α-tocopherol.

[m] Vitamin D may not be required by the adult but is by growing children; it is made from normal skin constituents by ultraviolet light or it may be fed in the diet. 1 IU of D = 0.025 μg D_3, activated 7-dehydrocholesterol.

[n] Although definitely required, it is difficult to establish amount because this nutrient is so common in foods.

[o] It is known that this trace element is an integral part of vital enzyme systems, but the daily requirement has not been established. Sufficient amounts are in many foods.

The nutrients in Table 5.1 are listed roughly in the order that they are critical in supporting life. Oxygen heads the list because it is impossible for a human to survive for more than a few minutes without it. Aside from the heart and kidney, the brain consumes more oxygen on a continuing basis than any other organ, and is quickly destroyed or damaged when oxygen is lacking. The ocean of air in which terrestrial life evolved has determined man's requirement for a constant supply of oxygen.

The next nutrient in terms of continuous critical need is water. We can survive only a few days without it. To be sure, the amount of water required varies with exercise and climate, but there is a minimum need for fresh water so that the water activity, a_w (see Chapter 4), of the tissues can be properly maintained. All life processes can only take place in water.

If oxygen and water are available, we are able to survive for some days without food, depending on nutrients stored in the body or consumption of our own tissues as a source of energy. We have emphasized repeatedly that energy is required for life, and it is indeed our next critical nutrient. Most of the energy we use comes from carbohydrate or fat oxidation. We can also use other substances for energy—for example, citric acid, acetic acid, lactic acid, even alcohol and protein or amino acids (see Fig. 3.4 in Chapter 3). However, most of the energy we require comes from a combination of carbohydrate and fat. In fact, this combination is necessary; fat combustion alone (as in inability to burn glucose in diabetes mellitus or in starvation since we store relatively little carbohydrate) can cause metabolic difficulties such as acidosis (accumulation of β-hydroxybutyric and acetoacetic acids) and acetone production.

Next in order come proteins, which we commonly associate with the structural portions of our bodies even though now we know that many vital enzymes are proteins. Other substances are associated with bone, muscle, etc., and here we would list besides protein (amino acids) sodium, chlorine, potassium, phosphorus, calcium, magnesium, and essential fatty acids.

Finally, there are a host of nutrients almost all of which are related not so much to structure as to function. In other words, these nutrients are involved directly in the dynamic chemical processes of life. Here we find the vitamins and the trace elements.[2] They are parts of enzyme systems or transporters of electrons (cytochromes, Chapter 3) or of oxygen and carbon dioxide, hemoglobin and myoglobin of blood and muscle, respectively. Although the molecules associated with these processes are proteins and we can make these proteins if the necessary amino acids are in the diet, we are not able to make the vitamins or mineral ions which are necessary for the enzymes or proteins to function properly.

In Table 5.1, the nutrients are shown in molar quantities or the number of moles of each nutrient required each day as well as in terms of grams. Furthermore, the arrangement is such that with a single exception, water, the nutrients are given in descending order of daily molar requirement. Those

---

[2] An interesting observation here is that boron is not listed even though it is present in the human body. As already mentioned, this element is necessary in many plants but as yet no function for it has been found in man.

molecules needed for fuel or energy are required in molar quantities; protein requirement, as total amino acids, is an order of magnitude less, tenths of moles; next come the nutrients making up other structural components of protoplasm, which are needed in hundreths of a mole per day; the remaining nutrients, vitamins, etc., are required in 10^{-4}, 10^{-5}, 10^{-6}, 10^{-7}, 10^{-8}, and 10^{-9} mole per day quantities. Indeed, when requirements are studied in terms of moles or numbers of molecules needed, we can get a very good idea of how a nutrient may be functioning in our bodies. Observe that approximately 25,000 more molecules of glucose than niacin are needed ($3.75 \div 1.5 \times 10^{-4}$). Figure 3.4 in Chapter 3 clearly shows why—niacin is used over and over again and glucose is eventually oxidized to carbon dioxide and water. Also in Fig. 3.4 you will see less frequent use of thiamine and riboflavin than of niacin. Only one molecule of each of these nutrients is required for every 35 niacin molecules ($1.5 \times 10^{-4} \div 4.2 \times 10^{-6}$). We know from other work that niacin is required by most hydrogen-carrying enzymes involving oxidation and reduction processes in living things. It is not necessary to continue this line of discussion further, for the point is obvious that nutrient requirements in terms of actual numbers of molecules give a broad overview of the types of roles various nutrients perform in our bodies.

For the sake of simplicity, Table 5.1 is based on adult requirements. There are, of course, individual variations with regard to body size (weight and height), as one might expect. There are also very significant and important variations with age or position in the life cycle. A newborn baby has different requirements than an adolescent or adult. The pregnant or lactating mother will have different requirements than a nonpregnant or nonlactating adult woman of the same age, weight, and height. There are differences in nutrition requirements between the sexes and between the old and the young. There are even some variations in nutritional needs of individuals of the same sex, age, weight, and height. It is beyond our purpose to discuss each nutrient in detail or the variations in requirements just mentioned. All of these facts may be found in books dealing strictly with nutrition.[3] However, there are some further considerations which are of importance as we study how the nutrients are made available to us through our food.

ENERGY

Generally speaking, nutrient requirements are tied to energy needs, and the total energy requirement of any individual is a function of the surface area

[3] For more detailed information see *Recommended Dietary Allowances*, eighth edition, 1974, Food and Nutrition Board, National Research Council, National Academy of Sciences, 2101 Constitution Avenue, Washington, D.C., 20418.

of his body. The energy expenditures for the man and woman of Table 5.1 are, respectively, 62.5 and 51.2 Calories per square meter of body surface per hour. These energies represent two separate and distinct categories—the energy required for maintenance and nothing more, called the *basal metabolism*, and the energy required for activities such as reading, eating, working, and playing. Let us take a look at these two energy parcels.

Basal metabolism can be determined by actually measuring the heat produced, oxygen consumed, and carbon dioxide produced under conditions of absolute rest. Such tests are usually begun when the subject awakens in the morning and while he is lying in bed. The energy required for any activity may be determined from the same type of measurements by subtracting the basal energy requirement from the total energy expenditure during the activity.

One way to look at the general concept of basal metabolic activity is in terms of energy liberated per square meter of body surface per hour for different ages and sexes. The data in Table 5.2 indicate clearly that energy requirements depend on age. The energy needs of growth are unmistakable,

Table 5.2. Effect of Age and Sex on Basal
Energy Requirements

Age (yr)	Basal metabolic rate (Cal/m^2 body surface/hr)	
	Male average[a]	Female average[a]
1–3	60.1	54.5
3–5	56.3	53.0
5–7	52.3	49.7
7–9	49.5	46.2
9–11	46.5	44.1
11–13	44.5	40.5
13–15	43.7	38.3
15–17	41.9	36.2
17–19	40.1	35.4
21–23	39.0	35.2
25–27	38.0	35.0
32–34	37.0	34.9
38–40	36.5	34.3
40–50	36.0	33.4
44–70	33.1	31.3

Data from the handbook *Metabolism*, Federation of American Societies for Experimental Biology, 1968.
[a] 95% of individuals tested are within about ±7% of this average.

as is the gradual decline in basal metabolism with age. However, data presented in this manner tend to mask the effect of size and to magnify sex differences. Actually, in terms of Calories per unit of active metabolic tissue (that is, assuming most fat is not metabolically active) there is actually little or no difference in energy requirements between the sexes at any certain age. The figures in Table 5.2 reflect the fact that females tend to have more fat—particularly subcutaneous fat—than males. Also, males tend to be larger than females and to be somewhat more active.

By making estimates of average but not strenuous activity, and using average sizes and basal metabolic rates for the two sexes at various ages, the Food and Nutrition Board of the National Research Council has recommended daily energy intakes as shown in Table 5.3.

When we look at these figures and at the same time realize that eating habits are generally established during the growing and adolescent years of greatest activity, we can readily appreciate why overweight is a common problem in cultures such as ours where there is an abundance of food and labor-saving devices are a way of life. The reason, of course, is that physiologically we, as animals, have a means of storing energy for use when needed—as fat. In general, in those biological systems requiring movement to complete

Table 5.3. Recommended Daily Energy Intake

Age in years	Calories per day		
0–½	117/kg body weight		
½–1	108/kg body weight		
1–3	1300		
4–6	1800		
7–10	2400		
		Male	Female
11–14		2800	2400
15–18		3000	2100
19–22		3000	2100
22–50		2700	2000
51 +		2400	1800
Pregnancy			Add 300
Lactation			Add 500[a]

From *Recommended Dietary Allowances*, Food and Nutrition Board, National Academy of Sciences–National Research Council, eighth edition 1974.

[a] Up to 2 months. If lactation is continued beyond this period, an additional 1000 Cal or more may be required.

all or even some phase of their life cycle, we find energy reserves in the form of fat because there is more energy per unit of weight. Furthermore, fat storage does not require the presence of water and so we generally find fats in seeds as well as fat deposits in animals.

Man, being a social animal as well as one of habit, has developed many of his customs around the enjoyment of food. So it is easy to understand why we are often preoccupied with our weight and "dieting." When we then look at all the nutrient requirements besides calories in Table 5.1, it is clear that willy-nilly restriction of food intake to reduce weight can create shortages of other dietary essentials. Therefore, a calorie-limited diet must be more carefully planned than a high-calorie regime associated with a more active or arduous life.

Let us now take a look at the energy needed above the basal requirement to perform certain activities. Table 5.4 gives only a very limited résumé of the energy needed to do certain things with which we are all familiar. Of course, the speed at which each activity is accomplished will affect the energy required

Table 5.4. Energy Cost of Various Activities

	Percent over basal requirement
Basal, completely relaxed, supine position	0
Sitting, eating	23
Sitting, reading or writing	66
Washing, shaving, dressing	204
Walking, 3 miles/hr	336
Walking, 4.2 miles/hr	678
Walking down stairs, normal pace	494
Walking up stairs, normal pace	1445
Driving car	139
Bicycling, 5.5 miles/hr	285
Playing football	675
Swimming, normal rate	832
Typing, 40 words/min	51
Assembly line work, medium	131
Washing clothes by hand	175
Sweeping floors	225
Brick laying	242
Sawing wood	500
Tending open-hearth furnace in steel mill	772

Data from the handbook *Metabolism*, Federation of American Societies for Experimental Biology, 1968.

per hour. In Table 5.2 basal data are given on an hourly basis. Notwithstanding some variations in how fast you might climb stairs or dress yourself, Table 5.4 adds a quantitative dimension to how much energy you need under different circumstances.

You will note that the energy needed to perform tasks varies widely. No ordinary activity requires more energy than climbing stairs. To do so at normal rate and no load except body weight requires 14.5 times normal basal energy. This is almost twice the energy required in very heavy labor such as tending a furnace in a steel mill. Of the common sports, swimming requires the most energy but little more than half that of climbing stairs. Many sedentary occupations require only one-third to two-thirds more energy than basal requirements, revealing why people who have such jobs are often overweight.

A relatively minor factor in energy expenditure relates to food itself. This is the energy required for ingestion and assimilation of food. When protein is consumed, the body produces about 30% more heat than accounted for by the caloric value of protein. For fat and carbohydrate, the excess energy is only 6% and 4%, respectively. Such energy expenditure, called the "specific dynamic action of foods," is not usable for other activities and appears as heat. It is apparently related to the energy cost of putting food in a form usable by the various tissues of the body.

How is the energy available in a portion of food or a complete diet determined? The answer is simple.

All of the body's energy needs are met by the oxidation (combustion) of the constituents of the food consumed. Carbohydrate and fat provide most of the energy, with proteins supplying a much smaller proportion. We have already noted in Chapter 3 that fat yields 9 Cal per gram and that carbohydrate and protein yield 4 Cal per gram. So if we know the amounts of these major food constituents which are eaten, we can easily calculate the total energy available in the food. For example, the number of Calories available in a glass of whole milk weighing 225 grams (approximately 8 ounces) is determined as follows: From the food tables of the Appendix, we find that whole milk contains 3.7% fat, 4.6% carbohydrate, and 3.2% protein.[4] We calculate the caloric value of the milk as follows:

$$225 \text{ g} \times 0.037 = 8.3 \text{ g fat} \qquad\qquad \times 9 \text{ Cal/g} = 74.7 \text{ Cal}$$
$$225 \text{ g} \times 0.046 = 10.3 \text{ g carbohydrate} \times 4 \text{ Cal/g} = 41.2 \text{ Cal}$$
$$225 \text{ g} \times 0.032 = 7.2 \text{ g protein} \qquad\quad \times 4 \text{ Cal/g} = 28.8 \text{ Cal}$$

$$\text{Total energy in 225 g milk} = 144.7 \text{ Cal}$$

Note that skim milk (no fat) contains only one-half the energy of whole milk.

[4] The food tables in the Appendix also give the food energy in 100 grams of each food listed.

You should familiarize yourself with this procedure by calculating the calories available in the foods you consume in a 24-hour period.

PROTEINS

Proteins in our food provide nitrogen in a readily usable form for the body to make all necessary proteins, nucleic acids, and many other miscellaneous nitrogenous compounds required for living. You will note the inference in this statement that man requires protein in his diet. Here is a fundamental difference between people and green plants as well as most microorganisms. These latter organisms can use simple nitrogen compounds such as ammonia, NH_3, or ammonium salts, NH_4^+, nitrate salts, NO_3^-, nitrite salts, NO_2^-, urea, $CO(NH_2)_2$, and even nitrogen gas from the air, N_2. Some can even live on cyanide salts, CN^-, very poisonous compounds for most higher organisms. But man must have protein. This is also true of most higher animals.

The specific proteins which we find in a given species are different from those of all other species. In fact, in higher animals—man particularly—some of the proteins appear to be unique in each individual. Protein incompatibility is the reason for blood typing for transfusion and is the major difficulty in surgical transplants of organs (heart, kidney, etc.) from one individual to another. Thus it seems that a fundamental biological necessity is that each organism, and this means each person, must synthesize his own protein, his own nucleic acid, and certain other nitrogen-containing substances. The proteins of the organisms that man consumes as food are the raw material for the complicated manufacture of these materials.

Proteins are high molecular weight polymers of amino acids. In other words, proteins are constructed of amino acids. We might think of the amino acids as the bricks and mortar of proteins. Since our proteins in sum are individually unique and since we require proteins in our diet, we must then have a means to disassemble the proteins of our food organisms and use their constituents to construct our own particular proteins and other needed nitrogenous compounds and a means to discard what is left. This topic will be discussed in Chapter 7; here we are concerned with the nutrients required in our diet. If proteins are required but we can use them only after disassembling them into their amino acids, it follows that the protein building blocks—the amino acids—are the real nutrients. Let us now look at the amino acids.

Fortunately, almost all proteins are made of only 18 amino acids. There are some minor additional amino acids in some proteins but they are of no

Table 5.5. Nutritional Classification
of Amino Acids

Essential	Nonessential
Tryptophan	Tyrosine
Lysine	Cystine
Threonine	Glycine
Leucine	Alanine
Isoleucine	Serine
Valine	Proline
Phenylalanine	Arginine
Methionine	Histidine
	Aspartic acid
	Glutamic acid

concern to us here. The critical reader of Chapter 3 will recall that a number of amino acids are readily synthesized or oxidized (burned) in association with carbohydrate synthesis or combustion. If this is true, and it is, why then must we have a source of amino acids in our diet? The answer is that we cannot make all 18 amino acids. We can synthesize only ten of the 18; the remaining eight must be present in our diets. These eight are the *essential* amino acids and are shown in Table 5.1. Here you will note that only 12.7 grams of essential amino acids is recommended to be included in 56 grams of protein. The difference is to provide for the necessary nitrogen for all other needs already mentioned. Furthermore, these recommended dietary levels are based on the consideration that there are rather wide individual quantitative variations in nutrient requirements and that not all proteins contain equal amounts of essential amino acids. The recommendations in Table 5.1 are designed to cover the needs and variability of more than 95% of the population.

Table 5.5 lists the 18 different amino acids commonly present in protein, divided into the essential ones and the nonessential ones that can be synthesized by man himself.

The newborn baby or a rapidly growing child may not be able to synthesize histidine rapidly enough. This is true to a lesser extent with arginine. However, normal adults can make all they need of these amino acids if sufficient nitrogen is available. In many proteins aspartic and glutamic acids occur in part as their amide derivatives which are known as asparagine and glutamine respectively. Table 5.6 lists the amino acid composition of a number of common foods. The animal protein foods tend to have a somewhat higher content of essential amino acids than other foods. Also, whereas in animal foods the amino acid compositions are relatively uniform within species, this is not true with vegetable foods. In wheat, corn, beans, etc., there are significant

Table 5.6. Amino Acids in the Proteins of Certain Common Foods
(Grams per 100 g of Proteins or per 16.0 g of Nitrogen)

Amino acids	Wheat	Corn	Rice	Beans	Soybeans	Peanuts	Potatoes	Cassava	Eggs	Meat	Cow's milk	Human milk
Essential												
Tryptophan	1.3	0.8	1.4	1.2	1.4	1.1	1.3	1.3	1.8	1.3	1.4	1.6
Lysine	2.7	3.0	4.2	9.3	6.4	3.5	5.0	4.1	6.7	8.2	7.8	6.2
Threonine	2.9	4.2	3.5	5.9	4.0	2.9	3.7	2.8	5.3	4.5	4.6	4.5
Leucine	6.4	12.0	8.2	7.7	7.7	6.2	4.6	4.1	9.0	7.8	9.9	9.4
Isoleucine	3.8	4.0	4.8	4.9	5.3	4.2	4.5	2.8	5.8	5.1	6.2	5.6
Valine	4.3	5.6	6.2	4.3	5.3	5.0	5.1	3.0	7.2	5.3	7.0	6.2
Phenylalanine	4.6	5.0	4.6	4.6	5.0	5.0	4.2	2.8	5.3	4.2	5.1	4.0
Methionine	1.6	2.1	2.1	0.9	1.3	1.0	1.6	0.6	3.0	2.4	2.4	2.1
Nonessential												
Tyrosine	3.2	3.8	5.8	2.5	3.7	3.0	2.9	1.8	4.3	3.4	5.6	4.8
Cystine	2.1	2.1	1.8	3.6	1.9	1.6	1.3	1.1	2.1	1.3	0.8	1.9
Glycine	3.8	3.0	6.6	4.9	4.5	5.4	1.9	2.8	3.8	4.5	1.9	2.2
Alanine	3.4	9.9	5.6	4.3	5.0	2.9	4.2	5.1	7.5	6.2	3.7	3.8
Serine	4.8	4.2	5.1	6.2	5.8	6.6	2.7	2.6	7.7	4.2	5.8	4.8
Proline	10.1	8.3	4.5	4.2	5.3	5.1	2.6	2.2	4.3	4.2	9.8	8.6
Arginine	4.3	5.0	8.5	7.8	7.4	10.6	5.3	10.0	6.4	6.6	3.7	3.4
Histidine	2.1	2.4	2.2	3.1	2.6	2.4	1.4	1.5	2.6	3.2	2.7	2.2
Aspartic acid	5.0	12.3	4.5	11.9	12.3	14.1	17.2	5.3	10.7	9.1	8.2	9.3
Glutamic acid	27.7	15.4	10.7	15.1	19.0	20.0	23.8	12.2	12.3	15.4	22.2	19.8

Compiled from *Scientific Tables: Seventh Edition* (Geigy) and other sources.

differences in total protein content and amino acid composition, with variety, for example, in navy, pinto, and black beans. Consequently, the values in Table 5.6 for the vegetable foods represent reasonable average values. The variations in the amino acid contents of plants are sufficiently great that plant breeders can alter them to some degree. Recently nutritionists have shown great interest in corn of significantly higher lysine content. (The reasons behind this interest will be apparent shortly.) In animals, the breed of chicken, for example, has relatively little to do with the amino acid content of eggs or meat, or the breed of cow with amino acid composition of milk or meat, or the race of human mother with the amino acid composition of breast milk. Furthermore, the differences in amino acid composition of meat from various species are not great.

Protein nutrition is much more complex than simply measuring the amount of protein needed for normal nutrition and identifying the essential amino acids and providing these willy-nilly in our food. Dietary protein is the raw material which the body uses for many diverse functions. Protein requirements are affected by such things as disease, trauma, activity, and even the kinds of food consumed. Then, too, the ease of chemical analysis and the mass of detailed chemical information available on protein foods tempt one to oversimplify the role of proteins in our food supply. It is beyond our purpose here to consider the detailed ramifications of menu planning or diet formulation to achieve optimum nutrition, but we must gain some appreciation of the nature of the problems involved.

Once the necessity for proteins was recognized, early nutritionists soon discovered that all proteins are not of equal nutritional value. It was found that it was one thing to consume protein and another thing to use it, because all proteins are not easily or completely digested and absorbed. In studying the efficiency of digestion of food in the gastrointestinal tract and absorption of the digested food into the bloodstream for use in the body, it was found that whereas most fats in foods are utilized to the extent of 90% or more of the amount ingested and most carbohydrates excluding cellulose 95% or more, this is not true for the protein portion of foods. The digestibility of foods can be fairly easily determined, and such studies of farm animals, which are much easier to study in this respect than humans, are of great economic importance. By measuring the amount of protein, fat, and carbohydrate consumed in a food and determining the amounts of these constituents eliminated in the feces, a digestibility coefficient for each major food constituent can be obtained that is of practical significance. The coefficient of digestibility for protein is defined as follows:

$$\text{Coefficient of digestibility} = \frac{\text{nitrogen in ingested food} - \text{nitrogen in feces}}{\text{nitrogen in ingested food}}$$

Little error is involved in using nitrogen, which is easily determined chemically because most of the nitrogen in foods is in the protein. Using similar relations, coefficients of digestion for fats and carbohydrates in particular foods can be determined. In the case of carbohydrates, cellulose is not included because man is unable to digest cellulose. Man cannot digest the proteins in hair or wool, but these are usually not in food. (A common problem in cats, dogs, and cattle, and at times in humans, is hair balls in the stomach becoming of such size as to form obstructions in the gastrointestinal tract.) Table 5.7 gives the coefficients of digestibility of the protein, fat, and carbohydrate exclusive of cellulose or total fiber, which is the common notation used in tables of food composition. It is clear from Table 5.7 that digestibility is a major consideration in dealing with human feeding problems. This is particularly true for proteins since they are in shorter supply, are harder to digest, and are more expensive than fats and carbohydrates.

If foods are not completely digested, does this concern not only the total available energy of the food discussed in the preceding section but also the protein, fat, and carbohydrate? The answer is yes. Then what is the justification for the generalization that 4, 9, and 4 Cal per gram are, respectively, the available energy in the protein, fat, and carbohydrate in food when not all these constituents are utilized completely? The actual heat produced by the combustion of most food proteins is 4.40–4.55 Cal per gram. So assuming that all dietary proteins on the average have a coefficient of digestibility of 88–89%, we get, with a small error, 4 Cal per gram of ingested protein. Similarly, for fats, which usually give 9.3–9.5 Cal per gram and have an average digestibility of about 95%, we get 9 Cal per gram.

The data in Table 5.7 reveal two factors of importance: (1) The refining of cereals results in a more digestible fraction of wheat, rice, and corn, although, as we know, this processing can decrease the content of some vitamins and minerals while making the cereal more palatable (see p. 68). (2) Excessive heat may actually destroy or render indigestible a large part of the protein and carbohydrate, as in chocolate or coffee. Some thermal degradation in other foods is desirable for flavor development, as in the crust formation on bread or in roasted, fried, or broiled foods. In such cases, however, the quantity of protein and carbohydrate converted to flavoring substances is very small.

Even when variations in digestibility were accounted for, pioneering investigators found that some proteins are more nutritious than others. This led to the concept of the biological value of proteins. The proteins of eggs, milk, and meat were found superior to proteins of many cereals, legumes, and other vegetables. Accurate determination of biological value for humans or other animals, although simple in theory, is somewhat involved practically. It was found that smaller amounts of human milk proteins would provide for

Table 5.7. Approximate Coefficients of Digestibility of Proteins, Fats, and Carbohydrates in Selected Foods

	Coefficients of digestibility (%)		
	Protein	Fat	Carbohydrate
Animal food products			
Eggs, meat, fish, milk	97	95	98
Fats			
Butter, margarine, animal and vegetable fats and oils		95	
Fruits			
All fruits	85	90	90
All fruit juices	85	90	98
Grain and grain products			
Wheat, whole wheat flour	79	90	90
Wheat, flour 85–93% extraction[a]	83	90	94
Wheat, flour 70–74% extraction[b]	89	90	98
Wheat, flaked, puffed, shredded, etc.	79	90	90
Wheat, bran	40	90	56
Macaroni, spaghetti, etc.	86	90	98
Corn, whole grain meal	60	90	96
Corn, degermed meal	76	90	99
Rice, unpolished, brown	75	90	98
Rice, polished, white	84	90	99
Barley, pearled	78	90	94
Oats, oatmeal or rolled oats	76	90	98
Rye, whole grain flour	67	90	92
Legumes, nuts			
Mature or immature peas, beans, lima beans, soybeans, soy flour, nuts	78	90	97
Vegetables			
Potatoes and roots such as beets and carrots	74	90	96
Mushrooms	70	90	85
Other vegetables	65	90	85
Sugars			
Cane or beet (sucrose), glucose			98
Miscellaneous			
Yeast	80	90	80
Chocolate[c]	42	90	32

Adapted from data in *Composition of Foods*, Agricultural Handbook No. 8, Bernice. K Watt and Anabelle L. Merrill, U.S. Department of Agriculture, 1963.

[a] Bread or all-purpose flour.
[b] Cake flours.
[c] Only these foods in this table are prepared by roasting at 150°C (302°F) or above, which makes much of the protein and carbohydrate rather poorly digestible.

good growth and nutrition than other proteins, although those of whole egg were not far behind in value, with cow's milk and meat next. It was also observed that some proteins of poor biological value when consumed together would have a much improved biological value, although never higher than that of human milk or egg proteins. This led to much research relating the biological value of proteins to the kinds and amounts of amino acids in proteins, with the resulting identification of the essential amino acids.

Identification and proof of the essentiality of the amino acids in proteins were formidable tasks spanning four decades of diligent research by many workers. Determination of how much of each amino acid is required by individuals is equally difficult and is currently under intensive investigation. Among the reasons for the difficulty are variations in individual physiology and the fact that dietary proteins and amino acids serve so many functions that absolute amino acid requirements vary according to many factors such as period in the life cycle of the test animal, its activity, time of feeding, and so on. For example, amino acids are used most efficiently when all are metabolically available at the same time since apparently there is no selective storage of amino acids in the body. All amino acids must be present in adequate amounts when the cells of the body make their own protein. Statements about absolute requirements of proteins and amino acids in an individual's diet or the average diet of a population require some knowledge of which individual or which population is meant. Nevertheless, we can get some understanding of the problems confronting certain populations of the world where food (calories), particularly protein, is in short supply. At the same time, we can gain considerable appreciation for certain dietary practices which have evolved in different cultures throughout the world. We are here today because our ancestors developed certain food habits that were nutritionally sound, largely based on the consumption of a variety of foods of animal and plant origin.

Let us now take a closer look at Table 5.6. Here are listed the amino acid composition of major foods used throughout the world: three cereals, wheat, rice, and corn; three leguminous seeds, beans, soybeans, and peanuts; a tuber, potatoes, consumed more in temperature climates; and a root, cassava, a common tropical food; and four animal foods, eggs, meat, cow's milk, and human milk. Although these foods may be refined, they are often consumed whole or after a minimum of processing and refining, particularly in underdeveloped areas of the world. Therefore, the data in Table 5.6 are average values for the entire food, for example, whole wheat. Bear in mind that there is more variability in foods of plant origin than of animal origin.

There is good evidence that the proteins of human milk are those most nutritious and most efficiently used by humans. As already mentioned, researchers have found that the proteins of whole chicken eggs are almost equal to human milk proteins. Since eggs are easier to obtain, they are often used as the standard proteins for human nutritional studies.

It can be shown that most 70-kilogram men (Table 5.1) can be maintained in adequate protein nutrition on only about 35 grams of proteins of optimum biological value, say human milk proteins. We can use the data in Table 5.6 to compare the proteins of each food with those of human milk as the standard. This has been done in Table 5.8, which shows the grams of the proteins of each food necessary to supply the same amount of each essential amino acid as is present in 35 grams of human milk proteins. It is quite simple to see the approximate relative nutritional value of the proteins of the various foods—the higher the number, the poorer the quality of the protein as a source of each amino acid. Further, for each food the limiting amino acids are at once apparent; they are those which occur in the lowest concentration relative to human milk proteins and thus set the limits of the nutritive quality of the proteins of the food. Consequently, in Table 5.8 the limiting amino acids are those of the highest numbers. For example, 80.4 grams of wheat proteins is necessary to give the same amount of lysine as 35 grams of human milk proteins. Whereas wheat is a poor source of lysine, corn is a poor source of both lysine and tryptophan. Of the cereals, rice has the best-quality protein from the standpoint of amino acids.

Beans are an excellent source of lysine and can compensate for the lysine deficiency of some cereals. Note that cow's milk and meat are about equivalent to egg—quite close to human milk. Although not as good as human milk, whole soybean proteins approach meat and milk. Potato proteins are fairly good except for leucine, and cassava proteins are the poorest of all.

But Table 5.8 is not the whole nutritional story by any means, even though it clarifies the significance of essential amino acids. It is not enough that the food contain the necessary quantities of amino acids; there are two additional factors of major concern: (1) Are the food and its proteins digested and utilized completely? (2) What else of nutritional significance does the food contain or not contain? Table 5.9 points up the importance of these considerations.

Let us use the digestibility coefficients of Table 5.7 and the data of Table 5.8 to determine the actual amount of each food necessary for our 70-kilogram man to consume in order to get the equivalent of 35 grams of human milk proteins in terms of essential amino acids. From this information and the protein content of the particular food, it is simple to compute the quantity that must be consumed in order to get the protein required. But since the foods contain fats and carbohydrates as well as protein and all contribute to the total energy of the diet, Table 5.9 also shows the amount of food energy in the food which contains the necessary protein. Since our 70-kilogram man needs only 2700 Cal, it is impossible to have both energy and

Table 5.8. Grams of Proteins of Various Foods Required to Give the Essential Amino Acids Present in 35 g of Human Milk Proteins

	Wheat	Corn	Rice	Beans	Soy-beans	Peanuts	Potatoes	Cassava	Eggs	Meat	Cow's milk	Human milk
Tryptophan	43.1	70.0	40.0	46.7	40.0	46.6	43.1	43.1	31.1	43.1	40.0	35.0
Lysine	80.4	72.4	51.7	23.2	33.9	62.0	43.4	53.0	32.4	26.5	27.8	35.0
Threonine	54.4	37.5	45.0	26.7	39.4	54.3	42.6	56.3	29.7	35.0	34.2	35.0
Leucine	51.5	27.4	40.2	42.8	42.8	53.1	71.6	80.2	36.6	42.3	33.3	35.0
Isoleucine	51.6	49.0	40.9	40.0	37.0	46.7	43.6	70.0	33.8	38.4	31.6	35.0
Valine	50.5	38.8	35.0	50.5	41.0	43.4	42.6	72.3	30.2	41.0	31.0	35.0
Phenylalanine + tyrosine	39.5	35.0	29.6	43.4	35.4	35.8	43.4	67.0	32.1	40.5	28.8	35.0
Methionine + cystine	37.8	33.3	35.9	31.1	43.8	58.3	48.3	82.4	27.5	37.8	43.8	35.0

Table 5.9. Quantities and Caloric Values of Certain Whole Foods Which Must Be Ingested in Order to Obtain the Nutritional Equivalent of 35 g of Human Milk Proteins (Limiting Amino Acids Are Given in Parentheses in the Column Heads)

	Wheat (lysine)	Corn (tryptophan + lysine)	Rice (lysine)	Beans (valine)	Soybeans (methionine + cystine)	Peanuts (lysine)	Potatoes (leucine)	Cassava (methionine + cystine)	Eggs (leucine)	Meat (tryptophan)	Cow's milk (methionine + cystine)
Grams of protein equivalent to 35 g human milk proteins	80.4	72.4	51.7	50.5	43.8	62.0	71.6	82.4	36.6	43.1	43.8
Digestibility coefficient	0.79	0.60	0.75	0.78	0.78	0.78	0.74	0.60	0.97	0.97	0.97
Grams of protein to be ingested	102	120	68.9	64.8	56.2	79.5	96.7	137	37.8	44.4	45.2
Percent protein in the food	13.3	7.8	7.5	24.0	34.0	26.2	2.1	1.1	12.8	21.5	3.2
Grams (pounds) of food to be ingested	767 (1.69)	1540 (3.39)	919 (2.02)	270 (0.594)	165 (0.364)	303 (0.667)	4600 (10.1)	12,500 (27.5)	295 (0.65)	206 (0.45)	1410 (3.1)
Calories per 100 g[a]	333	368	360	338	403	582	76	131	162	143	64
Calories to be consumed in order to get adequate protein	2560	5660	3310	913	665	1770	3500	16,400	477	295	903

[a] See Appendix.

protein requirements in balance if he eats only one cereal, root, or tuber as his diet. The only possible exception may be wheat. From Table 5.9 we see that it is impossible for a grown man to eat enough corn, rice, potatoes, or cassava to meet his protein requirements. Now we can understand a recent statement by the Food and Agricultural Organization of the United Nations that young children simply cannot ingest enough of these staples to meet both energy and protein requirements.

In some underdeveloped countries, corn and cassava make up the major part of the diet. Consequently, protein deficiency is common and is particularly acute and critical for babies at weaning time and for growing children. Kwashiorkor, acute protein deficiency, kills many children in large areas of the world. When corn or cassava replaces mother's milk, an adequate diet suddenly becomes deficient in protein because the child simply cannot consume enough to meet his requirements.

Tables 5.8 and 5.9 show quite clearly the importance of a diet of mixed foods to achieve good nutrition. Eggs and meat not only provide high-quality protein but also allow room in the diet for foods containing other necessary nutrients. Milk, beans, and soybeans also, if consumed in sufficient quantities to supply adequate protein, can permit other foods in the diet to supply the necessary calories and other nutrients listed in Table 5.1.

You will recall that on p. 88 we indicated that a mixture of proteins from different sources could have a greater biological value than the proteins of a single food. Tables 5.8 and 5.9 reveal why this is so. Note that when all of the proteins are supplied by rice 51.7 grams of rice proteins are necessary, and when all of the proteins come from beans 50.5 grams are required (Table 5.8). For rice lysine is limiting, and for beans valine is limiting. If rice and beans make up the diet and if each food supplies one-half of the dietary protein, only 43.3 grams of protein is necessary to equal 35 grams of human milk protein. Thus a 50/50 mixture of bean and rice proteins is more nutritious than either alone—approximately equal to soybean and meat proteins. This improved nutritional value will permit about a 16% reduction in the total amount of rice and beans consumed. Thus where 3310 calories of rice or 913 calories of beans will supply the necessary protein for our 70-kilogram man, only 1800 calories of a mixture of rice and beans is necessary to give equally nutritious proteins. This leaves 900 calories to come from other foods, which can supply nutrients not present in the mixture of 388 grams of rice and 113 grams of beans required to furnish the 43.3 grams of mixed proteins. You are encouraged to make similar computations for other combinations of foods in order to get a better understanding of the importance of a diet of a variety of foods. You will also appreciate even more the empirical foundation for the evolution of human food habits in different parts of the world.

After oxygen, water, and energy, proteins are the most critical nutrients.

Food proteins are in short supply worldwide, and much scientific and technological effort is being expended to try to increase protein supplies. Many potential sources of protein are either unpalatable or toxic. Whole soybeans have problems along this line. Food must be pleasing and palatable, because the most nutritious food is useless if no one will eat it. Custom, culture, and even snob appeal profoundly affect nutrition. New foods, even very nutritious and cheap foods, are very difficult to introduce into under-developed areas—particularly when such foods are introduced by persons who themselves do not eat the new foods.

There are a number of unpalatable and/or toxic potential sources of food proteins of high nutritional value. By proper refining and processing, these sources can yield excellent proteins for incorporation into food. Among these sources of protein are yeasts, fungi, and bacteria that can grow on organic matter inedible to man such as cellulose and petroleum; seeds from cotton, rape, vetch, etc., and leaves from various plants; certain fishes and other marine or terrestrial animals. For centuries, man has used refined sugar prepared from cane or beets, largely inedible. He has used refined fats and oils for almost as long. Protein refining is technically more complex, but today refining of protein for food use from organisms not usually considered

Table 5.10. Daily Protein Requirements

Age (yr)	Average weight (kg)	Grams of protein	
		Total	Per kg body weight
0–½	6	13.2	2.2
½–1	9	18	2.0
1–3	13	23	1.77
4–6	20	30	1.50
7–10	30	36	1.20

	Males	Females	Males	Females	Males	Females
11–14	44	44	44	44	1.00	1.00
15–18	61	54	54	48	0.89	0.89
19–22	67	58	54	46	0.78	0.79
22–50	70	58	56	46	0.80	0.79
51+	70	58	56	46	0.80	0.79
Pregnancy				76		
Lactation				66		

Compiled from *Recommended Dietary Allowances*, Food and Nutrition Board, National Academy of Sciences–National Research Council, Eighth Edition, 1974.

satisfactory for food is possible. Through this technology we will see refined food protein as an increasingly important part of our total supply of this critical nutrient. Fortunately, the science of nutrition itself has developed to the point that completely nutritious food supplies can be maintained using these refined nutrients heretofore unavailable to us. This will be discussed later.

In the foregoing discussion of protein nutrients, our reference point has been the adult 70-kilogram man. We could have used other reference points in the human life cycle as well. There are some variations in protein requirements with age, size, and sex. Table 5.10 shows the daily dietary protein allowances recommended by the Food and Nutrition Board of the National Research Council. By comparing Table 5.10 with Table 5.3 you will find that the ratio of protein to energy requirement is somewhat higher shortly after birth and during the period of rapid growth at puberty than in the intervening years. The ratio begins to increase in the later 30s and continues upward in the later years such that almost 50% more protein per 1000 Cal of dietary energy is recommended at age 75 than at age 10. This reflects differences in physical activity requiring energy.

FATS

All people like fat in their diets. The amount of fat in the diet not only reflects eating habits but also the economic level of the population. In some countries, almost 40% of the total calories consumed come from fat, and this is of concern to some nutritionists. In other countries, this figure may be as low as 10%. Semirefined or refined fats—butter, margarine, animal and vegetable fats used for frying and for incorporating into other foods such as cakes, cookies, sauces, salad dressings, and confections—are widely used. Also, many foods contain natural fat. Meat and dairy products, eggs, avocados, olives, nuts, etc., may contain from a few percent to almost 80% fat. Most other fruits, vegetables, and cereals contain relatively little fat (see Appendix).

Nutritionally, fats serve as a source of calories, as carriers (solvents) of fat-soluble vitamins, and as a source of so-called essential fatty acids (Table 5.1). Physiologically, fats and lipids, which are chemically related, have several functions, but we shall mention only two: energy storage, and as integral parts of cell membranes. These membranes include the external membrane, which encloses the cell and separates it from its immediate environment, and the intracellular membranes associated with the various organelles of the cell (see Fig. 3.2 of Chapter 3). Membranes are being intensely studied to determine how they perform many of their complex functions in living systems. Most appear to be a layer of fatty or lipid material sandwiched between two layers of protein molecules specific to the function of the particular membrane.

The body carries most of its excess energy reserve in the form of fat deposited in certain areas under the skin, particularly in women; around the intestines, kidneys, etc.; and between muscles near connective tissue. Body fat is derived directly from fat ingested in the diet or is made by the cells of the body from carbohydrate or even protein when a person consumes more energy in his diet than is required (see Fig. 3.3 of Chapter 3).

Fat is insoluble in water. In this respect fat differs from carbohydrate and protein, which in actively metabolizing tissue as in a living man or other animal are usually associated with almost four times their weight of water. Furthermore, fat yields 9 Cal per gram and carbohydrate only 4 Cal per gram. So it is apparent why an animal, which lives by moving about in his environment, carries its major energy reserve in the form of fat. For you to store your fat energy reserve in the form of carbohydrate or protein would require that your body weight be two or more times greater. Our 70-kilogram man, who is not overweight, contains about 16% fat. Let us assume that only 13% is actually energy reserve and the remaining 3% is for other physiological purposes: $70 \times 0.13 = 9.1$ kilograms of fat. This represents 81,900 Cal—enough to supply 2700 Cal daily for at least 30 days. In energy available to the body this is equivalent to 20.5 kilograms (45 pounds) of carbohydrate. In man such a carbohydrate is glycogen. But glycogen imbibes about four times its weight of water. So to carry around the energy of 9.1 kilograms of fat in the form of 20.5 kilograms of glycogen would mean a weight of 100 kilograms. So our 70-kilogram man would have to weigh $70 - 9 + 100$ or 161 kilograms (354 pounds) if his excess energy were stored as carbohydrate. You will surely agree that such a burdensome load would be too much to carry about during your day-to-day activities.

In Chapter 3, we noted that ATP is the immediate metabolic energy source for most body functions. But as ATP is used it is made again and again by the oxidation of glucose or glycogen to carbon dioxide and water. As this becomes used up, fat can enter the scheme and furnish energy (see Fig. 3.4 of Chapter 3). The economy of storing reserve energy as fat for organisms such as man which must constantly move to work for food is again clear. This is a rather important evolutionary accommodation.

The lipids making up membranes often contain what the chemist calls *unsaturated fatty acids*. Dietary fats and lipids supply some of these unsaturated fatty acids (see Chapter 6) because man is unable to make them himself. Most natural fats contain enough of these substances. In a diet of mixed foods containing a moderate amount of fat, a sufficient amount of these *essential* fatty acids—3 grams of linoleic plus linolenic acid—would be consumed.

Presently, many people are trying to relate the kind of fats consumed to certain degenerative diseases of the heart and circulatory system. As yet, no cause-and-effect relationship has been proved. Nevertheless, through press and TV there has been a big push by some fat and oil industries and some of the medical profession to have people consume less saturated fat and more unsaturated fat.

Unsaturated fat contains much more of the essential fatty acids. We shall see in Chapter 9 that such fats and fatty acids are much more susceptible to oxidative destruction by oxygen from the air. You observe this in rancid fatty foods—salad oils, potato chips, nuts, etc.—and it is a major type of food spoilage. In rancid fat, the fat-soluble vitamins have been destroyed and the fats are broken down to give a great number of compounds, some of which are toxic. In living plants, such fats are protected from oxidative destruction by naturally occurring antioxidants—preservatives to protect the fats from oxygen. You will recall that green plants not only live in oxygen but also produce it. One group of natural antioxidants in plant fats are the tocopherols—vitamin E. Man does not make antioxidants to protect his fat. But one of the several functions of fat-soluble vitamin E is protection of body fat from attack by oxygen—to act as a preservative so that the fat and the lipids can perform their proper function.

It has recently been observed that as the amount of unsaturated fat or fatty acids in the diet increases so does the requirement for vitamin E and, surprisingly enough, selenium. There seems to be a relation among these three nutrients but as yet just how they interact to perform their physiological functions is not known. Nevertheless, this interrelationship makes it quite clear that the quantities of some of the nutrients required by a person can be affected even by the quantities of other nutrients consumed.

SALT AND OTHER MINERALS

All of us have experienced a craving for salt when it is absent from our diet. So universal is the use of salt (sodium chloride) in food that many nations have used a tax on salt as a means of collecting revenue. Wild animals often have favorite "salt licks" where they can obtain salt and other needed minerals. We all know that sweat contains salt and that salt as well as water is needed to recover from periods of excessive sweating resulting from hot weather and hard physical activity. At the same time we all know that too much salt is lethal for man and most animals and plants. We cannot live by drinking seawater, although some other animals can. For example, many marine birds have special salt-secreting glands. Such common observations can give us a clue to the nutritional function of salt and some other minerals.

Sodium and potassium are particularly essential to the maintenance of the cellular environment for proper performance of all biological functions (Chapter 4). In man and other higher animals, sodium chloride is in much higher concentration than potassium salts in blood plasma and intercellular fluids but the reverse is true inside working cells. *In toto*, plants depend much more on potassium. Indeed, this element is one of the most critical in fertilizers for agricultural crops, which are usually formulated and sold on the basis of their content of available nitrogen, potassium, and phosphorus.

We have already given much attention to the role of phosphorus as an integral part of nucleic acids and the prominent role of phosphate esters in metabolic processes—ATP, ADP, AMP, the sugar phosphates, etc. The list of organic phosphate intermediates in the metabolic processes of all biological systems is almost endless. The toxic effect of arsenates is due to the fact that arsenates closely resemble phosphates chemically and thus they interfere with the biological use of phosphate because arsenate esters do not function precisely like phosphate esters. Also, some chemical derivatives of phosphoric acid such as fluorophosphoric acid are extremely poisonous because they are stable and interfere with normal phosphate function.

$$
\begin{array}{ccc}
O & O & O \\
\| & \| & \| \\
HO-As-OH & HO-P-OH & HO-P-F \\
| & | & | \\
OH & OH & OH \\
\end{array}
$$

| arsenic acid | phosphoric acid | fluorophosphoric acid |
| (arsenates) | (phosphates) | (fluorophosphates) |

Less toxic phosphate derivatives, many of which slowly decompose in the presence of water to ordinary usable phosphate, form the basis of a number of excellent pesticides that are necessary in efficient food production.

Phosphates function metabolically because there are *three* OH groups for connection to other biologically important molecules. Arsenates are very similar in molecular structure, but their OH groups differ slightly in reactivity. In small amounts, arsenates can stimulate biological activity but they are lethal in larger amounts. The shape of the fluorophosphoric acid molecule is almost identical to that of phosphoric acid because the fluorine atom, F, and the OH group are almost the same size. However, F does not react as OH and fluorophosphoric acid offers only two groups for metabolic function. H_2PO_3F is therefore extremely toxic. Fluoroacetic acid, FCH_2COOH, is also very poisonous. See if you can understand why by examining Fig. 3.4 of Chapter 3.

Since we are very conscious of our bones and teeth, we are often made aware of the importance of both calcium and phosphate in our diets. But, like phosphate, calcium and the other mineral ions have important metabolic functions, because proteins (enzymes) often require certain of these ions to perform their function. Magnesium is involved in muscular contraction and calcium in blood clotting. If calcium ions (Ca^{2+}) are removed from blood, it will not clot, which is why blood is collected in a buffer solution of sodium citrate for transfusion. The sodium citrate removes the Ca^{2+}. After transfusion, the citrate is metabolized (Fig. 3.4) and the transfused blood regains its normal ability to clot. The calcium ion content of human blood is rigorously controlled at between 9 and 11 milligrams of Ca^{2+} per 100 milliliters of plasma. (Blood plasma is blood excluding the cells.)

We have already indicated the importance of iron in the oxygen-carrying proteins of blood and muscle (hemoglobin and myoglobin, respectively) and in the electron-transporting enzymes so necessary in ATP production during the oxidation (combustion) of our food (Chapter 3).

Other minerals are known to be required for various vital enzymes to function. Zinc, copper, cobalt, and the other trace metal ions of Table 5.1 belong in this category.

Iodine and fluorine are the other halogen elements besides chlorine required in our diets. Fluorine is necessary for proper mineralization of bone and teeth. Iodine has a unique place in nutrition. It was discovered many years ago that one of the most common causes of goiter (oversized thyroid gland) was iodine deficiency. Iodine is necessary for the thyroid to make its iodine-containing hormones, which literally control the rate of metabolism of all of our tissues. So when there is not enough iodine in the diet, the metabolic rate drops and there is metabolic disorganization. In young children with iodine deficiency or in newborn infants of iodine-deficient mothers, the disorder is manifested as cretinism—a stunting of growth and an irreparable mental retardation. These problems occur in areas of the world where iodine is deficient in soils and so foods contain too little iodine. To solve the problem, iodine as 0.01% potassium iodide is added to table salt. This was perhaps the first nutrient intentionally added to our food supply. Iodine deficiency has been largely overcome but is on the increase again for two reasons: (1) Some people who need it will not buy and use iodized salt. (2) Iodized salt cannot be used in many processed foods which stand for extended periods of time because they may acquire the objectionable medicinal flavor of free iodine. (Iodide salts are tasteless in the quantities used but may yield free iodine of very strong odor and flavor if the food is oxidized to any extent on standing.) With the rapidly increasing use of ready-prepared food containing no iodized salt, the danger of iodine deficiency increases. Seafoods generally are good sources of iodine but in the United States as a whole, relatively little of these foods are consumed.

THE VITAMINS

No group of nutrients has captured the imagination of the public more than the vitamins. Because of television, press, and radio advertising, many people still seem to equate good nutrition with vitamins. How miraculous it seemed that such a tiny amount of a nutrient could cure beri-beri, scurvy, rickets, pellagra, etc., scourges known for centuries! No one wanted to be afflicted. Mothers wanted to protect their babies and children. Vitamin supplements were prescribed, often when not needed. It is little wonder then that vitamins received far more attention from the general population than any other nutrient.

The recognition, isolation, synthesis, and proof of nutritional necessity of all these organic compounds took 70 years—spanning the growth and maturation of the science of nutrition. As a number of vitamins were isolated and tested, it was discovered that some of the deficiency diseases were due to the lack of more than one vitamin. Then, too, as purified vitamins became available to nutritionists, new deficiency diseases were recognized. From such a situation, popular interest in vitamins was sustained for many years. Besides human disease factors, the economics of farm animal production provided interest and funds for nutrition research. Indeed, because of this, nutrition was a focal point around which the modern science of biochemistry developed.

To go into detail concerning the functions of vitamins would be a formidable exercise in the minutiae of biochemistry and physiology. However, perusal of the chemical formulas in Fig. 5.1 clearly reveals the complexity of learning what vitamins actually do and relating these biochemical functions to recognizable diseases. The quest for such detailed knowledge continues, for it is fundamentally important to the technologies of modern medicine and agriculture. Nevertheless, we can still appreciate the importance of the various vitamins by briefly relating the vitamins to their functions in general descriptive terms. This will once again make you aware of the importance of having a diet composed of a variety of food organisms properly processed and prepared. You will also gain some insight into the importance of fortifying some processed foods by the addition of vitamins prepared synthetically or obtained from organisms not generally used for food.

Fat-Soluble Vitamins

For convenience, vitamins are grouped into fat-soluble and water-soluble vitamins. Four vitamins are classified as fat soluble—A, D, E, and K—for they naturally occur in the fat or other fat-soluble fractions of biological

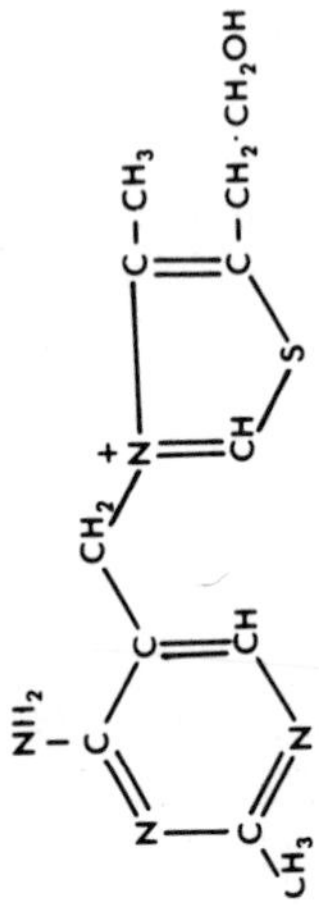

Thiamine, B_1

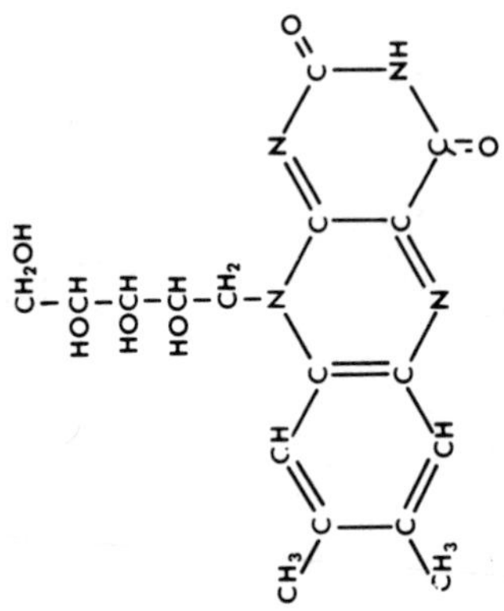

Riboflavin, B_2

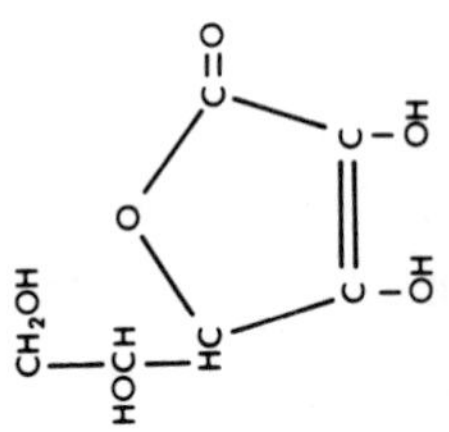

Vitamin C, L-ascorbic acid

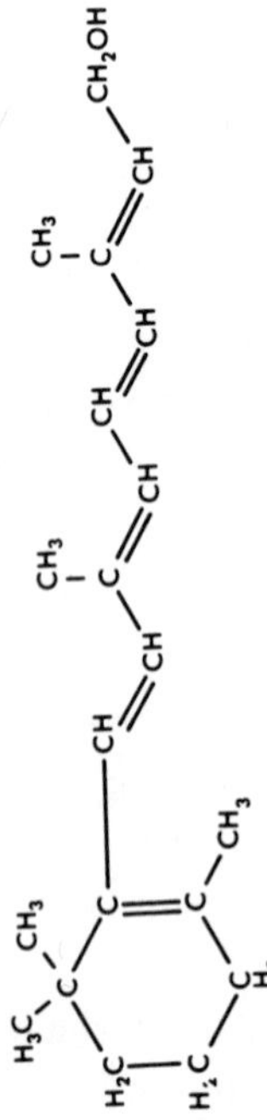

Vitamin A_2, retinol

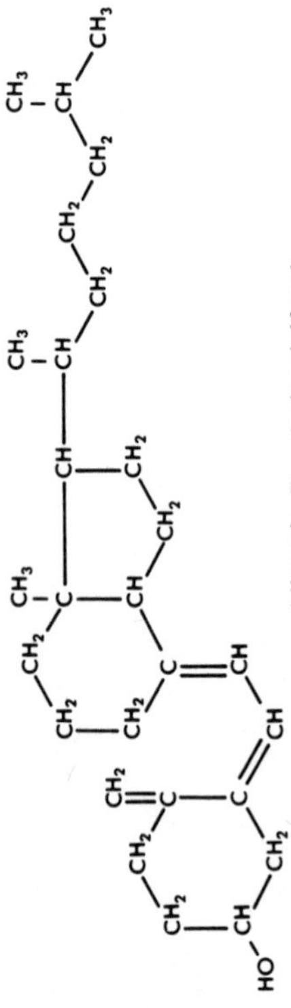

Vitamin D_3, cholecalciferol

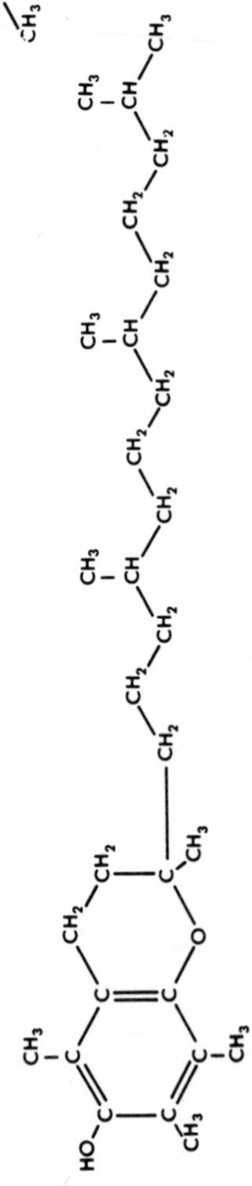

Vitamin E, α-tocopherol

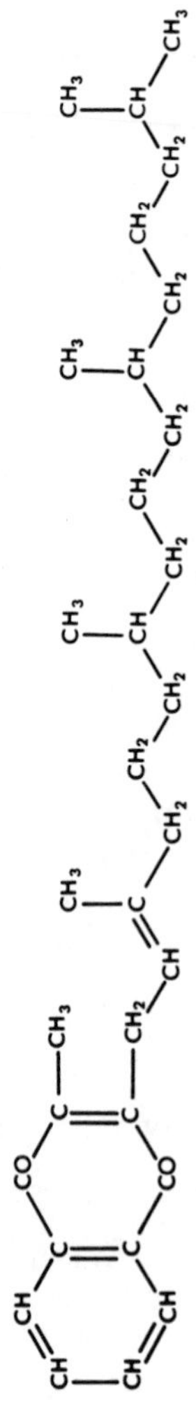

Vitamin K_1, phylloquinone

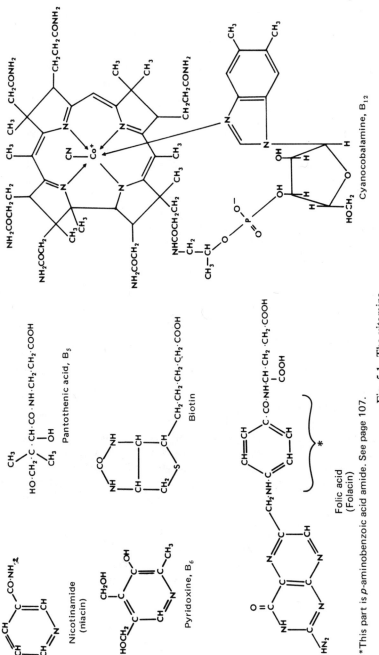

Cyanocobalamine, B_{12}

Pantothenic acid, B_5

Biotin

Folic acid
(Folacin)

Nicotinamide
(niacin)

Pyridoxine, B_6

*This part is p-aminobenzoic acid amide. See page 107.

Fig. 5.1. The vitamins.

systems (see Chapter 6). In each case, it has been found that a number of closely related molecules may serve the vitamin function.

Vitamin A appears to be required for the proper function of a number of tissues in the body. The tissues primarily affected by lack of vitamin A all appear to be derived from the ectodermal cells of the embryo—epidermis of the skin and mouth, sweat glands, nerve tissue, and receptor cells of the sense organs. Vitamin A definitely is an integral part of the photosensitive pigments in the eye. Impaired vision in dim light is an early symptom of vitamin A deficiency.

Vitamin A is synthesized in the livers of humans and other animals, including fishes, from the carotenoid pigments of plants. These pigments are some of the yellow, orange, and red oil-soluble colors in yellow, orange, and green plant tissues. The best of these pigments for producing vitamin A is β-carotene, a compound with a molecular formula of $C_{40}H_{64}$ (see Fig. 3.7 of Chapter 3). As you might guess, carrots get their color from the carotene pigments, which were first isolated in 1831. A century elapsed before their nutritional qualities were established. Liver is a good source of vitamin A, as this organ can store it. Fish liver is an excellent source of both vitamins A and D and codliver oil was one of the early sources of these vitamins, even though it may contain only 1/400 that in the liver oil of the swordfish or 1/1000 that in the liver of the black sea bass.

Vitamin D, the sunshine vitamin, prevents rickets, a crippling disease of bone formation and is necessary for the proper absorption of calcium from the intestinal tract. It is definitely needed for growing children who do not get enough exposure to ultraviolet light. Its necessity for adults is more difficult to demonstrate. Like A, D can be consumed as a vitamin from such animal foods as liver, eggs, and fish liver oil. However, it does not need to be fed if the skin is exposed periodically to ultraviolet light—full sunlight. (Glass absorbs ultraviolet light.) The human skin as well as exposed parts of other animals contains certain sterols—complex fatty alcohols—which are converted by the action of ultraviolet light to vitamin D, or calciferol. This in turn finds its way into the general circulation. The best provitamin D for humans is 7-dehydrocholesterol, which is synthesized in the body from carbohydrate or fats without difficulty, as is its close relative, cholesterol.

Vitamin E, or the tocopherols, was first discovered in relation to the inability of rats to reproduce. The embryos in pregnant rats deficient in vitamin E would die and be resorbed in the terminal stages of pregnancy. The tocopherols are a group of four closely related fatty alcohols abundant in plants, particularly in the oils of plants. It was discovered early that the tocopherols are preservatives for fats—they prevent action by oxygen to cause rancidity. Also, lack of vitamin E has been implicated in certain types of muscular dystrophy and paralysis of rapidly growing rats and rabbits.

Vitamin K is involved with proper coagulation of blood. Although it is not an integral part of the proteins responsible for blood coagulation, without it some of the necessary proteins for this important protective mechanism are not made by the body.

The history of vitamin K is fascinating. The Danish biochemist H. Dam was trying to find the vitamin whose deficiency was responsible for a bleeding disease in chickens at the same time that the American biochemist K. P. Link was trying to determine the toxic factor in spoiled clover which caused fatal hemorrhages in livestock. The two lines of research converged. One found vitamin K and the other found a closely related compound molecularly antagonistic to vitamin K. Both compounds led to important clinical drugs. The vitamin in water-soluble form is used for control of hemorrhagic disease of newborns and to treat biliary obstruction (lack of bile prevents fat and vitamin K absorption; see Chapter 7). Discovery of the antagonist led to the synthesis of coumadin, which is used to control blood coagulation in persons whose blood tends to clot too fast and form thrombi (clots in the heart, brain, extremities, etc.). The same chemical compound under the name warfarin is an extremely effective rat poison. Such are the fruits of the science of nutrition.

Water-Soluble Vitamins

The water-soluble vitamins include several known as the B complex as well as vitamin C. In the early years of nutrition research, the water-soluble factor B was found to contain a substance which cured scurvy but which did not cure diseases such as beri-beri. The antiscurvy factor was assigned the letter C. The B factor was found to contain a large group of nutritional factors. As the B vitamins were isolated, purified, and tested, the need for other substances was shown until finally B_{12} was isolated. Now we tend to use simple names derived from the more precise chemical names or biological terms rather than B_1, B_2, etc. The nutrients belonging to this group all appear to be related to specific enzymes responsible for certain biochemical processes.

Enzymes are protein molecules, but many of these proteins must have certain nonprotein molecules (or *coenzymes*) attached in order to function properly. The vitamins of the B complex are parts of many coenzymes and hence part of the total enzyme. In such cases, the protein part—or *apo-enzyme*—is not a catalyst. It is inactive until the coenzyme becomes attached:

Apoenzyme   + coenzyme $\longrightarrow$ enzyme
specific protein, inactive   inactive   active biochemical
                                                                            catalyst

The apoenzyme might be likened to the handle of a specific tool to do a specific job and the coenzyme to the tool head to fit the specific bolt, nut, or other fitting. Neither the handle nor the head of the tool can function alone, but when they are properly joined the complete tool can function. Since the vitamins of the B group are coenzymes necessary to perform specific biochemical reactions, scientists tend to think of these vitamins with regard to specific metabolic functions rather than certain diseases. This has already been apparent from Figs. 3.4 and 3.5 of Chapter 3.

The history of each vitamin is a fascinating story but beyond our purpose here. Nor can we relate in detail specific food sources for each. (Information of this type is given in the Appendix.) It is well to keep in mind, however, that since these water-soluble vitamins are involved in fundamental metabolic processes, we might expect actively metabolizing tissue to be a good source of these nutrients. This is precisely the case. Animal tissues—notably liver, eggs, muscle, and milk—are generally good sources. Actively growing tissues of vegetables and fruits are, also. The plant embryos within the seeds generally contain more vitamins per unit weight than the other seed parts. Some of the vitamins are commercially prepared from actively growing microorganisms. It is easier to make some of the others synthetically from nonbiological sources.

Ascorbic Acid, Vitamin C, is unique in that of the higher animals only man, other primates, and the guinea pig must have ascorbic acid in their diets. All others make their own. Scurvy is an often fatal disease that is usually associated with early sailors, explorers, etc., who had to live without fresh foods for long periods of time. As its name implies, ascorbic acid in the diet will prevent scurvy, which is caused by severe deficiency. Although it is required in fairly large amounts, the functions of vitamin C are not as precisely known as those of other vitamins. Ascorbic acid has been related to function of connective tissues, resistance to infection, elimination of potential toxins, etc.

Vitamin C in the presence of oxygen can be rather unstable at elevated temperatures or when certain enzymes are also present. Raw milk contains vitamin C, but there is very little in pasteurized homogenized milk. The layman thinks of ascorbic acid as something present in fruits, particularly oranges and lemons, but it is also present in many vegetables. In many areas of the world such fruits are rare, and potatoes, onions, and cabbage are consumed in large quantities and serve as vitamin C sources.

Thiamine, Vitamin B_1, was the first to be purified, crystallized, and synthesized, after more than 40 years of work. Beri-beri is the disease associated with B_1 deficiency. Biochemists relate thiamine to the removal (decarboxylation) of carbon dioxide from pyruvic acid, a key reaction in carbohydrate metabolism (Fig. 3.4 of Chapter 3). Thiamine is part of the coenzyme for this process—of greatest importance in brain and nerve tissue, which must use carbohydrate and particularly pyruvic or lactic acid as their energy source.

Riboflavin, Vitamin B_2, gives the greenish-yellow color to whey, the clear liquid remaining when the casein of milk is coagulated and removed to make cheese. Riboflavin is part of the coenzyme for a number of hydrogen-carrying enzymes.

Niacin, or *nicotinic acid amide,* forms part of the coenzymes for a large number of hydrogen-carrying enzymes. Figures 3.4 and 3.5 of Chapter 3 clearly show the key role of niacin and riboflavin in biological oxidation-reduction reactions. Table 5.1 shows that almost 40 molecules of niacin are required for every riboflavin molecule. Such is niacin's importance in biological oxidation and reduction processes. Niacin deficiency in man results in pellagra, a disease once fairly common in the southern part of the United States among poor people living on diets composed largely of corn.

Patothenic acid, was once christened the "gray-hair" vitamin by the popular press to glamorize nutrition. Its name, from Greek for *from anywhere,* indicates the widespread occurrence of this essential nutrient. Now that its function is rather well understood, we can understand why it is everywhere. Acetic acid, the acid of vinegar, is the primary intermediate in carbohydrate and fat metabolism and in the biosynthesis of sterols, carotenoid pigments, and even rubber. Acetic acid performs its biochemical functions through the use of a pantothenic acid–containing coenzyme. Many other organic acids also require pantothenic acid–containing enzymes.

Pyridoxine, Vitamin B_6, is found in the coenzymes for many of the enzymes involved in amino acid metabolism. In Fig. 3.4 of Chapter 3 it may be seen that whereas pantothenic acid–containing enzymes connect carbohydrate and fat metabolism, pyridoxine-containing enzymes connect carbohydrate with amino acid and protein metabolism.

Biotin gets its name from *bios,* a nutritional factor postulated for yeast as early as 1869 by Justus von Liebig. It is a required nutrient for many microorganisms as well as for man and other higher animals. We have already referred to the chicken, whose eggs contain a good deal of biotin in the yolk and an antibiotic in the white which reacts with any biotin in the white so that most bacteria cannot live there and kill the chick embryo. The amount of biotin required is relatively small. Biotin is found in enzymes which carboxylate, or put carbon dioxide (CO_2) in, certain compounds. We usually think of CO_2 as a waste product of metabolism. It is, for the most part, in animals, but even here some CO_2 is used in making certain essential intermediates.

Folic acid, Folacin, literally *leaf acid,* is indeed found in leaves. This essential nutrient also has a fascinating history. A human macrocytic anemia of nutritional origin was described in the early 1930s by Lucy Wills, a physician working in India. A few years later, Paul Day of the University of Arkansas reported that an additional vitamin was needed to prevent an anemia in monkeys fed all of the then known nutrients. He called it vitamin M. Others working on the nutrition of microorganisms reported that unknown factors were needed for certain of them. All of these studies ultimately led to the same molecule, folic acid. We now relate this nutrient to the enzymes capable of transfering one-carbon units, the formyl group (—CHO), in metabolic reactions. Because an anemia had been shown to be related to this vitamin and pernicious anemia is a dreaded human disease, much public attention was focused on folic acid. But, alas, this new-found nutrient was not the answer, although it helped in some cases.

Vitamin B$_{12}$, or *cobalamin*, was discovered soon after the isolation and proof of structure of folic acid. B$_{12}$, a very complex cobalt-containing molecule, seemed to pick up where folic acid left off in nutritional anemia problems. It is required in very minute amounts. Its isolation and proof of structure were triumphs for newly developed techniques of biochemistry and organic chemistry. This clinically valuable vitamin is the substance in liver shown in 1926 to have a beneficial effect in pernicious anemia. It is known that vitamin B$_{12}$ is part of the coenzyme related to several enzymes involved in metabolically important molecular transformations.

BENEFITS TO MAN FROM THE SCIENCE OF NUTRITION

Every animal, every plant, every microorganism has a specific set of nutrient requirements—molecules and atoms it must have to go through its complete life cycle and begin a new generation. What these molecules are and how they function is the science of nutrition. It is the basic core of knowledge on which all the technologies for feeding organisms rest.

Agronomists have developed a sophisticated technology for fertilizing (feeding) and cultivating needed plants. This has given more feed for animals and more food for man. Animal scientists have developed an efficient technology for growing animals to produce milk, meat, and eggs for our food supply. Food scientists and technologists must efficiently bring all of these foods to man in usable form while at the same time preserving all necessary nutrients. In developing new foods to meet the ever-expanding food requirements for our increasing propulation, the science of nutrition provides guidelines for the evaluation of foods.

Because of the development of the science of nutrition and its resultant technologies, the twentieth century has seen remarkable benefits for many people. The struggle for existence has been eased. Without the science of nutrition, starvation or the threat of it would hang like a pall over a larger proportion of the world than it does now. Whereas in many parts of the world production of milk per cow is still only about 1 liter (about a quart) per day, in the United States the *average* per cow is almost 12 liters per day, which means that a large proportion of cows far exceed this amount. Similar statistics could be given for meat, milk, grain, fruit, and vegetable production. Without these benefits in food, we could not support our highly developed economy. On the average, one farmer in the United States produces food for himself and 53 others. The science of nutrition has made much of this possible.

Good nutrition means better health. Well-nourished people are less apt to become victims of the parasites and infectious diseases which plague so many undernourished people of the world. The number of people afflicted with nutritional deficiency diseases has been reduced to the extent that in some areas physicians have difficulty recognizing scurvy or beri-beri. The direct contribution of the science of nutrition to the technology of modern medicine and health care is clear. The role of nutrition in public health education is even greater. People must learn to feed themselves properly.

The science of nutrition has made another more subtle but nonetheless important contribution to the treatment of disease itself. Treatment of disease with drugs dates from antiquity. Modern chemotherapy dates from the turn of this century with the imaginative Paul Erhlich and his "magic bullet." His idea was that it should be possible to find a chemical compound that would be specifically toxic to an invading organism but not to man. From his search he developed arsenic-containing drugs for the treatment of syphilis. But as chemotherapy slowly developed, newer concepts were needed.

In the early 1930s, a long-known dyestuff intermediate, sulfanilic acid amide or sulfanilamide, was found to be somewhat effective in certain bacterial infections, although the drug had many adverse effects on patients. During this time, numerous investigators were also working on nutritional problems. Many bacteria and other microorganisms were known to require folic acid or a part of the folic acid molecule, para-aminobenzoic acid. The molecular structures of para-aminobenzoic acid and sulfanilic acid (and their amides) are quite similar:

| p-Aminobenzoic acid amide | p-Aminobenzoic acid | Sulfanilic acid | Sulfanilamide |

Could it be that these two compounds, one a nutrient and the other a poison for bacteria, were antagonistic? Could the one interfere with the other? It was so! The toxic effect of sulfanilic acid or its amide could be overcome by feeding p-aminobenzoic acid or its amide to the affected organism. As this work developed, the vitamin K–coumarin story unfolded—one an essential nutrient, the other a toxic substance. The two compounds were very similar in molecular structure and were antagonistic to each other in biological activities. Thus a new concept of chemotherapy was born. The usefulness of this idea expanded far beyond the nutrients.

This concept has had a profound effect on biological and biochemical research generally. Essentially it is that metabolic processes can be affected by molecules structurally related to normal metabolites whether such metabolites be vitamins, hormones, amino acids, etc. Organic chemists are capable of making multitudes of compounds molecularly related to compounds having specific functions in biological systems. By testing such compounds on organisms, it has been possible to find substances effective for specific biological problems whether they be safe control of invading bacteria or parasites in the human body or control of insect pests in a field of soybeans. Thus the benefits of the science of nutrition are manifested in diverse ways. Similar interactions of the various fields of basic science and technology are common in our modern society.

As new ideas and concepts are translated into new technologies bringing changes benefiting a large society, new and unanticipated problems sometimes appear. So it is with the science of nutrition. During the life cycle of any organism, there seem to be certain periods in which certain events must take place. In humans and in many higher animals, growth and ultimate body size may be adversely affected by poor nutrition during certain periods. This is particularly so for brain growth and subsequent mental development. These problems can be overcome by good nutrition. Indeed, good nutrition throughout the life cycle can permit an organism to grow, develop, and reproduce to its fullest genetic potential. Vital statistics show not only that better nutrition means better health so that more people may live their natural life span but also that better nutrition has produced a population composed of taller and larger people than their forebears. Thus through better nutrition we are reaching our full growth potential. We are also reaching our full reproduction potential.

In 1850, the average age of puberty for girls was 17 years of age. Today it is 12.6. Boys reach puberty somewhat later than girls, but a similar drop has occurred in the age of puberty for boys. The problems of adolescence, the sex drive, and the ability to reproduce appear not in the terminal years of secondary school and the beginning of college but in the junior high years. Indeed, illegitimate children of 12- to 14-year-old mothers and fathers have created a real social problem. Yet our sophisticated economy has required an extension of the number of years we must spend in the process of education and at the same time remain dependent on our parents. For many young men and women, the first and most active decade of their reproductive life occurs while they are in school and still dependent and economically unable to support themselves or a family. Thus better nutrition has presented us with some major sociological and educational problems which as yet are unsolved. One is tempted to ask whether better nutrition advances the age of learning to the point that even elementary school should be speeded up.

One other adverse effect of better nutrition coupled with labor-saving devices of our modern cultures has been the greater tendency of what some have called "overnutrition"—overweight or obesity. The automobile and all manner of machinery to accomplish the tasks of home, farm, and factory with a minimum of manual labor have reduced per capita energy requirements. Our culture is more sedentary. More persons are consuming more calories than necessary and therefore gain weight to the point where the work of carrying the excess weight (fat) equalizes the energy of the food consumed. Thus weight control, or lack of it, is the topic of daily conversation. For many, the pleasure of eating conflicts the discipline needed to control what and how much to eat in order to assure good nutrition without obesity and its associated problems.

The foregoing pages summarize many of the salient features of the science of nutrition as they concern us. It should be clear that much of the basic knowledge is available to attack many of the problems of feeding ourselves. By using this knowledge widely, we should be able to feed ourselves for years to come. Similarly, the science of nutrition may point the way for those people living in food-deficient areas of the world to learn to feed themselves better.

WHAT IS OUR FOOD MADE OF?

> What are little girls made of?
> Sugar and spice and everything nice.
> What are little boys made of?
> Rats and snails and puppy dog tails.
> What are little plants made of?
> Water and air and sunlight fair.
> What are little animals made of?
> Plants and worms, anything that squirms.
> What are real people made of?
> Food and drink and brains that think.
> That's what we are made of.

The driving force which makes life possible is the sun. By photosynthesis, plants, the autotrophs, grow and start new generations. The heterotrophs, including man, eventually use all of the substance produced by photosynthesis and convert it back to the raw materials of photosynthesis—water, carbon dioxide, nitrogen, etc. By such processes the heterotrophs get their nutrients and energy to complete their life cycles and start new generations. The amount of material produced by biological systems is equal to the amount of such material consumed and decomposed by biological systems. Consequently, there must be systems for building protoplasmic material and systems for breaking it down. Otherwise, the world would have long since been buried in biological or organic waste.[1]

Viewing biological activity in this way tells us much about the unity and very little about the great diversity of biological organisms. No single organism by itself completes the cycle of building up and tearing down; some organisms are producers (autotrophs) and some are consumers (heterotrophs). Ecologists are much concerned about the various trophic (feeding) levels of organisms occupying any particular place or ecosystem. In terms of individual organisms,

---

[1] The fossil fuels, coal and petroleum, are accumulated biological material which has not yet been consumed.

biological systems often appear radically different from each other. The processes of evolution and the struggle for survival have resulted in organisms of such great diversity that biologists can fairly say that almost anything or any process can be found somewhere in the realm of biology.

When we study the organisms which we use for food, the diversity of biological systems is obvious. Perhaps this is one of the most important reasons for man's survival. We are omnivorous creatures using both plants and animals for food. As we identified our nutrients in Chapter 5, it became clear that eating both plant and animal foods appears almost essential; doing so offers the greatest margin of nutritional safety and latitude of food supply. It is impossible here to go into detail concerning the chemical makeup of individual food organisms, all of which contain chemical compounds that we can use and other compounds that we cannot use. Some of the latter may even be toxic. Rather, we will return to the unity of biology to discuss some of the common aspects of the chemical composition of the organisms we use for food.

Since the first product of photosynthesis is glucose (Chapter 3), substances chemically related to glucose occupy a major role in all organisms. These compounds are known as carbohydrates. The unique places of proteins and nucleic acids have already been repeatedly referred to in relation to enzymes and genetic material. Fats and other lipids are integral parts of protoplasmic membranes, and fats for energy storage are common to all organisms. In addition to these four major classes of molecules, there are numerous small molecules including metabolic intermediates for the multiplicity of the dynamic chemical processes which make life possible. Even though the chemistry of each of these five categories can be quite involved, we need not go into too many chemical details in order to get some idea of the nature of these compounds. It is these substances which make up our food. Although, as we know, water makes up the major part of living things, we shall not discuss it further here. Rather, most of our attention will be given to the other chemical entities making up organisms—the organic compounds, as they are called by the chemist.

Quantitatively, plant materials contain much more carbohydrate than substances of the other categories. Animal materials contain relatively little carbohydrate and much more protein and fat (lipid). Although nucleic acids are biologically very important, there is relatively little of them in foods of either plant or animal origin. Miscellaneous compounds also do not generally constitute a high percentage of the total organic compounds in foods. In animals the total mineral content is higher than in plants, a reflection of the fact that the skeletal structures (bones) in animals are composed largely of mineral matter whereas in plants similar structures are carbohydrate. By neglecting bone, which we do not eat, the amount of mineral matter is roughly

Table 6.1. Proximate Analysis in Percent for Typical Foods of Plant or Animal Origin

	Protein	Fat	Carbohydrate		Ash	Water
			Total	Fiber		
Sweet corn	3.5	1.0	22.1	0.7	0.7	72.0
Apple	0.3	0.5	15.0	0.9	0.3	84.0
Whole wheat	13.3	2.0	71.0	2.3	1.7	12.0
Whole egg	12.8	11.5	0.7	0.0	1.0	74.0
Lean sirloin steak	21.5	5.7	0.0	0.0	1.0	71.8
Cooked hamburger	24.2	20.5	0.0	0.0	1.1	54.2

about 1% in both plant and animal tissue. (See the Appendix for details concerning the percentages of the major food constituents in particular foods.)

Table 6.1 shows what is called the "proximate analysis"—total protein, fat, carbohydrate including fiber (cellulose), ash, and water—for a few common foods. In such analyses, water is measured by the loss of weight on drying. Ash or total mineral is determined by loss of weight on burning in air. Protein is determined by measuring the total nitrogen; in most foods, percent nitrogen × 6.25 = percent protein. The material dissolved by ether is considered fat. Actually, this fraction is more properly termed "lipids," but generally 95% or more of lipids is fat (see below). The sum of all of these analyses subtracted from 100% is generally considered total carbohydrate. If determination of fiber (indigestible cellulose) is required in the analysis, it is determined by weighing the dry residue after boiling the food separately with dilute sodium hydroxide and dilute hydrochloric acid.

The proximate analyses of other biological materials would be quite different from those for the foods listed in Table 6.1. For example, the woody portions of plants would contain little carbohydrate other than fiber. Furthermore, the analysis of the whole corn plant would be quite different from that for the kernels of sweet corn that are the edible portion of the plant. As the term "proximate" implies, the different categories include other substances (such as nucleic acid, citric or other fruit acids, and sterols) of great importance biologically but of relatively minor importance when the total quantity of each fraction is concerned. Shortly, we will take a closer look at each of these major groups of organic chemical constituents (the nonwater and nonmineral portions) of the biological organisms or parts of organisms that make up our food supply. Before we do, a few general comments are in order.

All organisms are composed of molecules ranging in molecular weight from less than 100 to several million. Remember that molecular weight also reflects the size of the particular molecule. The large molecules are constructed by combining smaller molecules, which in a very real sense may be considered the fundamental biological building blocks. We have referred to this fact many times already. The chemist calls the very large molecules *polymers*, and they are made up of small molecules or *monomers*. Whereas biologically important monomers are water soluble, many of their polymers are not, and with their associated water such polymers give biological systems a solid appearance.

Repeatedly we have emphasized that biochemical processes take place in water. We have already indicated on pp. 16 and 17 that a very basic chemical process of building large molecules from smaller ones is joining two molecules by what appears to be the removal of water—a hydrogen from one small molecule and a hydroxyl from another. We can represent this by

$$MOH + M'H \quad\quad M \cdot M' + H_2O$$

where MOH and M′H are the biological monomers. To cause a reaction to take place by removing water in a water solution is quite a chemical feat. It requires energy. We pointed out that this can be effectively done by converting at least one of the reactants to a phosphate ester.

Remember ATP, adenosine triphosphate?

$$\text{MOH} \quad + \text{adenine—ribose—O—} \overset{\overset{\textstyle O}{\|}}{\underset{\underset{\textstyle OH}{|}}{P}} \text{—O—} \overset{\overset{\textstyle O}{\|}}{\underset{\underset{\textstyle OH}{|}}{P}} \text{—O—} \overset{\overset{\textstyle O}{\|}}{\underset{\underset{\textstyle OH}{|}}{P}} \text{—OH} \longrightarrow$$

monomer   ATP

$$\text{MO—} \overset{\overset{\textstyle O}{\|}}{\underset{\underset{\textstyle OH}{|}}{P}} \text{—OH} + \text{adenine—ribose—O—} \overset{\overset{\textstyle O}{\|}}{\underset{\underset{\textstyle OH}{|}}{P}} \text{—O—} \overset{\overset{\textstyle O}{\|}}{\underset{\underset{\textstyle OH}{|}}{P}} \text{—OH}$$

monomer   ADP
phosphate

$$\text{MO—} \overset{\overset{\textstyle O}{\|}}{\underset{\underset{\textstyle OH}{|}}{P}} \text{—OH} + \text{M'H} \longrightarrow \text{MM'} + \text{HO—} \overset{\overset{\textstyle O}{\|}}{\underset{\underset{\textstyle OH}{|}}{P}} \text{—OH}$$

monomer   monomer   connected   phosphoric acid
phosphate   monomers

Thus by converting the metabolite or monomer to its phosphate, it is possible in the presence of the proper enzyme to join it with another monomer by eliminating a free phosphoric acid molecule rather than water. These reactions which take place in water have the effect of joining two small molecules to form a larger molecule by removal of a molecule of water. This is the most common way to join small molecules in living systems. In the subsequent discussion of the constituents of biological systems (and our food), we may, for the sake of simplicity, refer to the synthesis of molecules by removal of water, whereas actually the mechanism for doing it may be more complex. This simplification can facilitate our understanding of the chemical processes in organisms which produce our food.

In order to visualize the nature of food constituents, a few simple molecular formulas will be used. We may use a few statements you may have to take for granted since answering why certain things happen would require more discussion of the detailed principles of chemistry than is appropriate at the moment. In Chapter 4 (pp. 44–46), it was shown that the atoms of carbon, oxygen, hydrogen, and nitrogen which make up most of the compounds of biological significance join together by sharing electrons with each other. This is also true for all atoms combining to form molecules of all kinds, not just molecules of biological significance. Carbon atoms have the great facility of joining with other carbon atoms. The carbon atom can be thought of as having four points of attachment for other atoms. (Each point of attachment is a shared pair of electrons; see pp. 45–46.) Oxygen has two points of attachment, nitrogen three, and hydrogen one. Given any specific number of atoms, we can have a stable compound as long as all points of attachment of all the atoms are filled. For example, with two carbon atoms, six hydrogen atoms, and one oxygen atom, there are two possible arrangements:

ethyl alcohol   dimethyl ether

Ordinary ethyl alcohol and dimethyl ether, a rather useless gas, are said to be *isomers* because they are two different molecules resulting from a different arrangement of the same atoms. In this case, the isomers are of different chemical classes, one an alcohol characterized by the —OH and the other an ether characterized by —O— joined to two carbon atoms. It is not necessary that isomers be of a different class. We can have isomers of the same class, for example, *normal* propyl and *iso*propyl alcohol:

```
   H  H  H                                  H  H  H
   |  |  |                                  |  |  |
H—C—C—C—O—H                              H—C—C—C—H
   |  |  |                                  |  |  |
   H  H  H                                  H  O  H
                                               |
                                               H
   n-propyl alcohol                         isopropyl alcohol
```

In biological systems, isomerism is very important. Molecules are of course three-dimensional structures. The formulas we show on paper are two-dimensional conventions used for illustrative purposes only. Isomers then have different shapes. Some differences may indeed be subtle, but as these isomers are used to build complex molecular structures a slight difference in the shape of one molecular building block can have profound differences in biological structure and function. In the construction of a great building, each stone must be shaped to fit a certain place, and so it is with molecules—they too must fit.

You will recall that small molecules are joined by effective elimination of water. You have also already been exposed to the fact that certain arrangements of atoms relate to certain chemical and physical properties. Let us now bring these two ideas together to see how certain molecules can have the property of joining with another molecule by the elimination of two hydrogen atoms and an oxygen atom to form water. Let us take the simplest of all organic compounds methane, CH_4—a hydrocarbon since it contains only carbon and hydrogen—and stepwise oxidize each hydrogen:

```
    H            H             H              H
    |            |             |              |
 H—C—H        H—C—O—H       H—C—O—H       H—O—C—O—H
    |            |             |              |
    H            H             O              O
                               |              |
                               H              H
```

```
                   H
                   |
                   O
                   |
               H—O—C—O—H
                   |
                   O
                   |
                   H
```

When there is more than one —OH attached to a single carbon atom, there is a tendency to lose water. So the three compounds on the right lose water as follows:

$$CH_4 \qquad CH_3\underline{OH} \qquad \begin{matrix} H \\ | \\ H{-}C{=}O \\ \text{or} \\ HC\underline{HO} \end{matrix} \qquad \begin{matrix} O \\ \| \\ H{-}C{-}O{-}H \\ \text{or} \\ HCOO\underline{H} \end{matrix} \qquad \begin{matrix} OH \\ | \\ C{=}O \\ | \\ OH \end{matrix} + H_2O$$

methane	methyl alcohol	formaldehyde	formic acid	carbonic acid, H_2CO_3
		$+$	$+$	$CO_2 + H_2O$
		H_2O	H_2O	carbon dioxide

We have now made from the hydrocarbon three important classes of compounds—alcohols, characterized by the —OH group; aldehydes, characterized by —CHO; and acids, characterized by —COOH. Let us now take two molecules of methyl alcohol and remove a molecule of water:

$$CH_3OH + HOCH_3 \longrightarrow CH_3{-}\underline{O}{-}CH_3 + H_2O$$

methyl alcohol · · · · · · · · · · · · · · · · dimethyl <u>ether</u>

Similarly, if we react ethyl alcohol

$$CH_3CH_2OH + HOCH_2CH_3 \longrightarrow CH_3CH_2{-}\underline{O}{-}CH_2CH_3 + H_2O$$

ethyl alcohol · · · · · · · · · · · · · · · · diethyl <u>ether</u>

we get ordinary ether, which is used in anesthesia and also as an important solvent for fats and lipids in general. Let us now react one molecule of methyl alcohol and one molecule of formic acid:

$$HCOOH + HOCH_3 \longrightarrow HCOO\underline{CH_3} + H_2O$$

formic acid · · · · · methyl alcohol · · · · · methyl formate, an <u>ester</u>

When acids and alcohols react by losing a molecule of water, we get an ester. From ethyl alcohol (from yeast fermentation) and acetic acid (from vinegar), we can get ethyl acetate, a sweet-smelling liquid of a multiplicity of uses (among them as a solvent for lacquers and fingernail polish):

$$CH_3COOH + HOCH_2CH_3 \longrightarrow CH_3\underline{COOCH_2CH_3} + H_2O$$

acetic acid · · · · · ethyl alcohol · · · · · ethyl acetate

This compound also occurs naturally in a number of foods—for example, in tomatoes (see Table 6.5).

Ammonia, NH_3, also has useful hydrogens for forming water:

$$CH_3OH + NH_3 \longrightarrow CH_3NH_2 + H_2O$$

methyl   ammonia   methyl amine
alcohol

$$CH_3COOH + NH_3 \longrightarrow CH_3CONH_2 + H_2O$$

acetic acid   ammonia   acetamide

Acetamide is the nitrogen excretory product of mice. It is the odor of "I smell a mouse," but acetamide is also a normal constituent of some foods. In humans, the counterpart is urea as the odorless nitrogen excretory product:

$$\underset{\substack{\text{carbonic acid} \\ (CO_2 + H_2O)}}{\overset{\displaystyle OH}{\underset{\displaystyle OH}{C{=}O}}} + 2NH_3 \longrightarrow \underset{\text{urea}}{\overset{\displaystyle NH_2}{\underset{\displaystyle NH_2}{C{=}O}}} + 2H_2O$$

As the numbers of atoms in a molecule increase, so also does complexity. Indeed, it is common in biological compounds to have molecules with a number of different functional (reactive) groups. Lactic and citric acid molecules are alcohols as well as acids. Some amino acids (see below) are acids, alcohols, and amines at the same time. So unless you know your way chemically it is easy to get lost among the innumerable molecules found in biological systems. We do not propose to do that. But from this brief excursion into organic chemistry it is clear that under proper circumstances many compounds of biological significance have the ability to react to form new compounds by elimination of water. These are the building blocks of biological systems, the molecules that give biological systems their diversity as well as much of their commonality. These are the nutrients we seek in our food.

THE CARBOHYDRATES

Carbohydrates occupy a unique position in all biology. Glucose, a simple sugar, is among the primary products of photosynthesis. This carbohydrate containing six carbon atoms occupies a key position in all biological systems— animals as well as plants (Chapter 3). The name *carbohydrate* means literally "hydrated carbon" or "carbon and water," for pioneering chemists found that sugars contain carbon together with hydrogen and oxygen in the same ratio that these two elements exist in water. Today *carbohydrate* is a term more loosely applied to substances derived from and closely related to the sugars.

Monosaccharides

Monosaccharides are the simplest sugars and are the fundamental building blocks from which more complex carbohydrates are made. As you might expect, glucose is the most important. Its empirical formula is $C_6H_{12}O_6$, but such a formula tells very little about glucose and the other monosaccharides of the same formula. In Chapter 3, in our discussions of glucose, photosynthesis, and combustion, we found that glucose could be converted to other sugars in biological systems. These monosaccharides may contain 3, 4, 5, 6, or 7 carbon atoms. Glucose occurs in at least three structural forms, all of which are important biologically. Let us focus our attention on the two structures found most commonly in biological organisms:

α-glucose   β-glucose

For convenience in identification, the carbon atoms have been numbered. Note that the only difference is that the —OH on α-glucose is below C-1 and on β-glucose it is above C-1. This means that in three dimensions these —OH groups point in different directions. Thus if these groups are used in connecting glucose to another molecule, the shape of the larger molecule will be greatly affected. For example, plants make two very important polymers of glucose. One is starch, which we can use for food. Although it is insoluble in water, it imbibes much water and it forms viscous pastes with water. The other is cellulose, the water-insoluble fibrous material which forms the skeletal structures of plants, and we use cellulose as cotton, paper, wood, etc. Both starch and cellulose molecules are made up of hundreds of glucose molecules joined together between C-1 of one glucose and C-4 of the next by the elimination of water. In this way, long chains of glucose molecules are joined by oxygen atoms, between C-1 and C-4. Why then are starch and cellulose so different? Only because of the direction that the glucose units take in space. In β-linked cellulose the chain is linear and in α-linked starch it is helical or spiral in form. We may represent the difference by a series of lines that denote the C-1, C-4 axis of the glucose ring. In cellulose, the axis of the glucose rings forms a line •—•—•—•—•—•—• with each ring turned 90° with respect to the next ring. Viewed from the end, a cellulose fiber would be

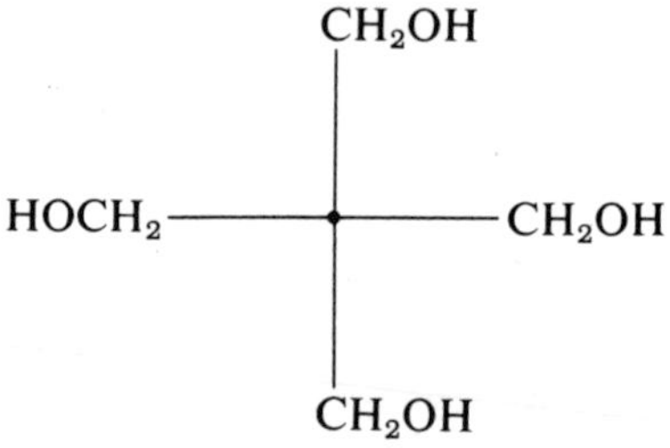

$$CH_2OH$$

$$HOCH_2———|———CH_2OH$$

$$CH_2OH$$

with the C-6 or CH_2OH groups at right angles. Furthermore, every fourth (first, fifth, ninth, etc.) glucose unit is in perfect alignment. When this micro-fiber is meshed with others along its axis, a larger fiber is formed. Such an arrangement of glucose units is not possible in starch, because they are arranged like a spring ᙁᙁᙁᙁᙁᙁᙁ or several springs tied together at different points.

Referring back to the structural formula of glucose, if the relative positions of the —H and —OH on carbons numbered 2, 3, 4, 5,[2] as well as 1 are changed, we will have different monosaccharides. Among the more important ones are galactose and mannose, which can be formed from glucose by most organisms. In addition, the ring can be closed in a different way to form a ring of five atoms (4C, 1O) instead of six (5C, 1O). Fructose has such a ring of five atoms and is also an important monosaccharide (see Fig. 3.4 of Chapter 3). All of these monosaccharides have the same empirical formula, $C_6H_{12}O_6$.

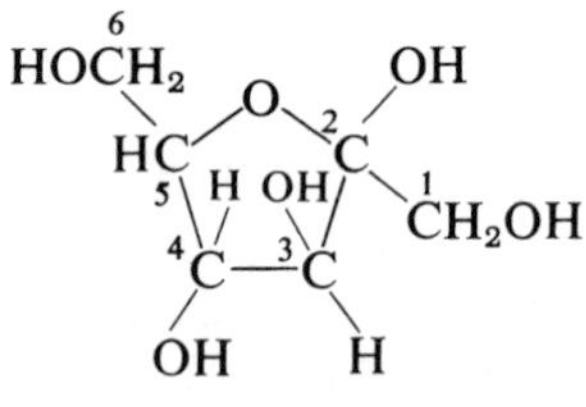

fructose

But these are not all of the monosaccharide building blocks. We have made a number of references to ribose, a sugar of five carbon atoms. There

___

[2] Carbon 5 as drawn does not contain an —OH but an —O—C— connection which itself can take either of two directions.

are some additional five-carbon sugars in nature such as xylose and arabinose, as well as monosaccharides with three-, four-, and seven-carbon atoms.

In addition to the simple monosaccharides we have discussed, there are a number of monosaccharide derivatives that are also important building blocks. Some of the more important are shown below:

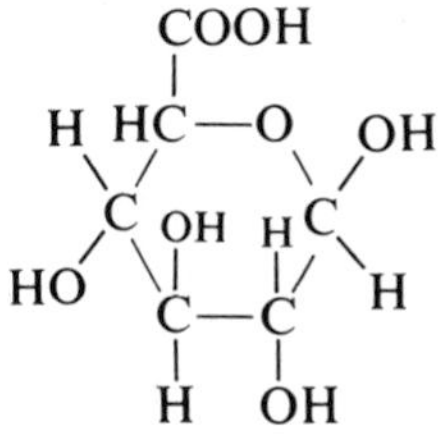

COOH
H HC—O OH
\ / \ /
C OH H C
HO \ | |/ \ H
C—C
| |
H OH
glucuronic acid

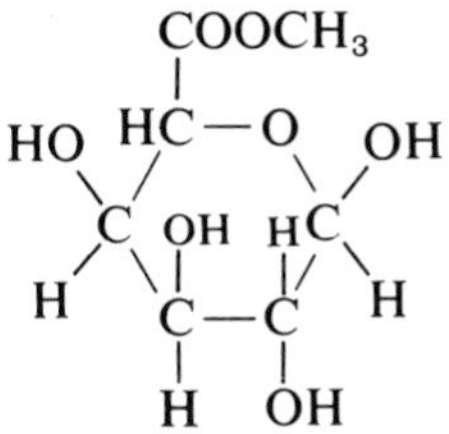

COOCH₃
HO HC—O OH
\ / \ /
C OH H C
H \ | |/ \ H
C—C
| |
H OH
galacturonic acid
methyl ester

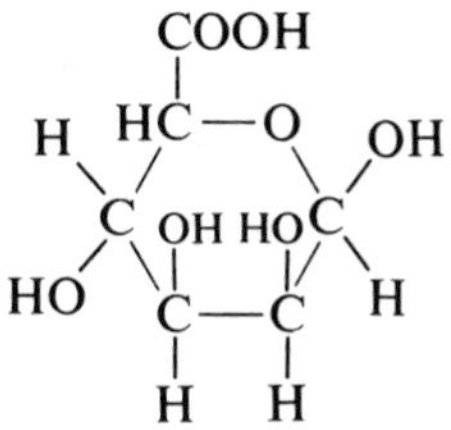

COOH
H HC—O OH
\ / \ /
C OH HO C
HO \ | |/ \ H
C—C
| |
H H
mannuronic acid

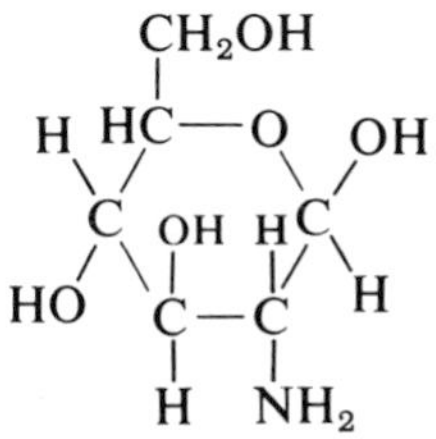

CH₂OH
H HC—O OH
\ / \ /
C OH H C
HO \ | |/ \ H
C—C
| |
H NH₂
glucosamine

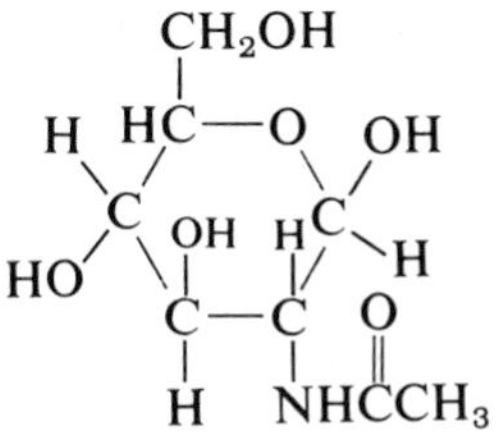

CH₂OH
H HC—O OH
\ / \ /
C OH H C H
HO \ | |/ \
C—C O
| | ‖
H NHCCH₃
acetyl glucosamine

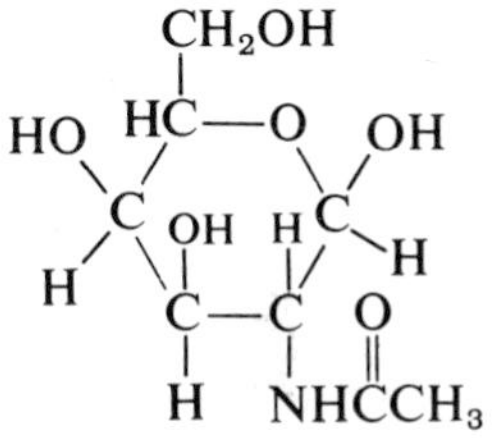

CH₂OH
HO HC—O OH
\ / \ /
C OH H C H
H \ | |/ \
C—C O
| | ‖
H NHCCH₃
acetyl galactosamine

These and other monosaccharides and their derivatives are the building blocks for carbohydrates of all kinds. Furthermore, these small molecules are also used in combination of building blocks of other kinds such as amino acids and fatty acids, and other metabolites such as lactic acid, phosphoric acid, sulfuric acid,[3] and even hydrocyanic acid, HCN. HCN in free form is toxic but it is widespread in many plants in combination with carbohydrates as cyanogenic glycosides. Cassava, lima beans, and the seeds of many fruits such as cherries, peaches, and almonds are notable in this respect. It is important to note that all of these building blocks of biological tissues of all kinds contain —OH and —H groups, which under proper conditions can be used to connect the building blocks by removal of water, H_2O. Also, we should not forget that such chemical processes take place in water and require energy, and that many times these syntheses require ATP and phosphate ester intermediates.

[3]
O
‖
HO—S—OH
‖
O

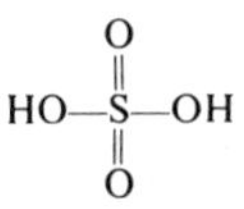

Of the monosaccharides and their derivatives, only glucose and fructose are present in significant quantity in the free form. Glucose, sometimes called *dextrose*, is common in grapes, sweet corn, etc. It is prepared commercially from starch and sold under the name *cerelose*, and is used in this form in many foods. *Fructose* literally means "fruit sugar," and it is the source of the sweetness of many fruits. Fructose is the sweetest sugar. Honey is a mixture of fructose and glucose.

Disaccharides

Disaccharides are sugars made by joining two monosaccharides together. They, like all the simple monosaccharides, are sweet to the taste. Three disaccharides are very important in foods. They are sucrose (cane or beet sugar), lactose (milk sugar), and maltose (malt sugar), and their constitution may be represented as follows:

$$C_6H_{12}O_6 + C_6H_{12}O_6 \xrightleftharpoons[\text{}]{\text{synthesis}} C_{12}H_{22}O_{11} + H_2O$$

$$\text{Glucose} + \text{Fructose} \xrightleftharpoons{} \text{Sucrose} + H_2O$$

$$\text{Glucose} + \text{Galactose} \xrightleftharpoons{} \text{Lactose} + H_2O$$

$$\text{Glucose} + \text{Glucose} \xrightleftharpoons[\text{hydrolysis}]{} \text{Maltose} + H_2O$$

Sucrose or ordinary sugar is present to some extent in many plants. In terms of tonnage, no other pure organic chemical crystalline compound is prepared in greater quantity. Chemically, sucrose is more stable than most simple monosaccharides and disaccharides. It is sweet, soluble in water, and used as food in most cultures throughout the world. In 1973, on the average each man, woman, and child in the United States consumed more than 102 pounds (46 kilograms) of sucrose. In some other countries, sugar consumption is even higher.

Lactose is the sugar found in the milk of mammals. Human milk contains about 7% and cow's milk about 4.6%. Lactose is not a particularly sweet sugar. If it were as sweet as sucrose, human milk would equal or surpass the sweetness of most soft drinks. Whereas other common disaccharides and the monosaccharides glucose and fructose are readily fermented by ordinary yeasts to give alcohol and carbon dioxide, this is not so for lactose.[4] It might be speculated that newborn animals are thus afforded some protection against yeasts, which are rather ubiquitous organisms.

[4] Some special organisms can ferment lactose. Kumys, a fermented mare's milk, is an alcoholic beverage consumed in some Asiatic cultures.

Maltose is the disaccharide most closely related to starch and is usually prepared from starch by starch-splitting enzymes. Maltose plays a role in many foods such as bread and in malt beverages such as beer.

Trisaccharides

Trisaccharides, formed from three monosaccharides, do occur in nature and are common in some plants and in some foods. Raffinose is made up of galactose, glucose, and fructose.

Sweetness

All of the simple sugars, mono-, di-, and trisaccharides, are sweet and contribute sweetness to our food. Table 6.2 indicates the relative sweetness of some of the sugars mentioned above.

Polysaccharides

The number of *polysaccharides* in nature is very great. We have already discussed starch and cellulose, the most common polysaccharides of glucose. The actual specific molecular structure of starch varies with the species of plant. The polysaccharides contain hundreds of monosaccharide units and are linear or branched chains. Furthermore, they may contain more than one type of monosaccharide. So you can see that it is impossible to go into detail about polysaccharides. Rather, we shall list a number of important ones present in our foods, indicate their source, and show the more important monosaccharide or monosaccharide derivatives from which they are made by the biological organism.

The polysaccharides of Table 6.3 are common in our foods. Most of them are quite completely utilized by man, a few are only partially utilized, and cellulose is completely indigestible by man. As food customs have developed in different cultures over the centuries, some of these plant materials have come to be used as food—edible seaweeds, exudates from plants injured mechanically or by insects, etc. It has long been known that certain fruits contain something that makes a jelly or viscous jam when sugar is used as a preservative. The chemical compound responsible is pectin. Edible seaweeds, interestingly enough, contain other carbohydrate polymers in their major structural parts. Table 6.3 lists three that have found considerable usefulness in foods—alginic acid, agar, and carrageenan (Irish moss). Gum arabic and

Table 6.2. Relative Sweetness by Weight:
Sucrose (Cane or Beet Sugar) = 100

Fructose	173
Invert sugar[a]	130
Sucrose	100
Glucose	74
Xylose	40
Maltose	32
Galactose	32
Raffinose	32
Lactose	16
Sodium cyclamate[b]	2500–3500
Aspartame[b,c]	20,000–30,000
Saccharin[b]	40,000–50,000

[a] Invert sugar is a mixture of equal parts of glucose and fructose somewhat like honey. It is made by hydrolysis of sucrose.
[b] These are not carbohydrates but are nonsugar sweeteners used in foods.
[c] The methyl ester of aspartyl phenylalanine, a dipeptide of two amino acids. See pp. 143–144.

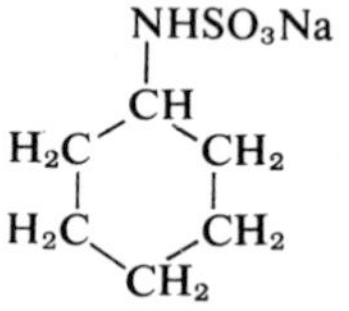

sodium cyclamate

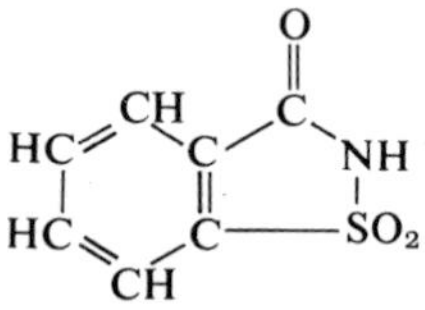

saccharin

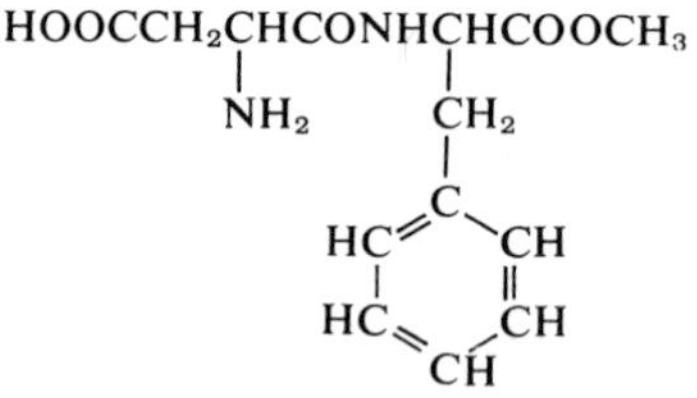

HOOCCH$_2$CHCONHCHCOOCH$_3$

Aspartame

gum tragacanth have been used in food since ancient times. These five polysaccharides have become even more useful in our modern food industry, which requires that food be prepared far in both distance and time from the person eating the food.

Most animal polysaccharides are consumed along with the food and are not used as added ingredients. Hyaluronic acid is widely distributed as a rather viscous or gelatinous coat around and between cells including the

Table 6.3. Biological Sources and Constituent Monosaccharides and Monosaccharide Derivatives of Selected Common Polysaccharides

Polysaccharide	Source	Constituent monosaccharides and monosaccharide derivatives
Starch	Wheat, corn, rice, potatoes, cassava, and many other seeds, tubers, fruits, and vegetables	Glucose
Cellulose	Most plants—cell walls and woody parts	Glucose
Glycogen (animal starch)	Muscle and liver	Glucose
Dextrans	Many species of bacteria, particularly when using sucrose	Glucose
Inulin	Many plants including some grasses and artichokes	Fructose
Xylans	Oat hulls, corn cobs, woody parts of plants, etc.	Xylose
Pectin	Oranges, apples, pears, gooseberries, and other fruits	Galacturonic acid, galacturonic acid methyl ester
Alginic acid	*Laminaria* species of seaweed	Mannuronic acid, guluronic acid
Agar	*Gelidium* species of seaweed	Galactose, galactose sulfate, anhydrogalactose
Carrageenan (Irish moss)	*Gigartina* species of seaweed	Galactose sulfate
Arabic gum	Exudate from trees of the genus *Acacia*	Galactose, rhamnose, arabinose, glucuronic acid
Tragacanth gum	Exudate from several species of *Astragalus*	Galactose, xylose, arabinose, galacturonic acid, galacturonic acid methyl ester
Xantham gum	Fermentation of carbohydrate by the bacterium *Xanthomonas campestris*	Glucuronic acid, glucose, mannose
Hyaluronic acid	Animal including human tissues	Glucuronic acid, acetyl glucosamine
Chondroitin sulfate	Animal including human tissues	Glucuronic acid, acetyl glucosamine sulfate
Heparin	Animal including human tissues	Glucuronic acid, acetyl glucosamine sulfate
Chitin	Exoskeletons of insects and crustaceans; fungi	Acetyl glucosamine

human egg cell; it also occurs in the vitreous humor of the eye and in joint lubricants. Chondroitin sulfuric acid is widely distributed in connective tissues, and heparin is found in blood, where it acts as a natural anticoagulant to keep normal circulating blood from clotting. Chitin, a polymer of acetyl glucosamine, is found in fungi and is also the major component of the exoskeleton of insects and crustaceans (lobsters, shrimps, crabs, etc.). Since it is a rather stable compound chemically, it can easily be detected in foods to determine insect infestation. From these several examples, it is clear that the great diversity of polysaccharides arising from the relatively few mono-saccharide building blocks serve many different functions. Since these compounds are in the organisms we consume, they are inevitable constituents of our food.

Glycosides

The monosaccharide building blocks are also found joined to other noncarbohydrate molecules. These substances are known as *glycosides*, and there are many subclasses of glycosides according to the nature of the non-sugar part. Many glycosides are quite nontoxic, and some are toxic if eaten. Many of the glycosides that are nontoxic when eaten are toxic if injected into the bloodstream. We have already mentioned the cyanogenic glycosides. The classic example is amygdalin of almonds, cherries, peaches, etc., which contain cyanide and benzaldehyde units. Benzaldehyde itself contributes to the flavor of these fruits. A group of glycosides are known as the saponins because they make foams (suds) when their water solutions are shaken. They are widely distributed in plants and are generally excellent emulsifiers. Some glycosides contain sterols, triterpenes, or other lipid parts. Among these are digitalis, a very useful heart stimulant. In biological systems, carbohydrates are also found attached to amino acids, proteins, and lipids of various types, but these compounds are usually classified as glycoproteins or glycolipids.

Lignins

Lignins are chemically not carbohydrates, even though they are made by plants from carbohydrate. They are usually grouped with carbohydrates because they are always associated with the cellulose. Quantitatively, there is about one-third to one-half as much lignin as cellulose in wood. In recovering cellulose fibers from wood for making paper or rayon, lignins are removed by strong alkalis or alkali sulfites. In general, lignins are amorphous polymers which cement cellulose fibers together in structural portions of plants. Lignins of different species of plants vary somewhat in composition. Like cellulose, they are generally indigestible by man. Cellulose and lignin do not contribute directly to our food supply, but in many foods of plant origin they affect digestibility and availability of other nutrients. The chemical structure of lignins has not been characterized completely. Coniferyl alcohol, shown below, is thought to be the monomer from which lignins are made by plants:

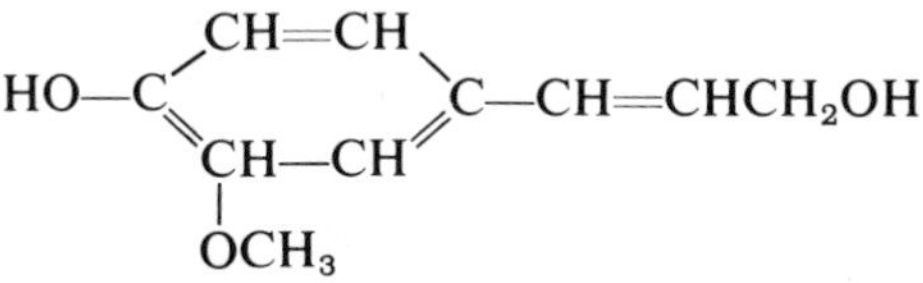

$$HO-C \overset{CH=CH}{\underset{CH-CH}{\diagdown\diagup}} C-CH=CHCH_2OH$$
$$\underset{OCH_3}{|}$$

You will find later that similar structures are found in two important amino acids and in a number of flavoring substances in essential oils—cinnamic aldehyde (cinnamon), eugenol (cloves), anethole (anise), etc.

LIPIDS

Lipids are customarily defined as those substances of biological origin which are soluble in ether (p. 117) and insoluble in water. Although such a definition includes fats, waxes, hydrocarbons, complex lipids containing carbohydrate and phosphate, etc., sterols, fat-soluble vitamins, and numerous substances classified as essential oils (flavoring substances), fats usually account for 90–98% or more of the total lipids in any particular tissue used for food. We have mentioned this before when we indicated how total fats are determined in foods or other biological material. Notwithstanding the heterogeneity of lipids, they do have some common smaller molecules or building blocks from which they are derived. As with carbohydrates and proteins, the smaller molecules are joined by the elimination of water.

Fats

Fats are what the chemist calls esters of glycerol and fatty acids. Glycerol comes from glucose via glyceraldehyde, and the fatty acids are built up from acetic acid (CH_3COOH), which also comes from glucose (see Fig. 3.4 of Chapter 3). The most common fatty acids contain 16 or 18 carbon atoms in a row. Although we find some shorter-chain fatty acids in the fats and oils of coconut and the milks of mammals, we also find longer-chain fatty acids in the fats and oils of fish, peanuts, etc. Note the terms *fats* and *oils*. *Oil* as we use it here does *not* mean petroleum or lubricating oil. Fats and oils are both esters of glycerol, but fats are solid at room temperature whereas oils are liquid. This reflects the makeup of the kinds of fatty acids combined with the glycerol.

Table 6.4 lists the names (which generally reflect their usual source) and

Table 6.4. Fatty Acids Common in Foods

Name	Usual source	Formula
Butyric	Butter, cow's milk	$CH_3(CH_2)_2COOH$
Caproic	Goat's and cow's milk	$CH_3(CH_2)_4COOH$
Caprylic	Goat's and cow's milk	$CH_3(CH_2)_6COOH$
Capric	Goat's and cow's milk	$CH_3(CH_2)_8COOH$
Lauric	Laurel, milk fat, coconut	$CH_3(CH_2)_{10}COOH$
Myristic	Nutmeg, milk fat, vegetable fats	$CH_3(CH_2)_{12}COOH$
Palmitic	Palm oil, most fats	$CH_3(CH_2)_{14}COOH$
Stearic	Beef stearin, most fats	$CH_3(CH_2)_{16}COOH$
Oleic[a]	Olive oil, most fats	$CH_3(CH_2)_7CH{=}CH(CH_2)_7COOH$
Linoleic[a,b]	Linseed oil	$CH_3(CH_2)_4CH{=}CHCH_2CH{=}CH(CH_2)_7COOH$
Linolenic[a,b]	Linseed oil	$CH_3CH_2CH{=}CHCH_2CH{=}CHCH_2CH{=}CH(CH_2)_7COOH$
Arachidic	Peanut oil	$CH_3(CH_2)_{18}COOH$
Arachidonic[a,b]	Animal tissues	$CH_3(CH_2)_4CH{=}CHCH_2CH{=}CHCH_2CH{=}CHCH_2CH{=}CH(CH_2)_3COOH$
Behenic	Rapeseed oil	$CH_3(CH_2)_{20}COOH$
Lignoceric	Brain lipids	$CH_3(CH_2)_{22}COOH$

[a] These fatty acids are called unsaturated because they contain less hydrogen than is possible. Compare these acids with stearic acid, containing 18 carbon atoms, or arachidic acid, containing 20 carbon atoms. Stearic and all of the other acids in Table 6.3 are saturated; they contain all the hydrogen it is possible for them to hold.
[b] Essential fatty acids. See Table 5.1.

formulas of the fatty acids commonly found in food. Saturated fatty acids in fats tend to make them solid at room temperature, and unsaturated fatty acids tend to make them liquid (oil) at room temperature. Furthermore, the more unsaturated fats or fatty acids are, the more susceptible they are to oxidation by atmospheric oxygen. This causes rancidity, which not only gives fats and fatty foods a bad taste but also destroys fat-soluble vitamins. This oxidative process may be retarded by naturally occurring (or added) anti-oxidants (see Chapter 5, pp. 96, 103, and Chapter 9).

The actual amount of uncombined fatty acids in the lipids of living organisms is small—usually less than 0.5%. The fatty acids are joined together (esterified) with glycerol. In any particular fat, there are usually several different fatty acids combined with glycerol. Most fats contain at least the five most common acids—palmitic, stearic, oleic, linoleic, and linolenic. But any one glycerol molecule can only hold three fatty acids, as indicated below:

$$CH_3(CH_2)_{14}COOH$$
palmitic acid

$$CH_3(CH_2)_{16}COOH$$
stearic acid

$$CH_3(CH_2)_7CH{=}CH(CH_2)_7COOH$$
oleic acid

$$HOCH_2$$
$$+\ HOCH \longrightarrow$$
$$HOCH_2$$
glycerol

$$CH_3(CH_2)_{14}COOCH_2$$
$$CH_3(CH_2)_{16}COOCH + 3H_2O$$
$$CH_3(CH_2)_7CH{=}CH(CH_2)_7COOCH_2$$
fat (triglyceride)

Consequently, since fats in general contain several fatty acids combined with glycerol, fats are a mixture of glycerol molecules combined with different fatty acids taken three at a time. In chemical parlance, fats are mixtures of triglycerides. This is why fats are not crystalline, have no sharp melting and freezing points, and indeed have melting points higher than their freezing points. Many food uses of fats and oils depend on these peculiar properties.

Plants make their fats from the carbohydrates produced by photosynthesis. Man and most higher animals also synthesize their fats mostly from carbohydrates but can directly incorporate fatty acids from dietary fats into their own fats. Warm-blooded animals—man and most of the animals he uses for food—tend to make more saturated fats from carbohydrates, and if much unsaturated fat is in the diet it tends to make body fat more unsaturated. For years, pigs fed whole peanuts and soybeans were discounted on the market, not because of inferior meat, but because they had oily fat and consumers wanted solid fat on their pork, ham, and bacon. Pigs fed corn (mostly carbohydrate) yield relatively hard fat.

Peripheral fats such as those found under the skin and in adipose tissues near the surface are generally more unsaturated (softer, lower melting and freezing points) than those of adipose tissues in the deep, warmer regions of the body such as around the kidneys and abdominal viscera. Fats are in liquid form in living warm-blooded animals. It is interesting to note that the internal fat of the lamb has a melting point 10°C or more higher than the lamb's body temperature but its freezing point is just below body temperature. These observations are related to the fact that fats serve as energy stores, and the more saturated a fat is the more hydrogen it contains and the more energy it contains per unit weight.

The relative amounts of different fatty acids in fats are generally characteristic of an organism, although absolute amounts do vary within limits. Only milk fats contain butyric, caproic, caprylic, and capric acids. Human milk contains only trace amounts of these except for capric acid, and even in this instance only about 1%. On the other hand, cow and goat milk fats contain up to 17% of these acids, which are quite objectionably odorous in free form. Of the other fatty acids in Table 6.4, only lauric acid has a characteristic odor or flavor. All others as well as pure fats (triglycerides) are without odor or flavor (see below). Fish oils contain a large amount of highly unsaturated acids of 18 to 24 carbon atoms. Their oxidation products, like those of linseed oil (for paint), are rather repulsive to the western palate. By and large, the fatty acids making up most of the common food fats are palmitic, stearic, oleic, linoleic, and linolenic acids. Many fats and oils are refined to remove objectionable substances. Then, too, in order to improve their stability to oxidation and to change them from liquid to solid, many oils are hydrogenated. This involves the addition to the fat of hydrogen gas (H_2), under pressure, in the presence of a catalyst (usually metallic nickel). In this process, unsaturated acids are changed to saturated acids. As commercially practiced, not all of the unsaturated acids are so converted. The process is controlled and stopped when the desired physical and chemical characteristics are attained. We may illustrate what happens as follows:

Linolenic acid $CH_3CH_2CH{=}CHCH_2CH{=}CHCH_2CH{=}CH(CH_2)_7COOH$

$H_2 \downarrow$

Linoleic acid $CH_3CH_2CH_2CH_2CH_2CH{=}CHCH_2CH{=}CH(CH_2)_7COOH$

$H_2 \downarrow$

Oleic acid $CH_2CH_2CH_2CH_2CH_2CH_2CH_2CH_2CH{=}CH(CH_2)_7COOH$

$H_2 \downarrow$

Stearic acid $CH_3(CH_2)_{16}COOH$

The rate of hydrogenation or receptiveness to hydrogen is in the order linolenic, linoleic, and oleic—the same order of their susceptibility to oxidation by air to cause rancidity. If the common food fats of either vegetable or animal origin (exclusive of milk fats) were completely hydrogenated, they would be essentially triglycerides of only palmitic and stearic acids. Such fats are quite hard and brittle.

We are all aware that fats and oils do not mix with or are insoluble in water. Yet in biological systems fats are intimately associated with water. Fats are often highly emulsified, and they are often associated with membranes. Lipids of the most complex sorts are parts of functioning organelles of cells. This apparent paradox of lipids and fats being insoluble in water but functioning in water results from the fact that molecules can have parts which try to dissolve in water and parts which do not. Such molecules (common soap is one) can emulsify fat or fat-soluble substances so finely that although they may be insoluble they can be dispersed in water. We all know that soap can remove grease from clothes or our hands. How?

Here is a molecule of soap—sodium palmitate or sodium oleate:

$$\underline{CH_3(CH_2)_{14}}COO^-Na^+ \qquad \underline{CH_3(CH_2)_7CH}\!=\!\underline{CH(CH_2)_7}COO^-Na^+$$

The hydrocarbon or nonpolar end (underlined) tends to dissolve in the dirt or grease and $—COO^-Na^+$ or the polar end tends to dissolve in water, with the result that the surface of the grease or dirt particle is literally covered with soap molecules with their hydrocarbon end in the dirt or grease and the $—COO^-Na^+$ in the water. The dirt and grease are therefore emulsified and "floated" away. Figure 6.1 illustrates the phenomenon.

There are many lipids in biological systems which have polar and nonpolar parts to their molecules. Such lipids serve many physiological functions, and many occur in the various membranes of cells of all kinds. They, like soap, are surface active and form interface layers between water and water-insoluble substances. For this reason, they can and do act as dispersing agents or emulsifiers in and out of functioning organisms.

Phospholipids

Like fats, *phospholipids* contain fatty acids and glycerol. However, they also contain phosphoric acid and usually a small nitrogen-containing alcohol. There are some phospholipids in brain and nerve tissue that also contain a fatty alcohol. Although we cannot discuss here the details of phospholipid chemistry, it might be of interest to show the makeup of the lecithins—phospholipids of similar structure which are found in all biological systems in varying amounts. Since the lecithins may contain more than one kind of

Fig. 6.1. Schematic representation of a particle of dirt or grease covered by molecules of soap.

fatty acid, they, like fats, represent a similar group of compounds. The following formula is illustrative:

$$\text{glycerol}\begin{cases} \overset{\displaystyle O}{\overset{\|}{\text{CH}_2\text{OC}}}(\text{CH}_2)_{14}\text{CH}_3 \quad \text{palmitic acid} \\[2mm] \overset{\displaystyle O}{\overset{\|}{\text{CHOC}}}(\text{CH}_2)_7\text{CH}{=}\text{CH}(\text{CH}_2)_7\text{CH}_3 \quad \text{oleic acid} \\[2mm] \overset{\displaystyle O}{\underset{\displaystyle O^-}{\overset{\|}{\text{CH}_2{-}\text{O}{-}\underset{|}{\text{P}}{-}\text{OCH}_2\text{CH}_2\overset{+}{\text{N}}{\equiv}(\text{CH}_3)_3}}} \end{cases}$$

phosphoric   choline
acid

lecithin

Note that it has an ionized polar part and a fat-soluble part. Choline itself is an important molecular building block for some other physiologically important substances besides lecithin. For example, acetylcholine, made of acetic acid and choline, is an extremely potent neural hormone and is essential for the transmission of impulses from one nerve to the next.

Egg yolks contain a large amount of lecithin and because of this they are used in food for emulsifying or dispersing fats and oils in aqueous mixtures in such things as mayonnaise, custards, ice cream, and cakes. The lecithin from soybeans also finds many food uses in chocolate, other confections, toppings, bakery items, etc., where fat dispersion is important.

Among other phospholipids in the foods we consume are the cephalins, sphingomyelins, cardiolipins, plasmologens, and phosphatidylinositol. All serve specialized biological functions and are made by the human organism from their simple building blocks.

Glycolipids

Glycolipids are a group of substances similar in some ways to the phospholipids, but they contain no phosphoric acid, rather having carbohydrate as the polar or water-soluble part of the molecule. Nerve and brain tissue are rich in these substances, which are sometimes known as *cerebrosides*. The carbohydrate usually present in these substances is galactose.

Waxes

Waxes are molecules in which a fatty acid is joined to a fatty alcohol (only one —OH where glycerol has three). Waxes are chemically quite stable. They serve as water repellants for the fur, feathers, and skin of animals, as protection against excessive evaporative water loss in plants (waxy covering of leaves, fruit, and some seeds), and as the natural packaging material for honey—beeswax. It is little wonder then that paraffin wax and waxed paper were among the early food-packaging materials. The paraffin waxes are not true waxes, although very similar chemically to some natural waxes; they are hydrocarbons and are refined from petroleum. The similarity can be illustrated as follows:

$$CH_3(CH_2)_{14}COOH + HOCH_2(CH_2)_{28}CH_3 \longrightarrow$$

palmitic acid   myricyl alcohol

$$CH_3(CH_2)_{14}COOCH_2(CH_2)_{28}CH_3 + H_2O$$

myricyl palmitate, $C_{46}H_{92}O_2$

This compound is a constituent of beeswax and some plant waxes. $CH_3(CH_2)_{14}CH_2CH_2(CH_2)_{28}CH_3$ or $CH_3(CH_2)_{44}CH_3$ is hexatetracontane, a paraffin hydrocarbon ($C_{46}H_{94}$) that is a constituent of paraffin wax. Both palmitic acid and myricyl alcohol are made biologically from the same simple molecule, CH_3COOH, acetic acid. Among the natural waxes, the fatty alcohol part of the wax may be a more complex molecule such as a sterol. For example, lanolin, wool wax, is a mixture of fatty acid esters of simple fatty alcohols as myricyl alcohol, sterols, and terpene alcohols (see below).

Steroids

All living things contain sterols and other closely related compounds, which are grouped under the name *steroids*. Although complex in molecular structure, they too are made from carbohydrate via acetic acid. Those sterols

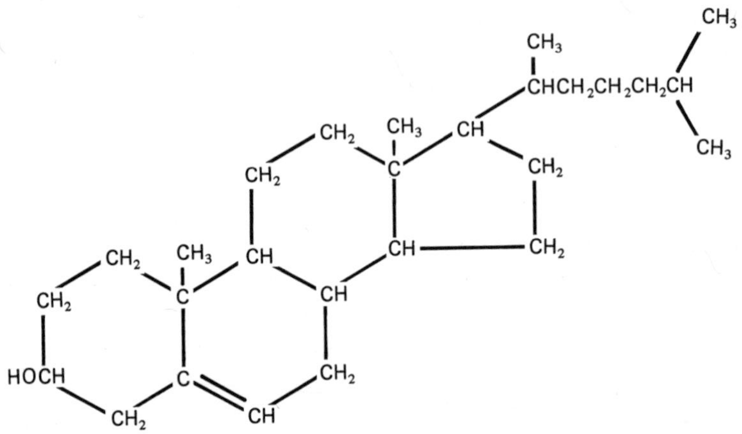

Fig. 6.2. Cholesterol, a steroid. Different steroids vary slightly in molecular structure, but subtle differences have profound physiological effects. For example, many hormones including the sex hormones are steroids. Hormones are metabolic regulators and control diverse biological activities. They are usually produced by specialized cells. In man, hormones produced by the pituitary and carried by the blood control growth, lactation, and the function of other endocrine glands, which in turn secrete hormones controlling other activities. To perform properly, hormones must be present in the proper amount at the proper time.

made by plants are sometimes called *phytosterols* (*phyto* = plant). One of the common sterols of humans and most higher animals is cholesterol (meaning "sterol of liver"). Each day, a human being synthesizes 1–2 grams of cholesterol. This substance is deposited in the walls of the arteries in certain diseases involving the circulatory system. When for some reason cholesterol along with other constituents of bile crystallizes in the gall bladder, gallstones result. Normally human bile contains about 0.6% cholesterol and human blood plasma contains about 0.25%. Figure 6.2 shows the molecular structure of cholesterol. Figure 6.3 shows the most potent of the several male and female sex hormones. You would not doubt that there are many profound as well as subtle differences between men and women, but the chemical differences between testosterone produced by the testes and estradiol produced by the ovaries are not great. Figure 6.3 also shows the structure of progesterone, another female hormone which, if present in sufficient amounts, inhibits ovulation at certain times during the menstrual cycle and during pregnancy. Some birth control pills are based on this fact. Since progesterone itself is inactivated when taken orally (see Chapter 7), the "pill" is based on chemical derivatives of progesterone which when ingested retain their ability to suppress ovulation and hence fertility.

In addition to the sex hormones, steroids include adrenal hormones, many other naturally occurring compounds used medically, and fungal

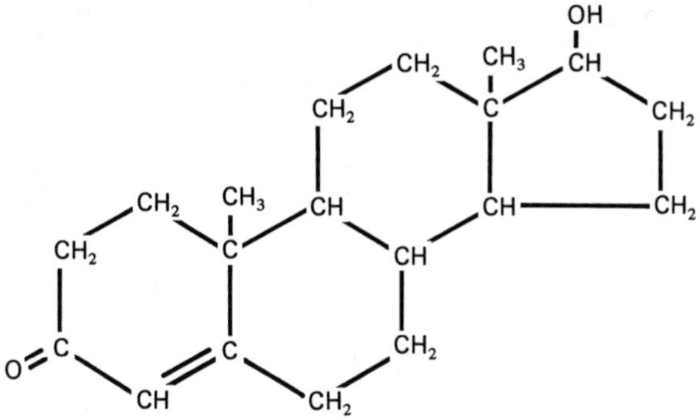

Testosterone

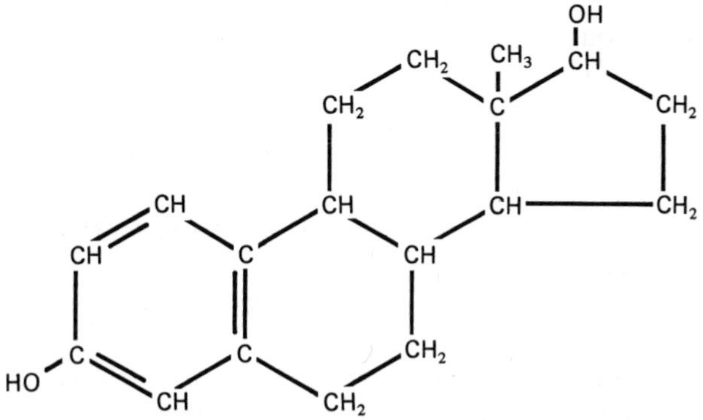

Estradiol

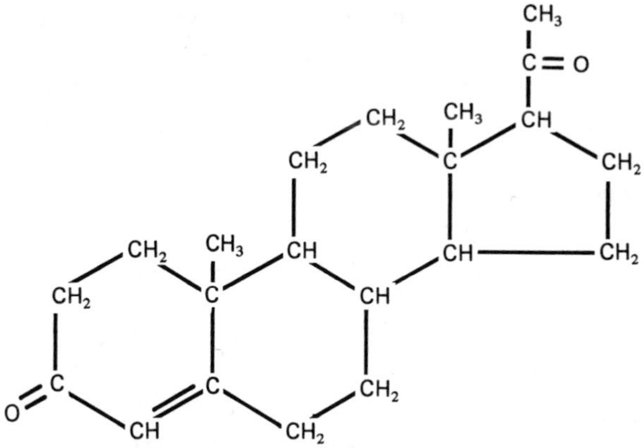

Progesterone

Fig. 6.3. Sex hormones.

(toadstool) toxins. Notwithstanding their diverse physiological activity, naturally present steroids account for a relatively insignificant amount of food lipids. In foods, the sterols are the most prominent. Of the common foods, eggs contain the most, 0.46% cholesterol. As you might expect, foods of animal origin may contain some cholesterol whereas plant foods contain various sterols classed as phytosterols.

Essential Oils

The odors and flavors associated with plant tissues (leaves, flowers, seeds, bark, fruit, pulp, peel, etc.) are found largely in the lipid fraction commonly called *essential oils* and *oleoresins*. We are not including here the sweet, sour, salty, and bitter tastes usually associated, respectively, with sugars, acids, salts, and other water-soluble substances, but rather the volatile and odorous substances which contribute significantly to the total flavor sensation so important and so sought after in our foods.

Although peoples of different cultures vary greatly in their food habits and preferences, people in general associate flavors (including both taste detected by the tastebuds on the tongue and odor detected by the olfactory organs of the nasal passages) with the desirability of food. This is a rather interesting biological phenomenon and evolutionary riddle because the major food constituents—proteins, fats, and most carbohydrates such as starches but excluding the simple sugars, are tasteless and odorless in pure form. Indeed, many of the other essential nutrients are without much flavor in the concentrations present in foods. Because the food flavors appreciated by man are so poorly correlated with nutritive value, good food habits must be learned. They are not instinctive. We shall have more to say about this later.

Let us now return to the natural food flavors which we include in essential oils and oleoresins. "Essential" as used here has quite a different meaning than the nutritional meaning in relation to amino acids, fatty acids, and other nutrients. *Essential oil* is a term antedating the science of nutrition and refers to the "oil" carrying the "essence" or characteristic odor or flavoring factors in biological materials.

If a plant material is extracted or percolated with ether, acetone, alcohol, or other suitable solvent and the solvent is then evaporated, the residue is called an oleoresin. Thus we might prepare an oleoresin of vanilla beans. True vanilla extract used in foods is essentially a solution of the oleoresin of vanilla beans. If, however, we take a plant tissue or even an oleoresin and distill it directly or with steam, the volatile flavoring oils may be removed and collected. Such flavoring oils are the essential oils. When peppermint leaves are boiled with water, the steam generated will carry with it the essential oil of peppermint. How many times have you smelled the odors coming from cooking food? You have little difficulty distinguishing tomatoes, onions, cabbage, and strawberries by their odors.

Spices and essential oils have been important to man from prehistoric times. Romantic tales of discovery, early trade, and even war involving these much sought after foods are prominent in early history. In the last century petroleum has become the international commodity of overriding concern, but even now spices are still very important economically for some nations. Black pepper is so important to American culture that when our supply was cut off during World War II much work was done on the chemistry of the oil of black pepper to see whether a satisfactory synthetic substitute could be developed. The early history of the science of chemistry is full of efforts to identify the compounds in essential oils. Early chemists found that major flavoring compounds of vanilla, anise, cinnamon, almonds, wintergreen, etc., were closely related in molecular structure, and these were named *aromatic* compounds. The organic chemist even today uses the term *aromatic* to denote a certain type of molecular structure and chemical behavior of a major class of substances.

From a closer examination of essential oils, we can get an unusual picture of the chemical complexity of the biological systems we use for food. Molecularly, these compounds are small enough to be volatile. They are, therefore, relatively simple chemically. Many of these compounds making up essential oils are repulsive or even quite toxic in highly concentrated form. Nevertheless, in small concentrations they are the stuff we want in our food— the stuff the gourmet seeks to thrill his palate. The chemistry of essential oils is complex and the physiological function of these substances in their natural state is often not well understood. We can only get a glimpse of the nature of these miscellaneous flavoring substances.

Most food flavors result from many compounds. More than 60 have been found in pineapple which contribute to its characteristic flavor. Over 180 compounds which contribute to the flavor of chocolate have been isolated and identified. In other foods, a single compound may predominate that accounts for most of the natural flavor. Figure 6.4 gives a few examples of principal flavoring compounds, showing the name of the natural food and the name and formula of the compound that gives it a characteristic flavor.

Table 6.5 has been prepared to give you some idea of the complexity of the chemical makeup of these food components, which are of very minor significance quantitatively but of major significance qualitatively. The marketing of food is profoundly affected by flavor and similar consumer attributes such as tenderness, softness, color, and other organoleptic[5] properties. For many foods, if flavor is unsatisfactory the food is unsaleable irrespective of its real nutritive value. Table 6.5 lists some of the known flavor

[5] Organoleptic properties are those which are detected by any sense organ.

Table 6.5. Some of the Compounds Naturally Present in Four Common Foods

Tomato (*Lycopersicum esculentum*)	Penta-3-enal	Butyl acetate
Decane	Hex-*trans*-2-enal	Pentyl acetate
Undecane	Hex-*cis*-3-enal	Hex-3-enyl acetate
Benzene	Hex-*trans*-3-enal	Methyl hexanoate
Toluene	Hept-*trans*-2-enal	Butyl nitrile
p-Xylene	Oct-*trans*-2-enal	Isobutyl nitrile
Isopropyl benzene	Non-*trans*-2-enal	Phenyl acetonitrile
Pseudocumene	Hexa-*trans-trans*-2,4-dienal	Hydrogen sulfide
Limonene	Hepta-*trans-cis*-2,4-dienal	Dimethyl sulfide
α-Pinene	Hepta-*trans-trans*-2,4-dienal	Dimethyl disulfide
Myrcene		2-Methyl mercaptoethanol
Δ³-Carene	Deca-*trans-trans*-2,4-dienal	3-Methyl mercaptopropanol
Methanol	Deca-*trans-cis*-2,4-dienal	2-Methyl mercaptoacetaldehyde
Ethanol	Benzaldehyde	3-Methyl mercaptopropanol
Propanol	Cinnamaldehyde	
Isopropanol	Hydrocinnamaldehyde	
Butanol	Phenylacetaldehyde	2-Ethyl furan
Isobutanol	Salicylaldehyde	2-Pentyl furan
Pentanol	Citral	2-Acetyl furan
2-Methyl butanol	Neral	Furfural
3-Methyl butanol	Geranial	5-Methyl furfural
Pent-1-en-3-ol	Acetone	2-2,4-Trimethyl-1,3-dioxalane
Hex-2-enol	Butanone	2-Isobutyl thiazole
Hex-*cis*-3-enol	Pentan-2-one	Acetophenone
Benzyl alcohol	Pentan-3-one	*o*-Hydroxyacetophenone
2-Phenyl ethanol	Pent-1-en-3-one	2-Methylhept-2-en-6-one
Phenol	Hexan-2-one	2-Methylhepta-2-*trans*-4-dien-6-one
Methyl salicylate	Non-*trans*-2-en-3-one	
o-Cresol	Diacetyl	β-Ionone
Guaiacol	Butan-2-ol-3-one	Pseudoionone
p-Ethyl phenol	Penta-2,3-dione	Epoxy-5,6-ionone
p-Vinyl guaiacol	Geranylacetone	2,2,6-Trimethyl-2-hydroxyhexanone
Eugenol	Farnesylacetone	Pentanoic acid
Linalool	2-Methyl butanal	
Linalool oxide	3-Methyl butanal	
2-Methylhept-2-en-6-ol	Acetic acid	
α-Terpineol	Propionic acid	Black pepper
Acetaldehyde	2-Methyl butyric acid	(*Piper nigrum*)
Propanal	γ-Butyrolactone	α-Pinene
Farnesal	γ-Hexalactone	β-Pinene
Glyoxal	γ-Octalactone	α-Phellandrene
Methyl glyoxal	γ-Nonalactone	β-Phellandrene
Hexanal	2,2,6-Trimethyl-2-hydroxycylcohexylidene acetic acid lactone	DL-Limonene
Heptanal		β-Caryophyllene
Penta-2-enal	Ethyl acetate	β-Elemene

Table 6.5—*Continued*

Black pepper—*Continued*
δ-Elemene
α-Cubebene
α-Copaene
α-*cis*-Bergamotene
α-*trans*-Bergamotene
α-Santalene
Hydrocaryophyllene
Isocaryophyllene
β-Farnesene
α-Humulene
β-Bisabolene
γ-Muurolene
α-Selinene
β-Selinene
δ-Cadinene
Calamanene
α-Thujene
Camphene
Sabinene
Δ^3-Carene
α-Terpinene
Myricene
p-Cymene
Terpinolene
Ocimene
Arcurcumene
Epoxydihydrocaryophyllene
Phenylacetic acid
Dihydrocarveol
Piperonal
Cryptone
cis-p-Menthen-1-ol
cis-p-2,8-Menthadien-1-ol
trans-Pinocarveol
Piperidine

Onion
(*Allium cepa* L.)
Propene
Propanal
Dimethylfuran
2-Methylpentanal
2-Methyl-pent-2-enal
Tridecan-2-one
5-Methyl-2-*n*-hexyl-2,3-
  dihydrofuran-3-one
Hydrogen sulfide
Methanethiol

Propanethiol
Allylthiol
Dimethyl sulfide
Allyl methyl sulfide
Methyl propenyl sulfide
Allyl propyl sulfide
Propenyl propyl sulfide
Dipropenyl sulfide
Dimethyl disulfide
Methyl propyl disulfide
Allyl methyl disulfide
Methyl *cis*-propenyl
  disulfide
Methyl *trans*-propenyl
  disulfide
Isopropyl propyl disulfide
Dipropyl disulfide
Allyl propyl disulfide
cis-Propenyl propyl
  disulfide
trans-Propenyl propyl
  disulfide
Diallyl disulfide
Allyl propenyl disulfide
Dipropenyl disulfide
Dimethyl trisulfide
Methyl propyl trisulfide
Allyl methyl trisulfide
Methyl *cis*-propenyl
  trisulfide
Methyl *trans*-propenyl
  trisulfide
Diisopropyl trisulfide
Isopropyl propyl trisulfide
Dipropyl trisulfide
cis-Propenyl propyl
  trisulfide
trans-Propenyl propyl
  trisulfide
Dimethyl tetrasulfide
2,5-Dimethylthiophene
2,4-Dimethylthiophene
3,4-Dimethylthiophene
3,4-Dimethyl-2,5-
  dihydrothiophene-2-one
Methyl methane-
  thiosulfonate
Propyl methane-
  thiosulfonate

Propyl propane
  thiosulfonate

Orange
(*Citrus sinensis*, Valencia)
Hexane
Isoprene
Methyl cyclopentane
Heptane
Octane
Nonane
α-Pinene
Sabinene
Myrcene
D-Limonene
β-Cubebene
β-Elemene
β-Copaene
Valencene
Ethanol
Linalool
Octanol
Nonanol
trans-2,8-*p*-Menthadien-1-ol
cis-2,8-*p*-Menthadien-1-ol
α-Terpineol
Citronellol
trans-Carveol
cis-Carveol
1,8-*p*-Menthadien-9-ol
8-*p*-Menthene-1,2-diol
Hexanal
Heptanal
Octanal
Nonanal
Decanal
Neral
Geranial
Dodecanal
Perillaldehyde
Acetone
Ethyl vinyl ketone
Carvone
Piperitenone
Diethyl acetal
Ethyl butyrate
1,8-*p*-Menthadien-9-yl
  acetate

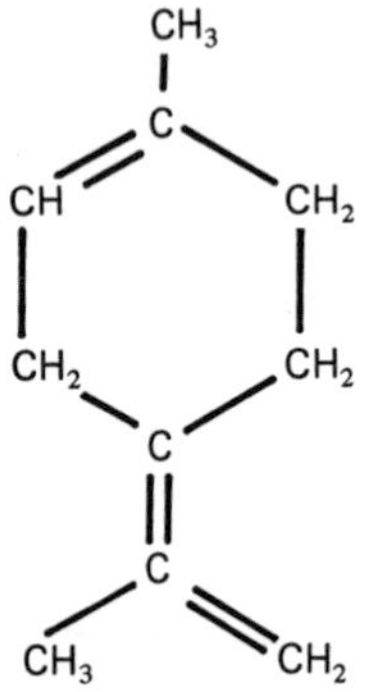

CH₂ = CHCH₂NCS

Allyl isothiocyanate
(Mustard)

CH₃COOCH₂CH₂CH₂CH₂CH₃

Amyl acetate
(Banana)

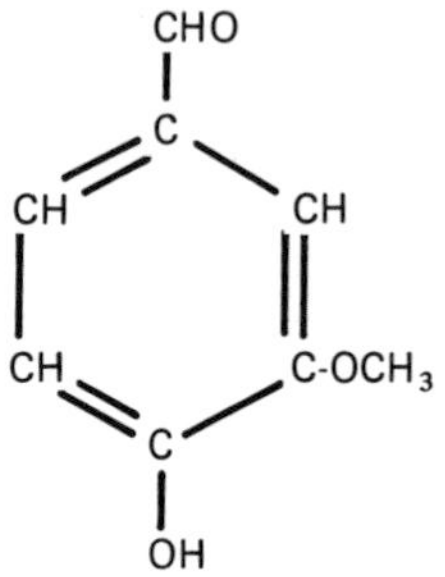

Vanillin
(Vanilla)

D-Limonene
(Orange)

Fig. 6.4. Flavoring compounds.

components in four common foods—tomatoes, black pepper, onions, and oranges. These foods are quite easily differentiated, and you would no doubt expect that the essential oils they contain would be quite different. However, some compounds occur in more than one food—for example, limonene is found in all except onions. In oranges, this terpene hydrocarbon is dominant—almost 95% of the volatile flavor of orange oil is limonene. It is of only minor importance in tomato, but the oil of black pepper contains about 25% of this compound and almost an equal amount of β-pinene and lesser amounts of β-caryophyllene, α-pinene, α-phellandrene, and β-phellandrene. These terpene hydrocarbons account for almost 85–90% of the essential oil of black pepper, but the flavor of natural black pepper is far more desirable than that of these five hydrocarbons alone. Onion flavor results from a number of complex sulfur-containing compounds. Fruity flavors can contain terpene hydrocarbons but also contain esters, alcohols, aldehydes, ketones, etc. In other words, each food organism has a unique group of compounds in its essential oil, but many compounds occur in a number of organisms.

Table 6.5 indicates the great diversity in biological systems and in the chemical components which make up these organisms. You may begin to wonder where is the order in these arrays of substances. To the trained chemist and biochemist, there are discernible patterns of compounds. In Fig. 6.4 we have illustrated what the chemist describes as three different classes: D-limonene is a terpene, vanillin is an aromatic compound, and allyl isothiocyanate and amyl acetate are aliphatic or linear compounds. We have already noted that monosaccharides are the unit compounds giving rise to all kinds of carbohydrate compounds and acetic acid is the basic metabolic building unit from which fats are synthesized. It can also be shown that acetic acid is the source compound for the sterols and for the aromatic and terpenoid components of biological systems. However, for the latter compounds there is a common metabolic precursor after acetic acid. In other words, acetic acid is used to make another parent compound—mevalonic acid:

$$\underset{\underset{OH}{|}}{\overset{\overset{CH_3}{|}}{HOCH_2CH_2CCH_2COOH}}$$

It is beyond our scope to go into the details of how this is accomplished, but the compound made from acetic acid via mevalonic acid which serves as the building unit for terpenes as well as for sterols, carotenoid pigments, aromatics and even rubber is the parent hydrocarbon isoprene:

$$\underset{\underset{CH_3}{|}}{CH_2=CH-C=CH_2}$$

Whereas natural rubber is a high molecular weight polymer of isoprene, C_5H_8, simple terpene hydrocarbons such as D-limonene, α- and β-pinene, etc., result from two isoprene units, $C_{10}H_{16}$, and a sesquiterpene results from three isoprenes, etc. β-Carotene, the precursor of vitamin A, is a quatroterpene, $C_{40}H_{56}$. Squalene, commonly synthesized by many organisms and in your body as an intermediate in making cholesterol, is a triterpene resulting from six isoprene units. Go back and look at the structures of the fat-soluble vitamins and of chlorophyll and note the chains of carbon atoms with the isoprene configuration:

$$\underset{\underset{C}{|}}{C-C-C-C}$$

It is clear again that metabolic common denominators exist in all biological systems and that these chemical compounds lead to the production of desirable or at times undesirable substances in the organisms we use for food.

PROTEINS

No biological organism can live and almost no biological process can take place without protein being involved. There is no biology without protein. Even the word *protein* connotes the importance of these molecules, for it is taken directly from the Greek word *proteios* meaning "primary." In Chapter 5, we discussed the role of proteins in human nutrition and noted that the nutritive value of proteins from different sources varies greatly. The information given there did not indicate the biological functions of the proteins in the various food organisms or in man himself. Here we shall get a glimpse of the diverse biological functions of proteins.

From time to time, we have emphasized that all biological systems are dynamic and require a continuing energy supply. Almost all of the highly organized chemical reactions of living things take place through the catalytic action of enzymes, which chemically are also proteins. For each reaction, a specific protein molecule or enzyme is required. In this book, whenever a chemical reaction taking place in a biological organism is noted, it must be understood that an enzyme, a protein, is involved whether or not this is explicitly stated. If we reflect on the thousands of chemical reactions taking place in our own bodies as well as in all other organisms, we must conclude that the array of specific proteins has to be enormous. The specificity can go even further—proteins serving the same function in different species may differ chemically. Indeed, hemoglobin, the important oxygen-carrying protein of blood, differs slightly from species to species. Even in the human there may be molecular differences, as seen in the abnormal hemoglobins involved in the inherited disease, sickle cell anemia. To get another view of the varied functions that proteins serve, reflect a bit on your own activities. There are contractile proteins in muscle which react and change shape as we walk about, blink an eye, play the piano, pump our blood, or move food through our gastrointestinal tract, etc. Proteins are involved in transmission of nerve stimuli; there are photosensitive proteins in your eye and in green plants; transparent protein in the lens of your eye; lubricating protein in saliva, the joints, etc.; protective proteins making up the hair and skin; and other protective proteins in our chemical defense systems against other organisms. These last include lysozyme in tears to protect the eyes and the immune proteins of our blood. There are specific proteins of the tissues that connect muscles to each other or hold organs in place, and even proteins to feed the young of the succeeding generation—milk and egg proteins. Such a list could be extended not only for the human but for every organism known. The proteins of our food are the proteins that serve the biological functions of the food organisms.

Proteins have a wide range of other properties. Some proteins are soluble in water and others are insoluble. Many are insoluble in water but soluble in dilute solutions of salts of various kinds. Our own blood contains sufficient salt to keep many vital proteins in functional solution. Many apparently insoluble proteins have the ability to imbibe or hold in their complex molecular structure varying amounts of water—remember that water is a very small molecule compared to proteins. Proteins may react readily with acids, bases, and salts, and indeed their molecular shape and biological function are determined by the pH of the water solution in which they occur. In Chapter 4, we noted that a major environmental parameter controlling life itself is pH. Proteins are also sensitive to heat, and at temperatures much above body temperatures most proteins are irreversibly changed, which shows how important this environmental factor is. Heat denatures and may coagulate proteins, i.e., make soluble proteins insoluble, which is what we observe in, for example, boiling or frying an egg.

All proteins are polymers of amino acids. In other words, amino acids are the building blocks (monomers) from which all proteins (polymers) are made. The number of amino acids is not great. Most proteins are made up entirely or almost entirely of only 18 amino acids (see Table 5.6 of Chapter 5). Although there are a few other amino acids in some proteins, they will not concern us here. In order to get an idea of how so few different types of building blocks are used in making up the enormous variety of proteins, we need only to consider the variation in molecular size of different proteins. They range in molecular weight from about 10,000 to 10,000,000 or more. Compared to a water molecule of molecular weight 18, protein molecules are 500–500,000 times larger. Putting it another way, 18 grams (about $\frac{2}{3}$ ounce or 4 teaspoons) of water contains the same number of molecules as 10–10,000 kilograms (22–22,000 pounds) of protein. Thus proteins, because of their size, must have a very complex molecular architecture. In terms of the numbers of building blocks (amino acids), a small protein of molecular weight 10,000 would contain approximately 90 amino acids while one of 10,000,000 would contain 90,000 amino acids. Since these building blocks (amino acids) are of 18 different kinds, all the permutations and combinations of arranging 90,000 or even 900 or 90 of them give a truly astronomical number of proteins. So you see that there is no problem in accounting for the great complexity and specificity of proteins.

The fundamental way in which amino acids are joined to make a protein molecule is similar to the manner that the carbohydrate building blocks or the lipid building blocks are joined to make more complex carbohydrates for lipids—i.e., they are joined by the effective elimination of a molecule of water. Again, since such a reaction requires energy and takes place in water, the biosynthetic steps in protein synthesis are complex. Suffice it to say that as with many other biosynthetic processes adenosine triphosphate, ATP, is involved and phosphate derivatives of amino acids are metabolic intermediates. The overall scheme of protein synthesis is indicated by the following equation, in which the —COOH group of one amino acid joins the next amino acid at the —NH$_2$ attached to the carbon atom adjacent to the —COOH:

$$
\underset{\overset{|}{NH_2}}{R\overset{\overset{O}{\|}}{CH}\!COH} + \underset{\overset{|}{NH_2}}{R'\overset{\overset{O}{\|}}{CH}\!COH} \longrightarrow R\underset{\overset{|}{H_2N}}{CH}\!\underset{\overset{|}{H}}{\overset{\overset{O}{\|}}{C}N}\!CHCOOH + H_2O
$$

peptide linkage

Here is shown a dipeptide—two amino acids joined by a peptide bond. In a protein molecule, many amino acids are joined by similar peptide linkages to form a primary repeating chain of two carbon atoms and one nitrogen atom, —C—C—N—. R and R' represent the unique portion of each of the amino acids, which are shown in Fig. 6.5.

The part of each molecule to the right of the dashed lines in Fig. 6.5 is the part that is involved in forming the peptide chain of the protein. The peptide chain occurs as a helix or coiled spiral or spring, with the various R groups exposed. Long chains assume positions in space according to the R groups. The chains may fold or turn in space to form linear or globular molecules. Also, the peptide chains may be cross-linked. Note in Fig. 6.5 that

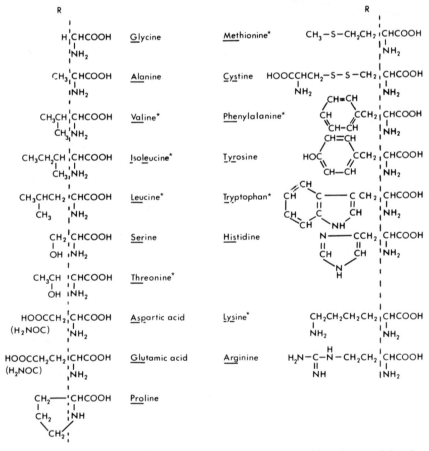

Fig. 6.5. The amino acids. Those with an asterisk are the nutritionally essential amino acids.

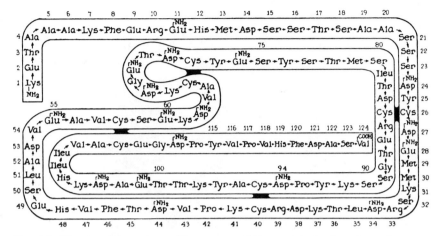

Fig. 6.6. The complete amino acid sequence of the protein bovine ribonuclease, an enzyme which digests ribonucleic acid. (From Symth *et al.*, *Journal of Biological Chemistry* **238**:227, 1963; reproduced by courtesy of that journal.)

cystine has two potential peptide groups, and this amino acid may serve to tie peptide chains together. The kinds and relative positions of R groups determine chemical reactivity as well as the shape of the protein molecule. So it is understandable that proteins are constructed by biological organisms for specific biochemical purposes at the molecular level.

In Fig. 6.6 a relatively simple, small protein of only 124 amino acids is shown. This protein is the enzyme ribonuclease, a biochemical catalyst involved in the digestion or hydrolysis of ribonucleic acid (see below) to its constituent building blocks or subunits. Because cystine holds chains of amino acids together, each half of cystine is counted and noted *Cys*. Figure 6.7 is a scale drawing of a protein molecule, pooling data of a great many kinds on the nature of myoglobin, a protein of muscle which picks up oxygen carried to the muscle by hemoglobin of the blood. Myoglobin is red in color, as is hemoglobin, and it or its derivatives give meat its color. The molecular structures of relatively few proteins are known in such detail as those illustrated here, but knowledge in this area is rapidly expanding as a result of some extraordinarily sophisticated chemical methodology. Even though the precise molecular architecture of most proteins is still not known, the biological functions of a great many different individual proteins are known.

The direct involvement of proteins in almost all biological processes means that proteins are associated with all active biological material. Metabolic processes cannot take place without them. This is significant with respect to our food supplies because proteins are usually associated with other essential nutrients, particularly vitamins and minerals in metabolizing tissues

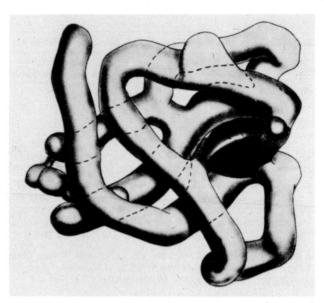

Fig. 6.7. Model of the myoglobin molecule, the oxygen-carrying protein of muscle which gives fresh meat its red color. (Reproduced through the courtesy of John C. Kendrew and the American Association for the Advancement of Science from *Science* **139**:1261, 1963).

(see Chapters 3 and 5). Consequently, food organisms containing large amounts of protein generally contain other nutrients as well. However, if proteins are isolated and refined and subsequently used in a formulated food, some of the other nutrients may be lost in the refining process. Thus the use in foods of such refined proteins as gelatin from animal tissues or proteins from seeds such as cotton, soya, corn, and wheat requires considerable sophisticated technology to assure that high nutritive levels are maintained in food products containing purified proteins as well as in those containing purified sugars and fats. If such factors are not taken into account in processing and distribution, then these high-protein foods will not generally be rich in the many other essential nutrients commonly thought to be associated with them (see Chapter 7).

A majority of proteins are polymers of amino acids only, and are some-times called "simple proteins" for that reason. However, there are proteins which have other non-amino acid groups in the biologically functional molecules. For instance, lipoproteins contain lipid subunits as well as amino acids. Glycoproteins contain carbohydrate, phosphoproteins contain phosphoric acid, nucleoproteins contain nucleic acid (see below), and so on. Lipoproteins are often found in cell membranes and nerve tissue; mucin, the lubricating protein of saliva and other body fluids, is a glycoprotein; casein of milk (and cheese) is a phosphoprotein. Hemoglobin of blood and myoglobin of muscle have an iron-containing porphyrin group, as do the cytochromes (Chapter 3). Chlorophyll, which contains a magnesium–porphyrin complex, is attached to protein as it functions in the photosynthetic process. There are many additional complex proteins wherein the amino acid polymers are combined with other non-amino acid parts to give a group of proteins roughly termed "complex proteins."

In foods, proteins are often altered by heat (cooking) or other processing techniques, such as beating an egg white, or by the digestion process itself (Chapter 7). These altered proteins are termed "denatured proteins," meaning simply that the proteins are chemically altered so that they no longer can perform their natural biological function. Proteins are such large molecules, with so many reactive groups, that they are easily changed by heat or other means. The protein molecules may uncoil or unfold in a different way or they may react with each other. These reactions are commonly seen when an egg or meat is cooked. If water is present in cooking, some peptide bonds may be hydrolyzed (broken by the insertion of water). In such cases, smaller polymers are made from larger amino acid polymers (proteins). All such changes in native proteins yield what are loosely called "derived proteins."

Native proteins and derived proteins of high molecular weight are tasteless.[6] But on heating proteins can react with other food ingredients such as sugars and other biologically important molecules to yield the characteristic flavorful substances of many foods. As hydrolysis (see Chapter 7) progresses, some proteoses and smaller peptides are formed, and many of these are bitter or have other flavors. The methyl ester of the dipeptide of aspartyl phenylalanine is very sweet and is now used in food under the trade name Aspartame (see Table 6.2). Some of the individual amino acids are bitter, but glycine is sweet and glutamic acid has a very desirable taste. Glutamic acid is the most abundant in food proteins (see Table 5.6 of Chapter 5) and monosodium glutamate prepared and refined from vegetable protein is used extensively in food as a flavor enhancer. Much of the characteristic flavor of soy sauce is from glutamic acid, some other amino acids, and peptides produced from the hydrolysis (depolymerization) of soy bean protein. So it is clear that proteins do play a role in flavor of foods quite different from that of lipids, essential oils, and carbohydrates.

Proteins play a very important role in quality attributes of food other than nutrition and flavor. The texture, toughness, tenderness, etc., of many

[6] Two fascinating exceptions are known. A protein isolated from the red West African serendipity berry (*Dioscoreophyllum cumminsii*) is 3000 times sweeter than sugar (sucrose) (J. A. Morris and R. H. Cagan, *Biochim. Biophys. Acta* **261**:114, 1972). A tasteless protein isolated from the African miracle fruit (*Synsepalum dulcificum*) has the unique property of changing sour taste to sweet when placed on the tongue. The effect lasts for about 2 hours (K. Kurihara and L. M. Beidler, *Science* **161**:1241, 1968).

foods are related to their unique proteins. Many of the characteristics of bread, cakes, macaroni, spaghetti, etc., are due to the proteins of wheat. Proteins relate directly to some of the eating qualities of other cereals such as rice and corn and the pulses such as beans. Milk proteins determine many characteristics of dairy products, and this is particularly true of the casein in cheeses. Tenderness and texture of meat and meat products including sausages are profoundly affected by the various proteins in animal tissues. It is clear that the chemistry of proteins of the biological organisms which we require for our food determines the nutritive value of our food and also in a large measure determines the qualities which whet our appetite and increase our desire to eat.

NUCLEIC ACIDS

The *nucleic acids* make up the fourth category of substances of major biological importance. As their name implies, these substances are most prominent in the nuclei of cells (see Chapter 3). To the nutritionist, food scientist, or technologist, the nucleic acids are of far less interest than they are to many other biologists. In the Appendix, the nucleic acid content of foods is not even listed. The nucleic acids contain no unique essential nutrients for man, since he is able to synthesize his nucleic acid needs from other nutrients. Quantitatively, nucleic acids and their monomers, the nucleotides, usually make up less than 1% of a particular food organism. But to many biological scientists nucleic acids are of great importance because the genetic material which passes information from one generation to the next is composed of nucleic acids. Viruses, the smallest of the infectious agents, are nucleic acids and/or nucleoproteins. Nucleic acids are involved in the transfer of information and instructions from the cell nucleus, which controls the metabolic activities of other parts of a functioning cell. Still other nucleic acids actually carry building blocks such as amino acids or monosaccharides, or even smaller metabolic units such as hydrogen atoms, as cells manufacture protoplasmic material. Adenosine triphosphate, ATP, which has been repeatedly mentioned as the carrier of energy for many metabolic processes, is itself a nucleotide derivative and nucleotides are the building blocks of nucleic acids.

Nucleic acids are polymers of nucleotides. Nucleotides are themselves composed of three relatively simple molecules—nitrogen bases, a five-carbon sugar, and phosphoric acid. As with other biological materials, synthesis of the three components of the individual nucleotides and the joining together of nucleotides are accomplished by the effective removal of water. There are two types of nucleic acids according to the two sugars which occur in their respective nucleotides—ribose and deoxyribose. Nucleotides containing deoxyribose form one type of nucleic acid called *deoxyribonucleic acid*—DNA. DNAs are found in cell nuclei and represent the genetic material of the cell. Chromosomes are composed mostly of DNAs. Nucleotides containing ribose form the other type of nucleic acid, known as *ribonucleic acid*—RNA. RNAs may occur within cell nuclei but also in other parts of the cell—notably the ribosomes, the sites of protein synthesis. Using information stored in the DNA of the genetic material, RNA is synthesized in the nuclei and transferred to the cytoplasm. The RNA carries coded information to the cell organelles where proteins are made and other metabolic activities occur.

Let us look first at RNA. Some RNAs are very large polymers and others are relatively small. There are only four kinds of nucleotides since all the nucleotides of RNAs contain ribose and phosphoric acid and only four different nitrogen bases are present. These bases are adenine and guanine, both classed chemically as purines, and cytosine and uracil, both pyrimidines. Figure 6.8 shows how a nucleotide is formed from adenine, ribose, and phosphoric acid. Adenylic acid is sometimes called *adenosine monophosphate*, AMP. When three phosphates are joined, ATP, *adenosine triphosphate*, is formed (see Chapter 3). The other nucleotides found in RNAs are shown in Fig. 6.9.

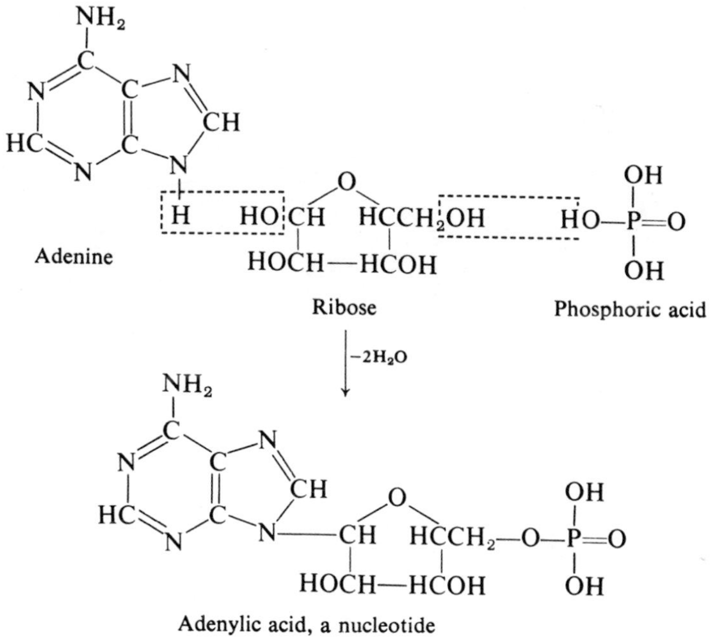

Fig. 6.8. Formation of adenylic acid.

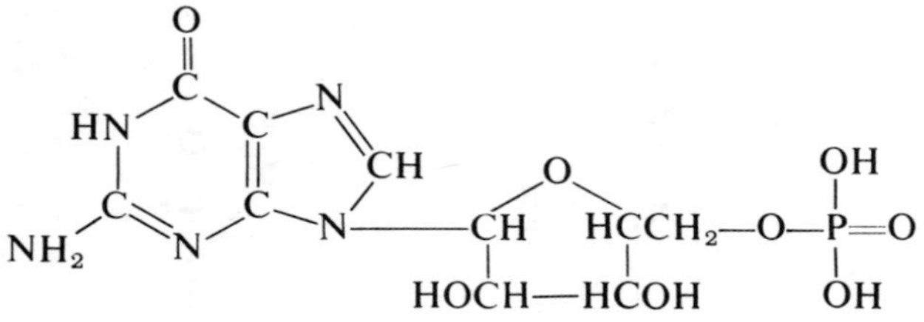

Guanine Ribose Phosphoric acid

Guanylic acid, a nucleotide

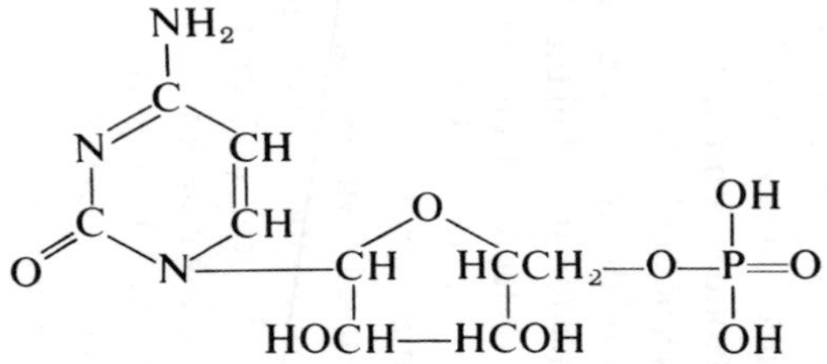

Cytosine Ribose Phosphoric acid

Cytidylic acid, a nucleotide

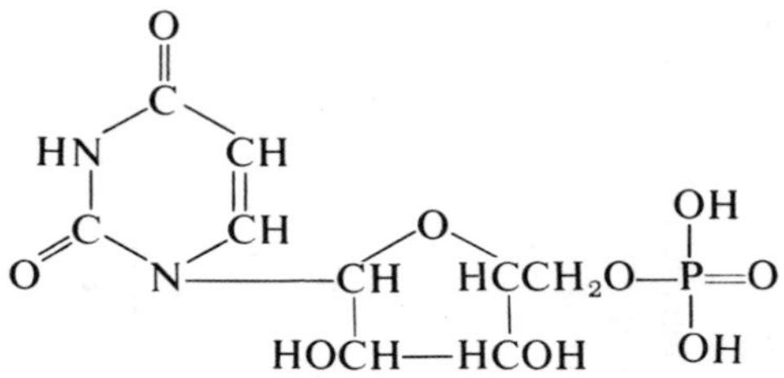

Uracil Ribose Phosphoric acid

Uridylic acid, a nucleotide

Fig. 6.9. Three of the RNA nucleotides.

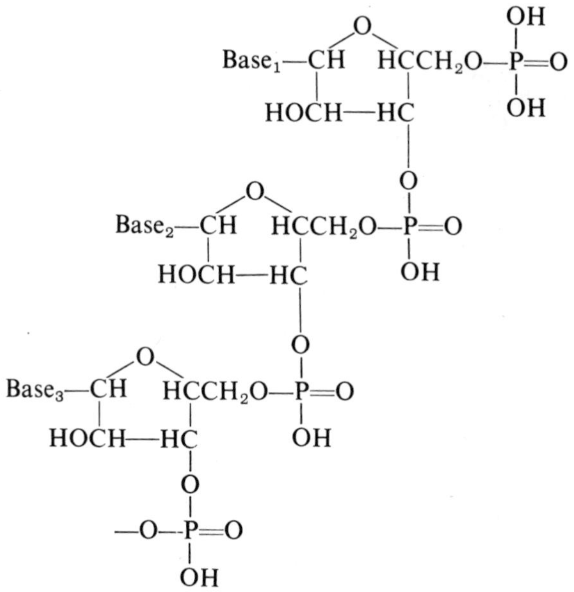
Base₁—CH HCCH₂O—P=O
HOCH—HC
OH
OH
O
Base₂—CH HCCH₂O—P=O
HOCH—HC
OH
O
Base₃—CH HCCH₂O—P=O
HOCH—HC
OH
O
—O—P=O
OH

Fig. 6.10. Formation of RNA chains.

To make ribonucleic acid, RNA, the nucleotides are joined by the effective elimination of a molecule of water[7] when another —OH of the phosphoric acid is linked to the —OH of ribose of another nucleotide as shown in Fig. 6.10. A relatively few or hundreds or thousands of nucleotides may be joined in this way. In essence, they consist of a linear chain of alternating ribose and phosphoric acids with different nitrogen bases attached to the ribose units. The uniqueness of the nucleic acid is determined by the order in which the four different nitrogen bases, adenine, guanine, cytosine, and uracil, appear on the chain, which assumes the shape of a helix (similar to a long coiled spring). The order of nitrogen bases is determined by the order of nitrogen bases on the directing deoxyribonucleic acid of the genetic material in the cell nucleus.

Deoxyribonucleic acids, DNAs, differ chemically from RNAs in only two ways. Rather than ribose, DNAs contain deoxyribose, a close chemical relative to ribose in which one of the hydroxyl groups (—OH) is replaced by a hydrogen atom (—H). The other difference is that the nitrogen base uracil does not appear but rather thymine. Otherwise, DNAs resemble RNAs. The nucleotides of DNAs are shown in Fig. 6.11.

[7] More accurately, in the living cell the analogous triphosphates, of which ATP is one, are the precursors, and in the presence of the proper enzymes and directing DNA the nucleotides polymerize by elimination of two phosphoric acids.

Deoxyadenylic acid, a nucleotide

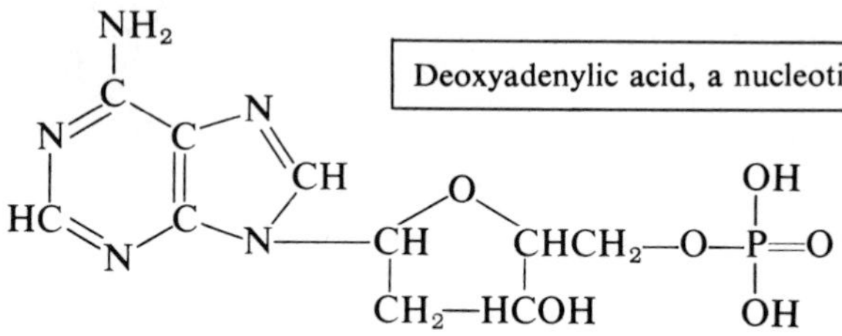

Adenine Deoxyribose Phosphoric acid

Deoxyguanylic acid, a nucleotide

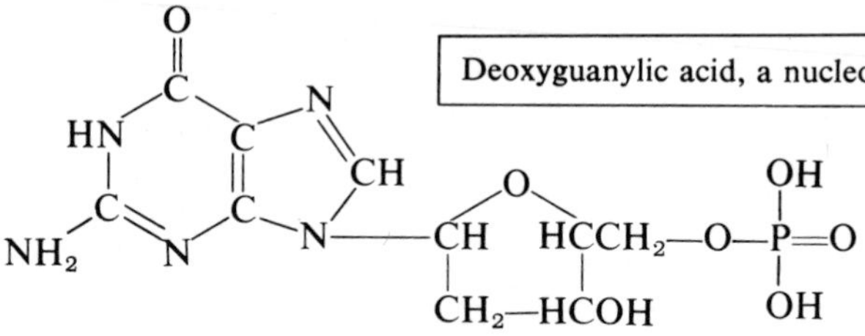

Guanine Deoxyribose Phosphoric acid

Deoxycytidylic acid, a nucleotide

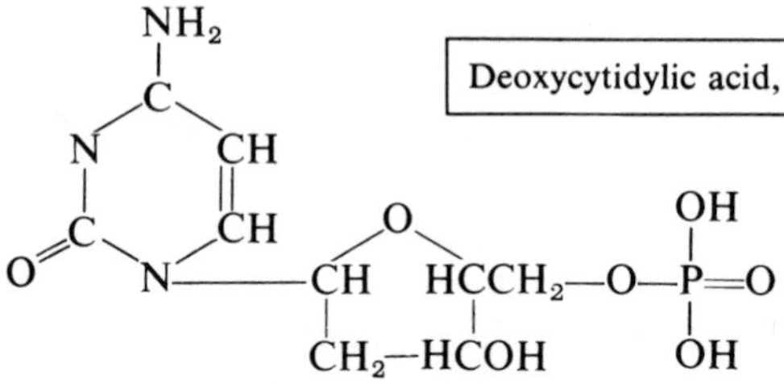

Cytosine Deoxyribose Phosphoric acid

Deoxythymidylic acid, a nucleotide

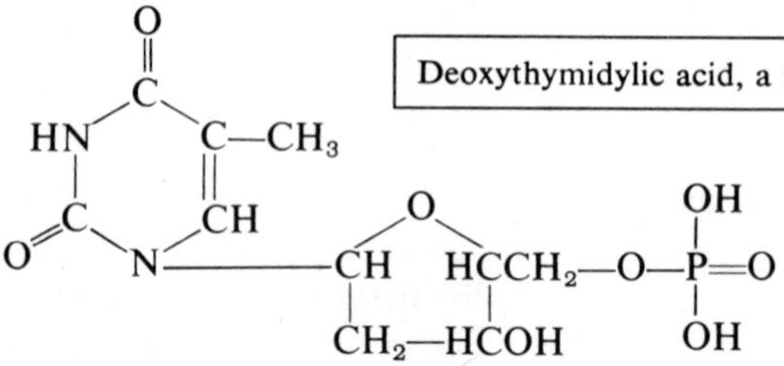

Thymine Deoxyribose Phosphoric acid

Fig. 6.11 DNA nucleotides.

The nucleotides of DNA are joined as in RNA to form a chain of alternating deoxyribose and phosphoric acid units with the nitrogen bases adenine, guanine, cytosine, and thymine extending from the deoxyribose units. The uniqueness and specificity of DNAs are determined by the order in which the four bases occur on the chain, which is also helical in shape. The DNAs carry all the information necessary to start a new organism of the next generation, and DNAs can reproduce themselves precisely in a living cell. They can also precisely program the synthesis of RNAs, which in turn control the precise sequence of amino acids in each protein making up the organism and further program the integrated and complex processes which take place during the growth, development, and reproduction of the organism. It is beyond our scope to go into detail here as to how this is done, but DNAs in the cell nucleus occur as a double helix like two springs intertwined. The two helices are so positioned that a purine base is paired with a pyrimidine base. In DNAs, adenine pairs with thymine and guanine with cytosine. The pairings are stabilized by hydrogen bonds. As shown in Fig. 6.12, the two helices may be thought of as a twisted ladder where the sides of the ladder are the chains of sugar (deoxyribose) and phosphoric acid and the rungs are

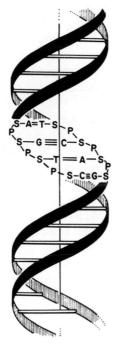

Fig. 6.12. A representation of the double helix of paired deoxyribonucleic acid (DNA) molecules.

the paired nitrogen bases of the nucleotides. As the base sequence is translated to RNA during its synthesis, guanine pairs with uracil.

Many of the nucleic acids, certainly the DNAs and many of the RNAs, are informational molecules in the true sense of the word. The transfer of "information" from one generation to the next has long been observed in man, animals, and plants. You need only to reflect on likenesses of children to their parents. How this is done through the biochemical processes of life has intrigued scientists of all persuasions, and it is a triumph of modern biochemical and genetic science that we now have the general rudiments of this fascinating process.

The DNAs of genetic material are long molecules whose overall shape is linear, and so they may be compared to recording tapes or the program tapes of a giant computer. The information necessary is coded in a four-letter alphabet—A, G, C, T—the *a*denine, *g*uanine, *c*ytosine, and *t*hymine attached to the sugar phosphate chains. It is now known that the sequence of the amino acids in proteins is determined by triplet groupings of the corresponding alphabet of RNA, i.e., A, G, C, U (U for uracil). For example, the UUU sequence in the RNA calls for phenylalanine and GCA for alanine. Not only protein structure but also the sequences in biological growth, development, maturation, and even reproduction itself are programmed in DNA. In the nucleic acids, we indeed have single molecules which influence biological systems profoundly. Compare this situation with that for the nutrients shown in Table 5.1 of Chapter 5, wherein the numbers of molecules required per day are enormous. For example, you require about 3.7×10^{-9} mole of vitamin B_{12} per day: $3.7 \times 10^{-9} \times 6.0 \times 10^{23}$, the number of molecules in a mole, equals 2.2×10^{15} or 2,200,000,000,000,000 molecules of B_{12} daily. Quite a difference in number of molecules when compared to DNA. But whereas you cannot make your own vitamin B_{12} you can make your own DNA and RNA from other nutrients—sugar, amino acids, phosphoric acid—because all the methods and tools for doing so are programmed in the minute amount of DNA you received from your parents.

An interesting digression is in order here. One human sperm contains only 2.5×10^{-12} gram of DNA. There is a similar amount from the human egg, so the initial fertilized human cell contains only 5×10^{-12} g of DNA. Let us say that the combined DNA is the final program for the human organism and it must be coded on those linear DNA molecules in only four letters—one for each nucleotide. The average molecular weight of a nucleotide is about 500, and so the 5.0×10^{-12} gram of DNA represents $5.0 \times 10^{-12} \div 500 = 10^{-14}$ mole of nucleotides in the DNA. This means that there are $10^{-14} \times 6 \times 10^{23}$ or 6×10^9 nucleotides in the 5.0×10^{-12} gram of DNA or 6×10^9 code letters in the DNA program for the human being, analogous to the taped instructions for a giant computer. If such instructions were to appear on a typed page, let us see how many pages would be required. Assume that it is possible to type 3000 letters on one side of a page. The DNA program in the original fertile human cell represents 1,000,000 pages typewritten on both sides—a formidable amount of information. It is this information that governs all life processes and responses to the environment, both external and internal.

Nucleic acids function essentially within a cell. Viruses, which are nucleic acids or nucleoproteins, may be of either the DNA or the RNA type, and they must get into a living cell in order to act. They may disrupt the cellular metabolic machinery and divert it to the production of virus molecules. When tissues grow seemingly out of control, the cell nuclei are generally involved and the tissues are cancerous. So in a sense cancers are related to disarranged nucleic acids. There is evidence that some invading viruses may initiate cancer. Some other simple chemical compounds can also initiate cancers, but how they do so is not known at present. Some of the carcinogenic or cancer-initiating substances are produced in biological organisms, whereas others are compounds produced in the external environment. Usually if DNA is too incorrectly made in the initial cell, the developing organism does not survive. Yet if the DNA is only slightly awry, the organism may develop with a genetic defect which may diminish its chances of survival or of normal life after it is born. Man is subject to a number of diseases of genetic origin such as sickle cell anemia, albinism, color blindness, and hemophilia.

To the food scientist, there are two important factors concerning nucleic acids—their flavor and the fact that too much nucleic acid in the diet cannot be tolerated by most people. Nucleic acids themselves have no flavor or odor but some of the nucleotides do; however, only two are apparently in sufficient concentration normally in food to affect flavor. We have already mentioned that the monosodium salt of glutamic acid has a flavor; like salt, it also has the ability to intensify other flavors in foods. Even more effective in this respect are the disodium salts of two nucleotides: guanylic acid (see Fig. 6.9) is about 300 times more potent than glutamic acid, and inosinic acid, which was found by the Japanese early in this century to be a major desirable flavor component of fish muscle, is about one-half as potent as guanylic acid. Inosinic acid is very closely related to adenylic acid and is a metabolic intermediate in the synthesis and degradation of adenylic acid, AMP, and adenosine triphosphate, ATP. Compare the structure of inosinic acid, shown below, with that of adenylic acid (Fig. 6.8):

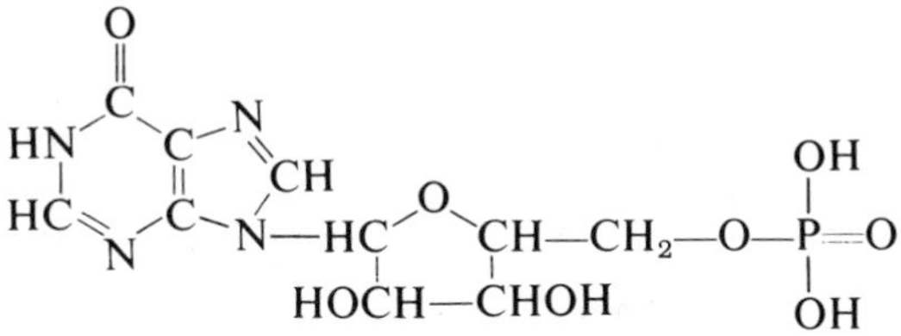

Both inosinic acid and guanylic acid are produced commercially and marketed as their disodium salts. Yeasts have a relatively high content of nucleic acids and can be easily grown. The nucleic acids are removed and treated with enzymes produced by other microorganisms so that the nucleic acids are hydrolyzed to the desired nucleotides. Then inosinic and guanylic acids are purified for use in food. So like sucrose (cane or beet sugar), these are chemical compounds purified from a biological raw material.

Because many microorganisms grow rapidly if properly cultured, much attention in recent years has been focused on algae, bacteria, and yeasts as a source of human food, particularly protein. The nucleic acid content of these organisms tends to be higher than in most food organisms consumed in quantity. Yeasts are particularly high in nucleic acids. When foods high in nucleic acids are consumed, the purines, i.e., adenine and guanine, of the ingested nucleic acids are converted to uric acid (see Chapter 7):

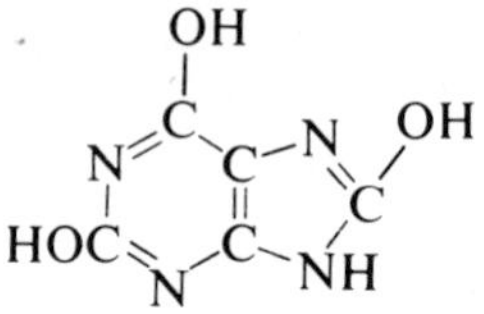

Even though uric acid is eliminated as a normal urinary constituent (about 0.5–0.6 grams per day), too much uric acid is not easily tolerated. Gout is characterized by relatively high blood uric acid and salts of uric acid or urates tend to crystallize into kidney and bladder stones. Consequently, too much dietary nucleic acids are to be avoided, but reasonable amounts cause no problem. If microorganisms are to be used as human food, they will likely require processing to remove some of the nucleic acids. Among mammals, only man, other primates, and the Dalmatian dog have difficulty in further metabolic degradation of uric acid. Yet uric acid is the major nitrogenous excretory product of birds and many reptiles. Such are the biochemical differences among the biota of the earth.

OTHER SUBSTANCES

Biological organisms produce thousands of chemical compounds other than those already mentioned. Generally, these do not account for a major portion of the nonwater substances, although in many instances they are important in our acceptance or rejection of a particular food. The diversity of these substances is generally greater in plants.

As might be expected, the intermediate compounds necessary to perform all of the biochemical transformations to make and use all of the foregoing compounds are present in different amounts in the organisms we use as food. Many of these are quite familiar—e.g., citric, lactic, and acetic acids, alcohol, caffeine, ammonia, urea, and nitrate. There are many other less familiar substances. The biological function of most of these is not understood, but nevertheless they are in our food. There is no doubt that some of these compounds are involved in the chemical defense of one organism against another. The antibiotics that are used medically and other such substances are naturally occurring materials, not only in microorganisms but in higher plants and animals as well. There are many similar compounds known which are not used commercially, for various reasons. The numerous tannins and other phenolic materials that are prevalent in plants may have defensive functions against predators, but quite probably they serve other functions as well for they are so common in many fruits and vegetables. The anthocyanins are common, also. Without chlorogenic, caffeic, and quinic acids coffee would not be coffee, yet these are common substances in many other plant tissues used as food. The gossypol of cottonseed has thwarted the use of this useful seed for human and animal food because it is toxic. Other potential food organisms are not used because of the chemical compounds they contain. Even in the organisms we use for food there are great numbers of substances naturally present which, if consumed in large quantities, might pose problems for us. Just to list them would be a task of enormous proportions. Nevertheless, toxicants which occur naturally in our foods are so common that the National Academy of Sciences–National Research Council published in 1966 and again in 1973 (2nd ed.) *Toxicants Naturally Occurring in Foods.* Under normal circumstances, usually no problems are encountered by the consumer. For example, serotonin, an extremely potent neurological agent, is found in bananas, pineapples, and tomatoes; vegetables of the Brassica family (cabbage, etc.) contain goitrogens or compounds capable of interfering with the thyroid gland and causing goiter. Estrogens, compounds having female sex hormone activity, have been found in soybeans and other plants, and meat, milk, and eggs contain estrogens as well as other hormones. Even carcinogenic compounds of microbial or fungal origin are present in many foods. The list of substances naturally present in our food goes on and on, a challenge to the imagination. In the following chapter, we will see how the body copes with the complex chemicals which are in our food.

Many chemical compounds which are not in food organisms as they grow nevertheless occur in our food. These are substances which are formed as we cook or otherwise process or store our food. Man has survived because he has learned to cultivate, process, and store his food until needed. Cooking antedates recorded history. Man discovered that cooking can make some otherwise inedible foods more digestible and useful and that cooking can increase acceptability by altering or enhancing the flavors and textures that add to the pleasure of eating. And since ordinary cooking requires temperatures above that which will permit life processes, it can kill all organisms in the food being cooked. Because some organisms are potentially pathogenic or parasitic to man, the hygienic value of cooking is obvious. This destruction is due no doubt to the fact that proteins and other heat-labile substances necessary for life of the potential invading organism are denatured, coagulated, and perhaps changed in other ways. Many denatured proteins are much more easily digested than undenatured proteins (see Chapter 7). At cooking temperatures, compounds in biological systems react with each other to produce other compounds not found in living systems. Proteins react with sugars to give substances of unique flavor and color. Simple sugars literally form hundreds of compounds as they are caramelized. Lipids may also be decomposed. You only need to observe the flavor development on cooking and the flavor degeneration on staling to know that new chemical compounds are formed and that many of these newly formed substances seem to be unstable. All of these changes are in reality changes in the chemistry of the food itself. Yet through the ages these hundreds of compounds formed on cooking appear to have caused no great problem for man.

Reliable techniques for identifying small amounts of the many compounds present in food have been developed in the last decade. For the first time in history, the chemist has been able to match the human nose as a detector of chemical compounds. Table 6.6 indicates many of the flavorful volatile components of cooked beef. To the trained biochemist, some of these compounds are recognizable as possible metabolic intermediates in the raw meat. However, most of these substances were probably formed during the boiling, roasting, or frying process. Although Table 6.6 is a formidable list, it does not contain many of the nonvolatile compounds which also play a major part in the flavor of cooked beef. Most of the compounds listed in Table 6.6 are present in very small amounts—on the order of parts per million or parts per billion—but even in such low concentrations many of them do affect flavor. It is the composite of all of these substances together with the added spices that makes up the flavor of the beef as eaten. Similar compilations could be made for other foods.

Many of the most popular foods are roasted or fried to the state of dryness at least in part, as evidenced by the crust of bread. Chocolate, coffee, nuts, cereals, etc., are dry-roasted. Snack foods such as potato and corn chips are examples of foods fried to dryness. Among the more reactive substances in raw foods are reducing sugars such as glucose, maltose, and lactose, the amino acids, and proteins. Sugars caramelize and also react with proteins and amino acids to give a great variety of compounds which characterize the food.

Table 6.6. Some Compounds Identified in Roasted, Fried, and Boiled Beef

n-Nonane	Phenol	2-Undecanone
n-Decane	Benzyl alcohol	2-Dodecanone
n-Undecane	Vinylguaiacol	2-Tridecanone
n-Dodecane	2-Methylbutanal	2-Pentadecanone
n-Tridecane	3-Methylbutanal	Acetyl furan
n-Tetradecane	n-Pentanal	2-Furfurylmethylketone
6-Methyltetradecane	n-Hexanal	Acetylpyrrole
n-Pentadecane	2-Hexenal	2-Methyl-acetyl pyrrole
n-Hexadecane	n-Heptanal	o-Hydroxyacetophenone
n-Heptadecane	2-Heptenal	4-Hydroxy-5-methyl-3(2H)-
n-Octadecane	n-Octanal	furanone
Toluene	2-Octenal	4-Hydroxy-3,5-dimethyl-3(2H)-
1,4-Dimethylbenzene	2,4-Octadienal	furanone
1,2-Dimethylbenzene	n-Nonanal	Acetic acid
Trimethylbenzene	2-Nonenal	Propionic acid
Methylethylbenzene	2,4-Nonadienal	Butyric acid
Diethylbenzene	n-Decanal	Isobutyric acid
n-Butylbenzene	2-Decenal	Valeric acid
2-n-Pentylfurane	2,4-Decadienal	Benzoic acid
2-n-Hexylfurane	n-Undecanal	Lactic acid
2-n-Octylfurane	2-Undecenal	Acetol acetate
Dimethylpyrazine	2,4-Undecadienal	γ-Butyrolactone
Trimethylpyrazine	n-Dodecanal	Methylbutyrolactone
Dimethylethylpyrazine	2-Dodecenal	γ-Hexalactone
1-Propanol	2,4-Dodecadienal	γ-Heptalactone
1-Butanol	n-Tridecanal	δ-Heptalactone
2,3-Butanediol	2-Tridecenal	γ-Octalactone
1-Pentanol	n-Tetradecanal	γ-Nonalactone
1-Hexanol	n-Pentadecanal	δ-Nonalactone
n-Butoxyethanol	n-Heptadecanal	γ-Decalactone
1-Heptanol	Benzaldehyde	2,4,5-Trimethyl-Δ³-oxaline
2-Heptanol	Phenylacetaldehyde	Thiophene-2-carboxaldehyde
3-Heptanol	Methylcinnamaldehyde	5-Methiofurfuraldehyde
4-Heptanol	Butan-2-ol-3-one	Benzothiazole
1-Octanol	2,3-Butadione	Dimethyldisulfide
3-Octanol	2,3-Pentadione	1-Methylthioethanethiol
4-Octanol	2-Heptanone	Methional
2-Octen-1-ol	3-Heptanone	Dimethylsulfone
1-Nonanol	2-Octanone	2,5-Dimethyl-1,3,4-trithiolane
1-Decanol	3-Octanone	
1-Undecanol	2-Nonanone	
1-Dodecanol	2-Decanone	

Data compiled from C. H. T. Tonsbeek, A. J. Plancken, and T. V. D. Weerdhof, *J. Agr. Food Chem.* **16**:1017 (1968); H. M. Liebich, D. R. Douglas, Albert Zlatkis, Francoise Muggler-Chavan, and A. Donzel, *J. Agr. Food Chem.* **20**:96 (1972); Kenji Watanabe and Yasushi Sato, *J. Agr. Food Chem.* **20**:174 (1972); H. W. Brinkman, Harald Copier, J. J. M. Deleuw, and Sing Boen Tjan, *J. Agr. Food Chem.* **20**:177 (1972); Stephen S. Chang, *Proceedings 26th Annual Conference*, Am. Meat Sci. Assoc. 76 (1973); Ira Katz, *Proceedings 26th Annual Conference*, Am. Meat Sci. Assoc. 102 (1973).

A group of chemical compounds that particularly contribute to flavor of cooked and roasted foods are the pyrazines. (Naturally occurring pyrazines also contribute to the characteristic flavors of some uncooked foods.) Pyrazine itself has the structure

$$
\begin{array}{c}
\text{HC} \overset{\text{N}}{\diagup\diagdown} \text{CH} \\
\text{HC} \diagdown_{\text{N}} \diagup \text{CH}
\end{array}
$$

and the pyrazines differ only in the substituent atoms or groups of atoms replacing the hydrogens (H). Compare the structure of pyrazines with that of the pyrimidines, the cytosine, thymine, and uracil of nucleic acids (Figs. 6.9 and 6.11).

Chocolate is a very popular food the world over. It is prepared essentially by grinding the roasted seeds of the tropical fruit *Theobroma cacao*. The seeds in the fresh pod are embedded in a rather tasty flesh or pulp not at all like chocolate in flavor. The pulp is usually removed from the seeds by fermentation and the seeds are air-dried at the site of production. The seeds are then roasted and ground to make chocolate. Thirty different pyrazines have been identified in chocolate. They result from the thermal reactions of the amino acids, sugars, and other compounds in the seed when it is roasted. Figure 6.13 indicates some of the chemical changes during the roasting of cocoa beans at 150°C (302°F). Most of the changes occur in the first 30 minutes. Note that whereas the unroasted beans contain no pyrazines, about 9 parts per million (ppm) is formed when they are roasted. Simultaneously, there are decreases in

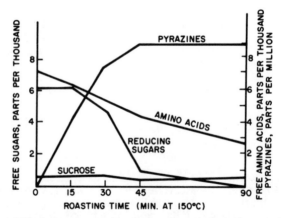

Fig. 6.13. Chemical changes occurring during the roasting of cocoa beans to make chocolate. (Reprinted with permission from G. A. Reineccius, P. G. Keeney, and Wendy Weissberger, *Journal of Agricultural and Food Chemistry* **20**:205, 1972. Copyright by the American Chemical Society.)

both free amino acids and reducing sugars while the sucrose content remains almost unchanged.

Beef and chocolate are only two of the many common foods which acquire characteristic flavors by chemical change on heating. Similar knowledge concerning other foods is rapidly accumulating. Now that we have methods for determining the minor constituents as well as the major constituents in the food we eat, a more meaningful perspective on the chemical nature of many of our foods is becoming possible.

CONCLUSION

In this chapter, the chemical components of our foods have been briefly described. In addition to the four major biochemical categories of substances present in all biological organisms—lipids, carbohydrates, proteins, and nucleic acids—there are many other miscellaneous compounds produced and found in microorganisms, plants, and animals. Also, many more substances not usually found in living things are produced when our food is prepared for use. There are other important chemical changes which occur on storage without heating, and some of these are both economically and nutritionally important, particularly when oxygen of the atmosphere is involved. Chemical knowledge about food will increase even more as we try to expand and improve the quality of our food supplies. People must eat every day, and always the public is concerned—even emotionally concerned at times—about food, over which they have little control in our modern sophisticated society.

Frederick Accum (1769–1838), an outstanding chemist and foremost popularizer and practitioner of the science in his time, would no doubt be amazed at our wealth of chemical knowledge about food. He was a gifted lecturer and demonstrator who wrote books on the science of beer, bread, and wine making (see Chapter 9). In 1821, in his book *Culinary Chemistry*, Acuum wrote:

> Cookery or the art of preparing good and wholesome food and preserving all sorts of alimentary substances in a state fit for human sustenance, of rendering that agreeable to the taste which is essential to the support of life and pleasing to the palate without injury to the systems, is, strictly speaking, a branch of chemistry. . . .

Now that we are acquiring so much concrete scientific information about what really makes up our food, we can better meet the challenge of feeding our expanding population with more wholesome food. To accomplish this goal, our detailed knowledge must be wisely applied to specific needs.

7

THE SEPARATION AND UTILIZATION OF NUTRIENTS FROM THE MULTIPLICITY OF SUBSTANCES NATURALLY PRESENT IN FOODS

Man, like every other animal, lives by consuming other organisms. In Chapter 6 it was pointed out that organisms differ from each other in the details of their chemical composition even though there are chemical similarities among all biological organisms. Thus man, like every other animal, must have mechanisms for separating substances which are usable and which he needs from the multiplicity of substances which are useless or perhaps even harmful. This is even more apparent when we reflect on the specialized functions of specific cells, tissues, and organs which make up the whole dynamic organism. In order for nerve cells, muscle cells, secretory cells, etc., to perform their specific tasks, they must depend on other cells, organs, and organ systems to maintain the proper biological (cellular) environment.

You will recall from Chapter 4, that the environmental factors controlling life are water, temperature, pH, osmotic pressure or water activity, and nutrients. Although the nutrients receive more detailed attention, the systems for maintaining all of the environmental parameters at the proper level to assure cell function are equally important. The food you consume presents your body with an almost infinite number of chemical challenges to the maintenance of rigidly controlled cellular environments. In addition to the environmental requirements of water balance, temperature, pH, and water activity, the immediate cellular environment must contain molecules which can be metabolized to furnish the energy necessary for life and the building blocks for making and maintaining the protoplasmic constituents of the functioning cells. It is also necessary to have a scheme for removing the materials the cell no longer needs. The constantly flowing blood serves to control all factors of the cellular environment. Through arteries, capillaries, and veins, blood is pumped continuously by the heart to all the cells and organ systems which work to provide the nutrients and maintain temperature, acid–base balance, and water balance as well as to all the cells and organ systems which make life possible. When all of these processes are correctly functioning, we are in good health but when any vital process is faulty we are unhealthy or we die.

The fundamental interactions of our bodies with our surroundings that keep us alive are with the air we breathe and the food and water we eat and drink. Our modern urban style of living is causing some concern about the quality of the air we breathe. Air pollutants, both suspended solids (dust) and gases (auto exhaust or industrial gases), are great problems confronting society, because they befoul our lungs and contaminate our blood and cellular environment. We do have some natural defenses against air pollutants, and assuming that the air is capable of supporting life, the chemical challenge to our physiological systems is rather constant in the air, which supplies us with only one nutrient, oxygen. The greater direct chemical challenge is from the food we must eat and the water we must drink. For our purposes, we shall consider the water pure and any organisms or other substances carried by the water as part of our food supply.

The compounds which constitute our food are many and varied because our food is a mixture of biological organisms. In Chapter 6, the chemical makeup of our food was discussed in some detail. From this raw material we must retrieve the molecules we can use to fuel each working cell of our bodies as well as all of the other nutrients that all of our cells require, which have been discussed in Chapter 5. Let us see how you, acting as a very efficient chemical processing plant, extract what you need and discard what you do not need so that you can have the fundamental raw materials in a usable form.

PRINCIPLES

Often the basic principles of a particular process are simple and rather easily comprehended, but the equipment for carrying out the process efficiently can be rather complex. So it is with the utilization of our food. Let us first state the principles involved and then describe the various pieces of equipment—the various organs of our bodies—and their functions.

First recall the nutrients you require each day (Table 5.1 of Chapter 5). Next consider the food you eat. In describing in Chapter 6 the lipids, carbohydrates, proteins, and nucleic acids which make up the protoplasmic material of our food, one aspect was common to all—that even though these molecules are large, complex, and often specific to the food organism and thus incompatible as such with our bodies, they are built up from small molecules common to all biological systems, including ourselves. Consequently, the problem is how to get these small molecules you can use from the food you eat.

You will remember that all the large molecules of lipids, carbohydrates, proteins, and nucleic acids are made biologically from their respective building blocks by the effective elimination of molecules of water. In this way, fats and other lipids are made from fatty acids and alcohols, complex carbohydrates are made from monosaccharides, proteins from amino acids, and nucleic acids from nucleotides, which themselves are made by joining phosphoric acid, monosaccharides, and nitrogen bases. Thus if the biosynthetic process of joining the small molecules to make large molecules by elimination of water is reversed, then fatty acids, alcohols, monosaccharides amino acids, and phosphoric acid may be made by hydrolysis—the insertion of water molecules to break the linkages in a large molecule to form smaller ones. So the fundamental building blocks or nutrients can be produced from the protoplasmic material of our food organisms. This is the process of *digestion* of food in the gastrointestinal tract (Fig. 7.1). Let us illustrate the digestion process by the following equations:

$$\text{Fats} + H_2O \longrightarrow \text{fatty acids} + \text{glycerol}$$
$$\text{Carbohydrates} + H_2O \longrightarrow \text{monosaccharides}$$
$$\text{Proteins} + H_2O \longrightarrow \text{amino acids}$$
$$\text{Nucleic acids} + H_2O \longrightarrow$$
$$\text{nitrogen bases} + \text{monosaccharides} + \text{phosphoric acid}$$

All of these reactions of hydrolysis are energy-liberating reactions and take place in water with relative ease provided that the necessary catalytic agents—enzymes—are present; the reverse or biosynthetic reactions, you will recall, require energy and so must be made through the intermediary phosphate complexes of the building blocks in the water medium of biological systems. To get a rough idea of the biosynthetic process versus the digestive process, consider the care, effort, and time required to build a building and the relative ease with which it is demolished. In both instances the same building blocks are present. A new building cannot be made from the material of an old building without demolition of the old structure and recovery of the building blocks for reuse. So it is with your food, you must reduce the food organism to its fundamental parts so that you can use these parts for your own needs.

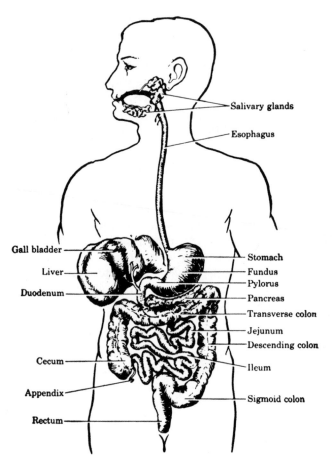

Fig. 7.1. The digestive tract. (From *Physiological Chemistry*, by Arthur K. Anderson. Courtesy of John Wiley and Sons, Inc.)

After you have digested or hydrolyzed your food as much as you can to the small molecular building blocks from which it is made, the result is a mixture of molecules you need and can use and other molecules you do not need and whose presence may even be deleterious. There will also be undigested food residue which, for a number of reasons, you cannot hydrolyze further. The next problem is separating the molecules you can use from those you must discard. You have remarkable chemical machinery for accomplishing this task so that you can supply the necessary nutrients to all the cells of your body without exposing them to undue amounts of harmful molecules.

Let us now proceed to the consideration of how your body accomplishes the two processes of digestion and utilization.

DIGESTION AND ABSORPTION

Although food is other biological organisms in whole or in part, it, of course, may have been cooked or otherwise preserved for use. Preservation of food requires that it be treated in a manner that will inhibit its consumption or destruction by other biological organisms competitive with us—e.g., microorganisms, insects, and other animals such as rats and mice. We can use Fig. 7.1 as a guide in outlining each stage of the digestion of food. As we do this, remember that the purpose of digestion is the hydrolysis of large molecules, many of which are insoluble in water, to small molecules that are soluble in water and can be carried to all of the cells of the body.

The digestive or gastrointestinal tract where digestion (hydrolysis) takes place in the presence of enzymes is a tubular structure extending from the mouth to the anus. Any substance that is eaten must pass through the walls of the gastrointestinal tract at some point in order to get to the blood and the body cells.

A few comments about enzymes are in order here. Several times it has been pointed out that chemically enzymes are proteins and they are the catalysts that facilitate the biochemical reactions which we recognize as the dynamic processes of biology. Since enzymes are proteins, pH affects their activity (see Chapter 6). For example, pepsin, a protein-hydrolyzing enzyme found in the stomach, is most active around pH 2.0 but is inactive at pH 8.0. On the other hand, trypsin, a protein-hydrolyzing enzyme found in the intestine, is active in the pH range of 6–8 but not around 2.

Enzymes are also specific for the types of reactions they can promote. Such specificity may relate to specific molecules or even specific linkages or arrangements among the atoms of a certain molecule or group of molecules. For instance, a protein-hydrolyzing enzyme may not be able to hydrolyze all the peptide linkages which connect amino acids in a protein molecule. In fact, it is more likely that a particular enzyme can hydrolyze only peptide linkages between specific amino acids or at certain positions in the protein molecules. You cannot digest hair, a protein, although you hydrolyze most food proteins in your digestive tract. The digestive enzymes may be likened to specific tools to do a specific job. So the protein-hydrolyzing enzymes hydrolyze or break some but not all peptide linkages between certain amino acids simply because the enzymes cannot fit particular peptide linkages. This type of enzyme specificity for certain molecules or molecular structures holds for all enzymes which promote biochemical reactions, including not only the digestion of proteins but also that of carbohydrates, lipids, nucleic acids, and the miscellaneous molecules present in biological systems and in our food.

Mouth

All food enters through the mouth, where the size of the food particles is considerably reduced by the cutting and grinding action of the teeth. This increases the surface area of the food greatly and facilitates digestion, because solid food must be attacked by the digestive enzymes at the surface. During the chewing process, saliva, a watery secretion from the salivary glands inside the mouth, is mixed with the food particles. From 1.0 to 1.5 liters (1.1–1.6 quarts) of saliva is secreted daily. Saliva contains mucin, a glycoprotein, which gives it its slippery character and serves to lubricate the food for easier passage to the stomach. In the mouth, the food comes in contact with the first digestive enzyme, amylase, which starts the hydrolysis (digestion) of starch. The end product of amylase digestion is maltose (see p. 122). Saliva also contains lysozyme, an enzyme capable of digesting bacterial cell walls, which kills some of the bacteria ingested with food. By movement of the tongue and cheeks, the chewed food is formed into a bolus which when swallowed passes into the esophagus, where a wave of contraction forces it down into the stomach. These waves are called *peristalsis*, and food is moved throughout the alimentary tract by peristaltic action.

Stomach

As the food reaches the stomach, where it will remain for 1.5–4.0 hours, it is continuously mixed with gastric juice by peristaltic contractions. The stomach, which has a capacity of 1–2 liters in the adult, is lined with cells which daily secrete about 2 liters of gastric juice containing hydrochloric acid, mucin, and a number of enzymes.

Hydrochloric acid, HCl, a strong acid sometimes called muriatic acid, is made and secreted by the cells in the stomach lining in sufficient amount to lower the pH of stomach contents to about 1.5–3.0. This serves several important functions. By lowering the pH of gastric contents below the level at which organisms can live, the HCl inhibits the growth of or kills outright most of the bacteria in the food. This is one of the major defensive mechanisms by which the human counteracts invading organisms, usually bacteria and small parasites, which may be in food. Achlorhydria, lack of gastric HCl, is a symptom of a number of diseases and is characterized by excessive bacterial growth and putrefaction of gastric contents. Another important function of the HCl is as a protein denaturant. Denatured proteins (see Chapter 6) are usually more easily hydrolyzed to amino acids than native proteins. Many proteins, of course, have already been denatured by cooking, so it is the proteins of raw or incompletely cooked foods that the HCl makes easier to digest. A third function of the HCl is adjustment of the pH of stomach contents to a level more favorable for the action of the enzymes of gastric juice.

The enzymes found in gastric juice are pepsin, rennin, and lipase. Pepsin is capable of hydrolyzing many large protein molecules to smaller polypeptide fragments, but it does not degrade proteins to amino acids. Other enzymes of the intestines do that. Rennin, which is only found in infants, denatures and initiates the hydrolysis of milk proteins, and rennin extracted from stomachs of other animals (lambs particularly) has been used for coagulating milk to make cheese for hundreds of years. Lipases are fat- or lipid-hydrolyzing enzymes and begin the digestion of fat to glycerol and fatty acids.

Stomach contents are gradually passed into the small intestine through the pyloric valve located at the base of the stomach. This valve is opened and closed periodically as determined by the relative acidic condition of gastric and intestinal contents. There is little or no absorption of food constituents into the blood from the stomach.

The Small Intestine

The small intestine is a long tubular organ having three distinct sections—duodenum, jejunum, and ileum, in order from the stomach. It is the place where the digestion initiated in the mouth and continued in the stomach is essentially completed and from where the end products of digestion—monosaccharides, fatty acids, amino acids, etc.—and other small molecules are absorbed into the blood.

As portions of the gastric contents are passed into the small intestine by the periodic opening of the pyloric valve, the acidified and partially digested food is immediately mixed with two other secretions—juice from the pancreas and bile from the liver—which together contain sufficient sodium bicarbonate and other somewhat alkaline components to neutralize the hydrochloric acid introduced by the stomach so that the pH of the whole mixture becomes about neutral, pH 6–8.

Bile, secreted by the liver at a rate of about 1 liter per day, is stored in the gallbladder, where it may be concentrated about 4–5 times by reabsorption of water, until it is needed to facilitate digestion. In addition to its function of facilitating digestion and absorption, bile serves as a vehicle for the excretion of certain substances from the body by way of the feces. The most obvious of these substances are the bile pigments, which result from the metabolic breakdown of hemoglobin (normally coming from worn-out red cells) and which give the usual greenish color to feces. Bile contains a relatively large amount of cholesterol and some of its metabolic derivatives, called bile acids, in the form of their sodium salts. Some of these bile constituents may be reabsorbed from the intestines. If cholesterol crystallizes from concentrated bile in the gallbladder, gallstones of cholesterol may form to plug the small duct carrying the bile to the intestine, a painful disorder often requiring surgery. The sodium salts of bile acids, like the sodium salts of fatty acids (soap), are extremely surface-active or soapy types of compounds in solution (see p. 131) and are excellent emulsifying agents—i.e., they disperse large fat globules into small ones and at the same time emulsify the fat which occurs in larger food particles. When fats and other lipids are so dispersed, they are more easily attacked by the enzymes which hydrolyze food to its fundamental constituents. Furthermore, this process exposes fat-coated food particles to digestive enzymes in much the same way that soap removes grease from our hands.

The juice secreted by the pancreas, in the amount of about 0.8–1.2 liters (0.9–1.3 quarts) per day, is a major source of digestive enzymes. Like bile, the pancreatic secretion is mixed with the partially digested food as it leaves the stomach and the whole mixture is slowly mixed and moved down by the peristalsis of the small intestine. The intestines themselves also secrete digestive fluids—2.5–3.5 liters (2.7–3.7 quarts) per day. There are a large number of different digestive enzymes acting on the food in the small intestine which continue the hydrolysis of the complex molecules of carbohydrate, proteins, lipids, nucleic acids, and other molecules of food to smaller and smaller molecules—carbohydrates to fragments containing a few monosaccharides, then two, and ultimately only one; proteins to fragments containing a few amino acids, then three, two, and ultimately only one; fats to diglycerides, monoglycerides, fatty acids, and glycerol; and so on. As these products of digestion approach or reach their ultimate small size, they, along with water, some of the bile acids, and the small molecules of our food not requiring digestion such as citric acid and acetic acid, are absorbed.

The small intestine besides being the site for the completion of digestion is a remarkable organ of absorption. The intestines are lined with *villi*, small finger-like protrusions that add enormously to the surface through which materials may pass (Fig. 7.2). It is estimated that the small intestine of the adult has about 7–8 square meters (8–9 square yards) of surface, which is not a passive membrane but rather lined with very active specialized cells. It is at this point that the selection process begins of sorting molecules which you need and can use from those you do not want.

Each of the villi extending into the contents of the intestines has a layer of cells which are capable of absorbing small molecules from the digesting mixture—water and very small carbohydrate and protein fragments. The

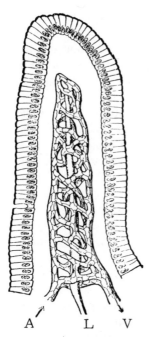

A   L   V

Fig. 7.2. Diagram of a villus. On the outside are the cells lining the intestine. Blood enters the network of capillaries by way of the arteries, A, from the heart and leaves by way of the portal vein, V, to pass through the liver before returning to the heart and general circulation. L is a lymphatic vessel embedded in the network. Loose tissue within the villus is not shown. (From *Physiological Chemistry*, by Arthur K. Anderson. Courtesy of John Wiley and Sons, Inc., and W. B. Saunders Co.)

absorbed tri- and disaccharides and very small peptides are hydrolyzed within the villi to their ultimate units—monosaccharides and amino acids. These and all other small metabolic molecules in your food, such as acetic and citric acids and vitamins, and water pass on through and are absorbed by the blood capillaries, which lead to the portal vein, which in turn carries all of the absorbed molecules to the liver, a most remarkable organ which will be discussed shortly.

Fats and lipids present a different problem in absorption since they are essentially insoluble in water, as are the fatty acids from which they are made. It was pointed out previously that fats become finely emulsified during the digestive process. Extremely small globules of mixtures of fats, fatty acids, mono- and diglycerides, and bile acids may pass the membranes of the cells of the villi, where the fatty acids and mono- and diglycerides are resynthesized to fats because too much of these compounds, which can serve as emulsifying agents, is not desirable in circulating blood. Finally the tiny globules of fat are passed on and eventually reach the blood by means of the lymphatic ducts.[1]

A very important aspect of the absorption of the products of digestion is that only the elementary building blocks should pass into the general circulation. This is particularly true for amino acids and monosaccharides, the building blocks for proteins and complex carbohydrates. The functioning cells of the body react against many of the molecules or fragments of molecules characteristic of another species. So if there is some defect in absorption such that a complex carbohydrate or protein or even a sizable molecular fragment of these larger molecules specific to another organism reaches the bloodstream, a complex series of other chemical defenses comes into play. If these defensive mechanisms fail, the individual affected experiences a series of responses loosely termed an *allergic reaction*. These may be manifested as sneezing, swelling of and excessive secretion by the mucous membranes of the nose or other respiratory passages, hives or other itching skin irritation, or even more general malaise. Persons so afflicted must learn to avoid the foods to which they are sensitive or allergic. It is estimated that at least one person in ten is allergic to one more or less common food, and when that food is one as commonly used as milk, eggs, wheat, or yeast a very real problem exists. Foreign molecules causing allergic reactions may enter the body in ways other than through the alimentary tract—through the mucous membranes of the nose, throat, and lungs, or by injection by natural (insect sting) or artificial means into various tissues or the bloodstream. The molecule causing such allergic responses is called an *antigen*, and the chemical defense mechanisms of the body can make a specific protein, called an *antibody*, to inactivate the foreign substance or antigen. A person may become very sensitive to foreign molecules if for some reason enough antibodies are not produced by the body, but when sufficient antibodies are produced the individual may become less sensitive or immune. It is roughly in this manner that a person becomes immune to reinfection after having certain diseases.

It seems almost incredible that the stomach can produce strong hydrochloric acid and that the stomach, pancreas, and intestines can synthesize so many digestive enzymes without themselves being digested. There are a number of ingenious mechanisms by which these organs protect themselves from digesting agents. There is some erosion of cells from these organs as a result of the friction of food passage and the action of digestive juices; generally such tissues are replaced and the sloughed cells may themselves be

---

[1] Lymph is the fluid which bathes the functioning cells of the body. It facilitates transfer for all essential nutrients and metabolic products between the circulating blood and tissue cells. Although lymph does not circulate in the same manner as blood, there is movement of the lymph and it is returned to the bloodstream via the lymphatic ducts.

digested with the food. When some protective devices do not function, ulcers can develop in the stomach and/or intestines. One defense mechanism of the stomach and intestines is that, although the digestive organs are themselves largely composed of proteins, the protein-hydrolyzing enzymes do not digest them because the secreting cells produce these enzymes in inactive form. The enzymes become active only when some other component of the digestion mixture is present. Thus the pepsinogen secreted by some of the cells of the stomach lining is inactive until converted to pepsin by the hydrochloric acid produced by other cells in the stomach lining. So also the pancreas secretes trypsinogen, which is inactive but becomes trypsin, a very active protein-hydrolyzing enzyme, when other enzymes start hydrolyzing trypsinogen. Enzymes themselves are proteins, and as digestion of food proceeds digestive enzymes are continuously produced and destroyed.

Not all of the food we eat can be digested completely. Some proteins such as those of hair cannot be digested, and neither can cellulose, the structural carbohydrate of plants. Other compounds of biological origin cannot be digested by man because he does not possess the necessary enzymes in his gastrointestinal tract. Indigestible fragments such as particles of sand, stones, metals, and plastics are also present, because the best of cleaning methods cannot remove all foreign materials.

Almost 80–90% of the water consumed with our food and the water introduced by the digestive secretions of the mouth, stomach, pancreas, liver, and intestines themselves is absorbed by the small intestines. The remaining water and undigested and unabsorbed substances are passed on to the large intestine, the colon, which has three sections—the ascending colon or cecum, the transverse colon, and the descending colon. In the colon, much of the remaining water is absorbed but not much else. In this way the feces are formed and eventually eliminated periodically from the body by way of the rectum through the anus.

You will recall that one of the fundamental environmental parameters of water solutions which permit cells to grow is pH. The hydrochloric acid produced in the stomach lowers the pH of its contents below that which most microorganisms consumed with the food can tolerate, and most of them are killed. However, the pH of the digestion mixture is changed to near the neutral point as stomach contents enter the small intestine, so that it is restored to a level permitting many bacteria to grow, which they do because the intestinal contents contain the necessary nutrients for them. These organisms also produce certain metabolic products which are absorbed. (In Chapter 8, we will see that bacteria and protozoa residing in the gastrointestinal tract are much more necessary in other species than in the human.) Although some of these bacteria are killed and digested, many remain viable, are retained in the intestine, and reproduce, particularly in the colon. Indeed, normal feces is composed to a large extent of microorganisms whose normal habitat is the human intestinal tract. The activity and relative numbers of various species of bacteria can be influenced by the types of food eaten. Among some of the metabolic products produced by the intestinal bacteria are the gases hydrogen, methane, carbon dioxide, and hydrogen sulfide and a number of odorous amines commonly associated with putrefaction.

Feces may be considered to be all the undigested material of the food, nonbiological insoluble matter consumed incidentally with food, and bacteria which grow on those residues not absorbed by the small intestine, plus water, which usually amounts to about 75% of the weight of the feces. The small intestines and the colon serve as very important organs for the conservation of body water. Food may contain certain chemical compounds or bacteria that stimulate peristalsis to the extent that materials pass through the alimentary tract too rapidly to permit proper water reabsorption. This condition is called *diarrhea*. In a way, this can be considered a defensive mechanism whereby the body rejects certain substances, but in doing so body water is lost via the feces. Dehydration of the body can at times be severe or even fatal as in such dreaded diseases as cholera and dysentery in infants.

SEPARATION OF NUTRIENTS FROM UNNEEDED SUBSTANCES IN DIGESTED FOOD

All of the organs of the alimentary tract function as an integrated system for performing two basic processes—digestion or hydrolysis of foods to their fundamental building blocks, most of which are soluble in water, and absorption of these substances into the bloodstream. However, all of the compounds so absorbed may not be usable when absorbed, or may not be usable at all, or may even be harmful. How then does the body protect its cells from unneeded or undesirable substances and yet provide all of its cells with all of the nutrients required to provide energy (fuel) and molecules for growth and repair?

The body must maintain for all cells at the proper level the environmental factors of water, temperature, pH, osmotic pressure, and nutrients. To accomplish these tasks, the liver, kidneys, lungs, and circulatory system (heart and blood vessels) work in a highly integrated fashion. To get some insight into to how all of these things are accomplished, we must return to the grand mixture of molecules absorbed by the intestines. This mixture must be sorted, so to speak, into good things (nutrients) and bad things (useless or harmful molecules).

The Liver

The substances absorbed from the intestines and passed into the capillaries of the villi do not go into the general body circulation. Rather, the capillaries of the villi lead to the portal vein, which carries the products of digestion directly to the liver—a most remarkable vital organ. Man or other animal quickly dies without a liver, and of all the so-called vital organs it is the only one that can regenerate itself if part is removed. In a very real way, the liver is a complex chemical factory, a minor storage facility, and a regulator of nutrient levels in the blood circulating to all cells of the body. Let us keep these functions in mind as we take a brief look at some of the chemical processes which the liver performs on the substances absorbed from the intestines.

All of the carbohydrates digested in our food yield monosaccharides. These may be glucose, fructose, galactose, ribose, and some of their derivatives (see Chapter 6). The liver converts all of these to glucose, because it is this sugar that must be circulated to the cells of the body. A glimpse into the nature of these transformations is found in Chapter 2. The energy for these chemical transformations is largely furnished by the oxidation of some of the glucose itself. The liver passes some glucose into the bloodstream up to a certain level, and the excess is made into glycogen—animal starch—from which the glucose can be regenerated and put into the circulating blood as needed.

Many of the common metabolic intermediates in our food such as acetic, lactic, and citric acids and the amino acids fit equally well into the chemical machinery of liver cells and may be either oxidized to give energy, converted to glucose, protein, fat, or other substances, or permitted to circulate at a prescribed level in the blood to be used by cells in other parts of the body.

One of the most effective ways to eliminate unneeded or even harmful molecules which occur naturally in food is by oxidizing (burning) them completely to carbon dioxide and water. Many substances made by other organisms but not needed by man can be effectively oxidized in the liver, which has within its own cells the necessary enzymes to accomplish this task. It is entirely reasonable that this is so because through the millennia of evolution man has survived by consuming other organisms.

One example of preferential oxidation of a molecule produced by other common organisms is ethyl alcohol. Produced by yeasts and thus a common ingredient of fermented beverages and yeast-leavened breads, alcohol is also present in small amounts in many foods, particularly fruit. When alcohol is consumed in quantity in beer, wines, and stronger distilled beverages, the liver preferentially oxidizes it to carbon dioxide and water, even though some of the alcohol may get into the general body circulation. About 98% of ethyl alcohol consumed is soon oxidized by the liver and other body cells, giving 7.2 Cal of energy per gram. The remaining 2% is eliminated in the urine and in the breath. Alcohol used in this way temporarily replaces carbohydrates, fats, and other food constituents as a fuel source.

Most of the fatty acids, alcohols, aldehydes, ketones, and even hydrocarbons mentioned in Chapter 6 are oxidized quite completely to carbon dioxide and water or are converted to normal metabolites by the liver. Evidence for this is that the short-chain fatty acids of milk fat such as butyric, caproic, caprylic, capric, and lauric acids are not usually present in human fat. Also, many of the odorous and flavorful substances in our food are quickly changed to odorless metabolic products. Thus the liver either oxidizes these substances completely or converts them to metabolic intermediates usable by the liver or other cells of the body.

In addition to the above chemical transformations, the liver can convert certain substances in our food to other compounds which are then eliminated from the body, usually by the kidney. These processes were termed *detoxification* by many earlier scientists, a word with a rather restricted meaning. To be sure, many potentially harmful compounds are converted to other substances which are apparently easier to eliminate. However, in a few cases the liver may change a relatively nontoxic molecule to a more toxic one. At times, some of the vitamins may be converted to ineffective molecules by the liver.

Some substances in food may not be oxidized completely but only partially. For example, benzene

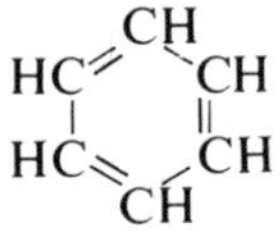

is converted to phenol

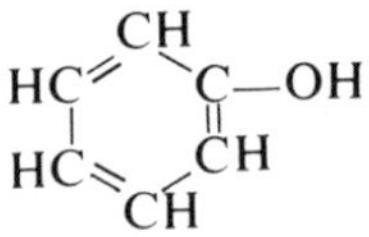

and vanillin (of vanilla)

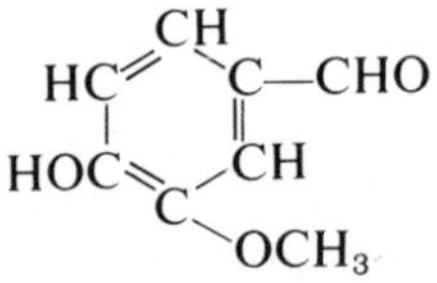

to vanillic acid

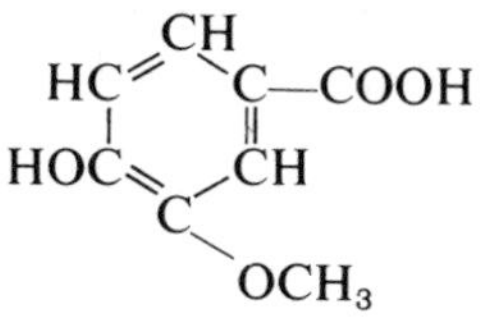

$$
\begin{array}{c}
\text{CH} \\
\text{HC} \diagup \diagdown \text{C—COOH} \\
\text{HOC} \diagdown \diagup \text{CH} \\
\text{C} \\
\text{OCH}_3
\end{array}
$$

Reductive types of chemical changes take place as well. Many nitro compounds (nitrates and organic derivatives of nitric acid) are converted to amino derivatives which themselves may be further modified (see below).

Another type of transformation brought about by the liver is *conjugation* or linking of a normal metabolite such as glycine or acetic acid wth a substance absorbed from digested food. Benzoic acid, a natural constituent of such foods as cranberries and plums that is used extensively in foods as a preservative against yeasts, molds, or bacteria, is conjugated by the liver to hippuric acid:

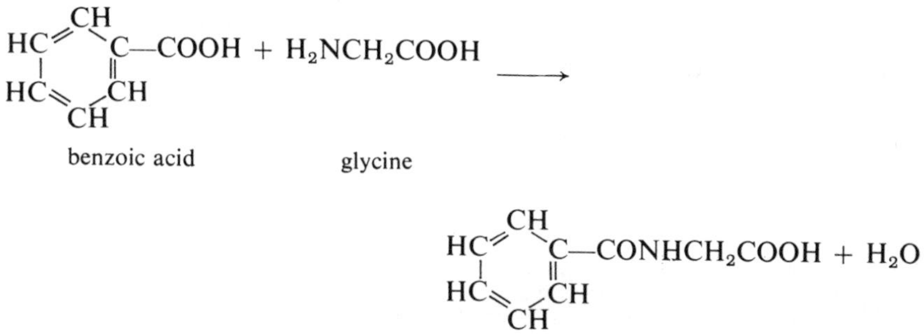

$$
\begin{array}{c}
\text{CH} \\
\text{HC} \diagup \diagdown \text{C—COOH} + H_2NCH_2COOH \\
\text{HC} \diagdown \diagup \text{CH} \\
\text{CH}
\end{array}
\longrightarrow
$$

benzoic acid   glycine

$$
\begin{array}{c}
\text{CH} \\
\text{HC} \diagup \diagdown \text{C—CONHCH}_2\text{COOH} + H_2O \\
\text{HC} \diagdown \diagup \text{CH} \\
\text{CH}
\end{array}
$$

hippuric acid

Salicylic acid, used commonly in medicine often in the form of aspirin, acetylsalicylic acid, is similarly conjugated, as is nicotinic acid, which is produced by the liver from the vitamin niacin:

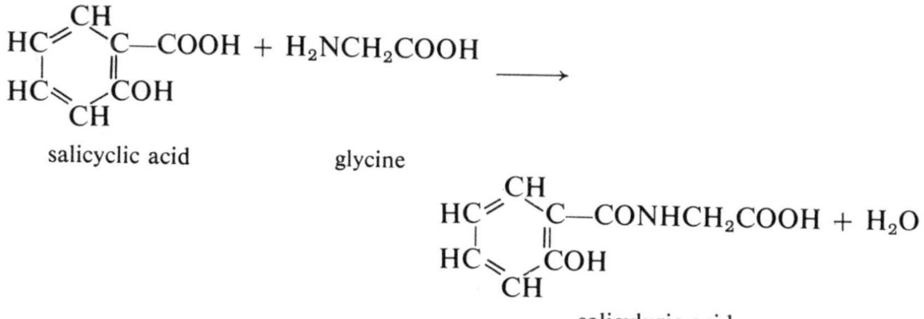

$$
\begin{array}{c}
\text{CH} \\
\text{HC} \diagup \diagdown \text{C—COOH} + H_2NCH_2COOH \\
\text{HC} \diagdown \diagup \text{COH} \\
\text{CH}
\end{array}
\longrightarrow
$$

salicyclic acid   glycine

$$
\begin{array}{c}
\text{CH} \\
\text{HC} \diagup \diagdown \text{C—CONHCH}_2\text{COOH} + H_2O \\
\text{HC} \diagdown \diagup \text{COH} \\
\text{CH}
\end{array}
$$

salicyluric acid

$$\underset{\text{niacin}}{\text{HC}\overset{\text{CH}}{=}\text{C—CONH}_2} \xrightarrow{\text{H}_2\text{O}} \underset{\substack{\text{nicotinic acid} \\ + \text{ NH}_3 \\ \text{ammonia}}}{\text{HC}\overset{\text{CH}}{=}\text{C—COOH}} \xrightarrow[\text{(glycine)}]{\text{H}_2\text{NCH}_2\text{COOH}}$$

$$\underset{\text{nicotinuric acid}}{\text{HC}\overset{\text{CH}}{=}\text{CCONHCH}_2\text{COOH} + \text{H}_2\text{O}}$$

Glutamic acid is another amino acid used by the liver for conjugation with other molecules.

Glucuronic acid (see p. 121) is conjugated by the liver with a number of different molecules. Among them are benzoic acid (showing that substances may be handled in more than one way), some of the terpenoid compounds, and steroids, particularly the steroid hormones including sex hormones. Acetic acid is joined with some aromatic amines including para-aminobenzoic acid, which is a part of the vitamin folic acid, and with the sulfa drugs used in medicine:

$$\underset{\textit{p}\text{-Aminobenzoic acid}}{\text{COOH}} \qquad , \qquad \underset{\text{sulfanilamide}}{\text{SO}_2\text{NH}_2} \xrightarrow[\text{(acetic acid)}]{\text{CH}_3\text{COOH}}$$

$$\underset{\text{acetyl } \textit{p}\text{-aminobenzoic acid}}{\text{COOH}} \qquad , \qquad \underset{\text{acetyl sulfanilamide}}{\text{SO}_2\text{NH}_2}$$

There are still other metabolic transformations of interest. The liver is capable of taking the sulfur from the sulfur-containing amino acids, oxidizing the sulfur to sulfuric acid, H_2SO_4, and hooking this sulfate to phenol and chemically related compounds to form organic sulfates which are excreted in the urine. Cyanide is found in many plant foods as cyanogenic glycosides (see p. 121, 126). Tobacco smoke also contains free hydrocyanic acid, HCN. HCN derived from either source is extremely poisonous in excessive amounts, but the liver has no trouble using a sulfur atom and converting small amounts of HCN to thiocyanic acid, HCNS, which is relatively nontoxic and can easily be detected in the saliva as well as in the urine of humans—particularly smokers.

The detailed information now available on the metabolism of thousands of compounds in our food is enormous and it is being expanded. This includes metabolism in other tissues as well as in liver. The need for more such information is obvious. Advances in medicine and agriculture to assure wholesome food supplies and to maintain health are possible only through understanding of how the chemicals which make up our food and medicines as well as those used to produce them actually are metabolized and used by ourselves, the organisms we use for food, and the organisms that compete with us. The rational use of such information is the only way we can use both foods and medicines with confidence.

There is one more major chemical transformation in the liver to consider. The liver, in conjunction with the kidneys, controls the amounts in the circulating blood of the many nutrients needed by all body cells. The body itself has no great capacity for the storage of amino acids, proteins, vitamins, and some other nutrients, although it can store small amounts of carbohydrate in the form of glycogen and relatively large amounts of fat as energy reserves. For this reason, consumption of foods to supply proteins, vitamins, minerals, etc., is a continuing requirement. The primary role of protein in our diet has been emphasized a number of times, so it is appropriate to look at how the liver handles the amino acids coming to it from the digestion of food proteins.

The individual amino acids entering the liver via the portal vein from the intestines are permitted to enter the general circulation only at specific levels, so that each of the amino acids may be available to other cells of the body as required to make the individual proteins necessary for growth or protoplasmic repair. If individual amino acids tend to rise above the prescribed level in the blood, they are removed and oxidized to give energy. The carbon and hydrogen of the amino acids are converted to water and carbon dioxide, but the nitrogen is converted to urea.

Observe the formulas for the amino acids in Fig. 6.5, Chapter 6 and note that the nitrogen is present as a derivative of ammonia, NH_3—that is, one or more of the hydrogens is replaced so that the nitrogen is connected to a carbon atom. As the hydrogen atoms and carbon skeleton of the amino acid are converted to water and carbon dioxide, the residual nitrogen would appear as ammonia. But ammonia, NH_3, and ammonium salts, NH_4^+, are not tolerated in blood to any extent. So the potential ammonia is converted by a series of reactions to urea. The overall scheme may be simply illustrated as follows:

$$CO_2 + H_2O \longrightarrow \overset{OH}{\underset{OH}{C}}\!\!=\!\!O \xrightarrow[\text{(ammonia)}]{+2NH_3} \overset{NH_2}{\underset{NH_2}{C}}\!\!=\!\!O + 2H_2O$$

carbon   water   carbonic   urea
dioxide   acid

As the preceding paragraphs have shown, the liver is truly a remarkable organ that performs a fantastic array of chemical processes. The liver takes the raw products that result from the digestion of food to its fundamental building blocks and chemically refines them by effective combustion and removal of some constituents or by modification so that unwanted substances can be eliminated from the body. The refined residue contains the essential nutrients for the body, and the liver controls the transfer of these cellular nutrients into the general circulation where they are carried to the various cells of the body to be used as needed.

The Kidneys

Water-soluble substances not needed by the body are eliminated in the urine, which is secreted by the kidneys. These substances include metabolic end products of all of the cells of the body and substances absorbed from the gastrointestinal tract which require disposal. Of course, among the metabolic end products are substances made by the liver from ingested material as described in the above paragraphs that do not serve the body cells in any way.

The kidneys perform their function by the process of filtration and reabsorption. The kidneys contain a great number of individually functioning units—one of which is diagrammed in Fig. 7.3. Blood is pumped from the heart by way of the aorta and renal arteries to the capillaries of Bowman's capsule, where an ultrafiltrate of blood plasma is made. This ultrafiltrate has the same composition as plasma except for the proteins, which are molecules too large to pass the walls of the blood vessel. This filtrate then passes through a long tubule around which there is a bed of small capillaries carrying the blood from Bowman's capsule. The cells making up the walls of the tubules are very active and require energy to effect the selective reabsorption into the blood of water and other nutrient molecules needed by the body. By this process, blood is reconstituted and flows back into the general circulation and the concentrated urine containing dissolved substances to be eliminated is formed. In an average adult, the kidneys make about 170 liters (45 gallons) of

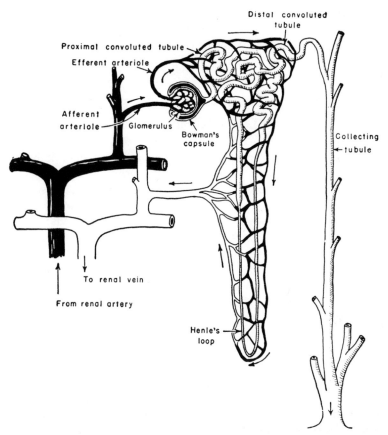

Fig. 7.3. Diagram of a single functioning unit of the kidney. (From *College Zoology*, by Francis R. Hunter and George W. Hunter, III. Courtesy of Francis R. Hunter.)

ultrafiltrate per day and reabsorb all except about 1.5 liters (1.6 quarts), which is the volume of urine excreted per day. The cells of the tubules are highly specialized and are capable of recognizing specific molecules such as glucose, amino acids, vitamins, and various mineral ions, absorbing them from the filtrate, and transferring them to the blood. Furthermore, these cells will allow excess blood constituents to pass into the urine, and during periods of deficiency they will conserve molecules needed by the body. In this way the kidneys serve to level out the surpluses and deficiencies of useful molecules contained in periodically consumed meals.

From this brief description of kidney function, it is clear not only that the kidneys rid the body of unneeded and even potentially harmful substances but also that they are organs which help to control the amounts of nutrients in the circulating blood. Consequently, the liver and the kidneys functioning in tandem with the gastrointestinal tract serve to maintain the proper environment for efficient function of all the cells and organ systems that make life possible.

MAINTENANCE OF THE CELLULAR ENVIRONMENT IN THE BODY

Maintaining the proper cellular environment is more than simply sorting and retaining or discarding molecules resulting from the digestion of food. These activities must be performed in such a manner that each aspect of the cellular environment is controlled within narrow limits while all of the varieties of functioning cells are discharging the end products of their metabolism into the same environment. Let us now briefly consider each of the five environmental factors—water, temperature, pH, osmotic pressure, or water activity and nutrients—from the point of view of how our food ultimately functions to maintain life processes.

Up to this point, we have discussed mainly the nutrients from our food and how they reach the bloodstream. Table 5.1 of Chapter 5 shows that oxygen and water head the list of human nutritional requirements. Water is consumed as such and as an integral part of our food and is also produced by the energy-giving oxidative processes of the body. The proper amount of water in the body is maintained by a balance of these inputs with the water outputs, which include loss in the urine, feces, and sweat and evaporation from the skin and particularly from the respiratory passages. Since we live in an ocean of air approximately one-fifth of which is oxygen, this nutrient is available for the taking. The oxygen of air is, of course, a gas, and the lungs are special organs for taking it into the body. They are composed of minute sacs in which blood capillaries are very close to the membrane surfaces that are exposed directly to the air. This membrane is permeable to oxygen, as are the capillary walls. Oxygen passes through these membranes and dissolves in the tissue fluids and blood plasma, but the solubility of oxygen in plasma is insufficient to carry the amount of oxygen required by the body. However, the blood carries red cells containing hemoglobin, a special protein which picks up oxygen by chemically combining with it as it passes into the bloodstream. As the blood is carried to the cells, the reverse processes take place and the oxygen passes to the cells where it is used. In this way, the critical nutrient oxygen reaches all body cells.

The lungs are also an organ of excretion. Remember that the energy used by the body results largely from the oxidation of food to carbon dioxide and water. Like oxygen, carbon dioxide is a gas not very soluble in water. Since air itself contains relatively little carbon dioxide, 0.03% by volume, the tremendous surface area of the lungs exposed to the atmosphere offers an escape route. Carbon dioxide produced in the cells of the body passes into the blood, in large measure into the red cells where it can also be carried by the hemoglobin to the lungs. Here the CO_2 passes from the hemoglobin of the red cells across the cell membrane and escapes as a gas to the air at the surface of the lungs. So the lung is another organ that plays a significant role in providing nutrients and eliminating a major metabolic by-product.

Assuming that all of the nutrients listed in Table 5.1 reach the cells, how are the remaining three cellular environmental parameters of temperature, pH, and osmotic pressure controlled and affected by the food consumed? Heat is produced by the combustion of food and it is this heat that maintains body temperature. Cellular metabolic rates are subject to both involuntary and voluntary controls. The involuntary rate of metabolism, called the *basal metabolism*, is regulated by a number of physiological mechanisms which need not be elaborated here except to say that the body thermostat in the hypothalamus of the brain serves to control body heat production, as do the hormones produced by the thyroid. The voluntary metabolic rate, determined by exercise or work of various kinds, also produces heat. The circulating blood serves to carry heat away from the tissues producing it to the skin and respiratory passages, where it can be dissipated to the outside. Heat leaves the body for the most part only through its surface, and you will recall from Chapter 5 that energy requirement of the body is a direct function of body surface. If we tend to get too hot because of exercise or because the temperature of the outside environment is too high, we sweat to increase heat loss by evaporation. One gram of water evaporating at body temperature consumes 575 calories or 0.575 Calories. To put it another way, the evaporation of 7 grams of sweat will dissipate the heat produced by the combustion of 1 gram of sugar. To facilitate heat loss, we may take off clothes or seek cooler places. If too much heat is being lost, we feel cold and put on more clothes or seek warmer places. Consequently, body temperature is controlled physiologically by combustion of nutrients in conjunction with behavioral patterns.

pH, or hydrogen ion concentration, is controlled within very narrow limits. This must be accomplished even though there is a relatively wide variation in the pH of foods as consumed. Most foods have acidic or approximately neutral pHs, although a few foods are slightly alkaline. Strongly alkaline foods are usually distasteful. It is not the pH of the food as eaten that has the greatest impact on the acid–base balance, the determinant of pH, but rather the acidic or basic nature of the food after its ultimate combustion in the body. This is so important that the food tables in the Appendix indicate whether particular foods are acid (A) or base (B) forming. Note that fruits and vegetables are base-forming foods even though they may be quite acidic as eaten—lemons, apples, spinach, etc. Many high-protein foods, particularly of animal origin, are acid forming even though as eaten they are close to neutral in pH. Why is this so?

Remember that foods are largely oxidized to carbon dioxide and water. The residue after oxidation is the factor that determines whether a food is acid forming or base forming. The analogous situation outside the body is simple burning in air, like the burning of wood. The resulting ash contains the mineral matter, which may be alkaline, neutral, or acidic. Wood ashes, composed largely of potassium carbonate, are quite alkaline and in earlier times were the major source of alkali for soap making. Orange juice tastes acidic because of the relatively high content of citric acid, some of which occurs as the potassium salt since orange juice is about 0.19% potassium. After being eaten, the citric acid along with carbohydrates is converted to CO_2 and water by combustion in oxygen. There is a residue of potassium which in the body occurs as potassium bicarbonate, an alkaline salt. In foods such as meat and whole eggs which contain relatively large amounts of sulfur and phosphorus, oxidation leaves these elements in the form of sulfuric and phosphoric acids so that the residue or ash is quite acidic. From these simple examples, it is clear that the body must somehow cope with this residual effect.

Fluctuations in the pH of the urine reflect the acid or basic character of the foods that have been eaten. The kidneys also are capable of making necessary adjustments in a positive way. You will recall that the kidneys serve to retain needed nutrients in the blood. The basic sodium and potassium ions and the acidic phosphate and chloride ions are major inorganic or mineral nutrients, and so the kidney must excrete only the excesses of these substances. If the food has an excess of residual acid, the kidneys are able to excrete some of it in the form of ammonium salts rather than sodium or potassium salts; thus potassium and sodium ions are conserved. But it was mentioned above that ammonia or ammonium ions are not tolerated well in blood and that nitrogen in the blood is usually in the form of urea. The kidneys have the necessary enzymes to convert some of the urea to ammonium ions as needed to combine with the excess acid in order to maintain the proper acid–base balance in the blood:

$$\begin{array}{c} NH_2 \\ / \\ C{=}O \quad + CO_2 + 3H_2O \longrightarrow 2NH_4{}^+HCO_3{}^- \\ \backslash \\ NH_2 \end{array}$$

| urea | carbon dioxide | water | ammonium bicarbonate |

If the food has an excess of residual base, meaning an excess of potassium and sodium ions, and a relative deficiency of chloride and phosphate ions, excess Na^+ and K^+ are excreted as bicarbonate salts $(HCO_3{}^-)$.

The body as a whole is much better protected against acidic residues, for the acid end product of food oxidation is largely carbon dioxide, which in water solution is only weakly acidic:

$$CO_2 + H_2O \rightleftharpoons H^+ + HCO_3^-$$

It is clear therefore that the kidneys not only excrete unneeded molecules from the body but in doing so must maintain the proper pH or acid-base balance in the blood in spite of the varying residual acidic and basic composition of food.

Osmotic pressure, or water activity, is the remaining critical environmental parameter of cells which must be controlled within very close limits in order for all parts of the body to function properly. Osmotic pressure is the pressure developed across membranes due to differences in diffusion rates caused by different concentrations of particles that are dissolved in water but are incapable of passing through the membranes as the water molecules do. Osmotic pressure plays a critical role in the exchange of fluids between various cells of the body and the blood, in the absorption of substances from the gut, and in the secretion of urine, sweat, and digestive juices. Without going into detail, it is the relationship between the osmotic pressure of the blood and the functioning cells and the hydrostatic pressure of the blood at the capillaries surrounding the cells that determines fluid transfer between cells and organs. It is this interrelationship between osmotic pressure and hydrostatic pressure at the capillary level that determines the force exerted by the heart. If for any reason blood flow is restricted, e.g., by increased friction or partial obstruction in the arteries, the heart pumps faster to raise the blood pressure and overcome the difficulties at the capillary level.

Osmotic pressure, determined only by the concentration of water relative to dissolved substances, must be also controlled, irrespective of food intake. The control of osmotic pressure in the circulating blood and thus in the functioning cells of the body is largely a function of the kidneys responding to variations of water intake and metabolic production as well as water loss in urine, feces, and sweat and other evaporation from the body. If we eat salty, dried foods or sugar in concentrated form as candy or other confections, or if we work hard and sweat, we become thirsty and drink water. If we consume excessive water in beverages or food, the kidneys respond quickly by increasing the volume of urine.

CONCLUSION

From the foregoing brief discussion it is clear that some of the organs involved in retrieving nutrients from the consumed food and air also serve to control the precise environment within the body that permits all of the diverse types of cells of the body to function properly.

As we study in the following chapter some of the organisms man uses for food, we can do so with a degree of understanding of the processes by which he utilizes the food he consumes. In short, man does it by digesting (hydrolyzing) the food to its fundamental molecular building blocks, absorbing them from the gastrointestinal tract, into his system, separating usable substances from harmful or unneeded substances, using the nutrients to meet metabolic requirements, and finally discarding and eliminating waste products—and while doing all of these things he must also maintain the internal environment of all of the cells in his body so that they can function properly and permit him to survive.

8

THE BIOLOGY OF SOME FOOD ORGANISMS

Man's ability to feed himself determines his survival. The ability of groups of people to form a viable community depends on their collective ability to provide food for themselves. The growth of civilizations depends primarily on their food supplies. So irrespective of level of social organization, survival of the human race is absolutely dependent on a continuing supply of food organisms. But what organisms? Of all of the hundreds of thousands of species of biological organisms on the earth relatively few are consumed by man and still fewer make up the major portion of his present-day food supply.

Early man obtained his food simply by gathering and consuming those organisms—plant or animal—available to him in the ecosystem of which he was a part. Probably by trial and error he learned which organisms he liked, made him feel good, and permitted him to live and which organisms had an adverse effect on his well being. Consequently, he selected some organisms and avoided others and passed this information to subsequent generations. As he learned to hunt and fish to enhance his food supply, these skills likewise were passed on to subsequent generations, thereby increasing the food supplies for the survival of more individuals. About 50,000 years ago the systematic use of fire in food preparation was discovered and cooking was found to make certain otherwise indigestible or noxious plants usable as nutritious food. This discovery expanded man's food supply immensely, and it is understandable that the mystical power of fire had a great cultural impact in the early recorded history of man. More organisms usable for food meant more people to feed.

More people required still more food and the next major step in the development of civilized man was that he learned to husband and to protect the animals and to cultivate and protect the plants he used for food, as well as the plants his animals needed for their growth. In this way agriculture came into existence. All of the activities related to the cultivation of the land to produce more food and other useful materials such as fibers for clothing have been the life work of the majority of people since the beginning of recorded history.

As agricultural practices became more effective so that some segments of the population could be freed from the labors of food production, other human endeavors were developed which contributed to the welfare of the community or society. Thus schemes of division of labor were developed among men, which led to the development of the sophisticated societies, cultures, and nations of today. Even now, in underdeveloped countries the majority of people are directly involved in cultivating the land and fishing the waters to feed themselves, and a minority of their populations are involved in other pursuits. In highly sophisticated economies, as in the United States, only a few percent of the population are farmers, who provide food for all. Indeed, so specialized have we become that, if left to our individual resources, most of us would not survive if we had to revert to the gathering and hunting and the agricultural skills of our not too distant ancestors. So the production of enough food to support all of the people is of prime importance—whether a society be primitive or highly organized and economically and politically complex.

To look at the primacy of food to the survival of civilizations in a somewhat different light, the food that can be produced is the major part of the only renewable resource on earth. Petroleum, coal, etc., are consumable resources of the earth, and cannot be renewed, but all of biological organisms of the earth are renewable because in their life cycles they use the constant source of energy from the sun and use and reuse the water, air, and minerals in the land and sea to grow and reproduce. Man's survival depends on how well he understands how this biological cosmos operates. It is within these limits that the human population is expanding and it is within these limits that each and every human being must procure in his lifetime the 20–30 tons of consumable biological material required to support him.

To provide some idea of the number and diversity of biological organisms required to supply the average American, Table 8.1 shows the daily per capita consumption of a few major foods and what each food represents in terms of numbers of organisms required per year to yield it. The number of organisms represents the number of complete life cycles of food organisms necessary to maintain an individual for a year, only a small fraction of the normal human life cycle of about 70 years. It is sobering merely to ponder the number of people necessary to care for all these diverse organisms. Some food organisms such as bacteria and yeasts have life cycles of only one-half to a few hours. Potatoes, tomatoes, small grains, and chickens require months to mature, and a year is the minimum for a hog and at least 2 years for cattle.

Table 8.1. Per Capita Consumption in the United States of
Selected Foods and the Number of Organisms Required
Annually to Supply These Foods

Food	Daily consumption per capita	Number of organisms (life cycles) per year
Milk	1.0 kg	0.10–0.12 cow[a]
Eggs	1.0 egg	2.0 hens
Meat	0.225 kg	
As beef		0.8 cattle
As pork		2.0 hogs
As chicken		100 broilers
Fruit (as apples, etc.)	0.300 kg	0.2 tree
Tomatoes	0.090 kg	4.0 plants
Potatoes	0.150 kg	50.0 plants
Small grain (as wheat, etc.)	0.225 kg	10^5 plants
Bacteria and yeasts	0.003 kg	10^{12} organisms

[a] At least 2 years are required for a cow to reach maturity, produce
milk, and calve to yield a future animal for meat or milk.

It must also be kept in mind that climatic conditions suitable for growth
of food may exist for only part of a year. Even though very important food
crops such as wheat, corn, and rice may require only months to complete
their life cycles to yield a harvest of food, there are few areas of the world
suitable for growing more than one crop a year. When we consider the number
of people on the earth, the vast numbers and varieties of food organisms
needed to feed these people, and the dynamic character of people and their
food organisms and then realize all are part of the limited biological cosmos
of the earth, we begin to understand the impact that the world's population
has on the entire biomass or ecosystem of the earth. It is also clear that, if
man is to survive, he must understand the limitations of the biosphere and
learn to function within them. With these considerations in mind, let us very
briefly consider some fundamental principles relating to man and the organisms that he cultures for food.

CHARACTERISTICS OF ECOSYSTEMS

An *ecosystem* may be defined as the sum of all physical and chemical
features and all biological organisms interacting in a given space. Hence an
ecosystem could be the ultracomplex one of the whole world or a relatively

simple one of a small lake, forest, desert area, wheat field, rice paddy, orange grove, or cattle meadow in any part of the world. Let us also remember (1) that all living things are dynamic, (2) that complex multicellular organisms such as higher plants and animals live only where they can modify the environment of their habitats to the environment required by their cells, (3) that all living cells making up more complex organisms must have liquid water and a source of nutrients, and (4) that an external energy source is required. So any ecosystem is dynamic and is kept that way by continuing inputs of energy. The light of the sun is the primary energy source for the world.

You will recall that all organisms can be divided into two classes—autotrophs, which can use light as their energy source, and heterotrophs, which must get their energy from the consumption of other organisms and the oxidation of the consumed protoplasm. Heterotrophs may consume autotrophs or other heterotrophs. Thus the autotrophs are the primary producers. They include phytoplankton of the sea, photosynthetic bacteria, algae, and all green plants from the smallest to the giant trees. Among the food organisms which are autotrophic are fruits, vegetables, and cereals. The heterotrophs include most bacteria, fungi, insects, and higher animals including man. Among the heterotrophic food organisms are cattle, pigs, chickens, yeasts, certain bacteria, mushrooms, fish, and lobsters. In a real sense the autotrophs, or light users, are the producers of biological substance and the heterotrophs are the consumers and decomposers of biological substances. Overall in any stable ecosystem there is competition between producing and consuming organisms coupled with dynamic interactions or symbiotic relationships. If the ecosystem is not disturbed, an ultimate balance of organisms is achieved.

As energy flows through an ecosystem, there is a loss at each step. The autotrophs by no means trap all of the energy of the sun to store as chemical energy in their protoplasmic substance. Much of the sun's energy that reaches the earth is lost by reflection to outer space or shines on areas incapable of supporting life. Only a small fraction of the sun's energy is stored by the photosynthetic process whereby water plus carbon dioxide, nitrogen, and mineral elements are converted to biological materials plus oxygen. When this biological material is consumed, most of it is oxidized back to carbon dioxide, water, nitrogen, and minerals. In the process heat is dissipated. Only a small fraction of the total energy consumed by the heterotrophs resides in their protoplasmic material. If the primary heterotroph is consumed by another heterotroph, most of the energy is dissipated as before and a still smaller fraction of energy is stored in the protoplasmic material of the secondary, heterotroph. So the process goes until all the energy originally stored in the protoplasm of the producers appears as heat and all the organic material is converted back to carbon dioxide, water, nitrogen, and minerals to be used again by the producers (autotrophs).

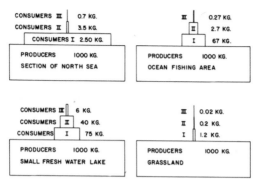

Fig. 8.1. Relative amounts of biological material (biomass) among the producer organisms and the first-, second-, and third-order consumer organisms in four different ecosystems.

Figure 8.1 illustrates the energy trapped at a particular point in time by the biomass in four simple ecosystems. Note the pyramid shape of the relationship between the producers and the rapidly decreasing amounts of biomass in the primary, secondary, and tertiary consumers. Figure 8.2 indicates how the types of organisms living in each organismic layer differ. The primary consumers are obligatory herbivorous species of animals, birds, insects, fishes, etc. Common animals used for human food which are at the consumer I level are cattle, sheep, goats, horses, chickens, pheasant, carp, some shrimps, etc. The consumer II level includes those carnivores that live by eating first-level consumers. Obligatory carnivores are not commonly used for human food and when they are they tend to be expensive and in limited supply. A number of fishes, such as tuna and bass, some crustaceans such as lobsters and crabs, some marine mammals such as walruses, seals, and whales, and frogs and even snakes could be considered as belonging in consumer levels II, III, IV, etc. However, species in the upper levels may actually consume organisms of several lower levels. Also among the various levels of consumers are found certain omnivorous animals—those that can live on both animal and vegetable tissues, eating not only autotrophic producers, the green plants, but also consumers of all levels. Man is the central character among the omnivorous organisms. The pig is also omnivorous, as are a few other familiar animals including the domesticated dog. However, in the complexity of human evolution, a few cultures have developed where only plants, primary producers, are used for food and a few, such as the Eskimo culture, where only animals are eaten.

Schemes that trace protoplasmic material through various levels of organisms are generally called the *food chains* or *webs* within a particular ecosystem. Figure 8.3 is a much oversimplified food web showing the inter-relationship of producers and consumers as well as decomposers, which

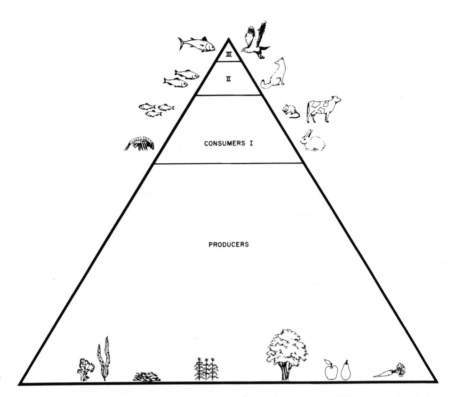

Fig. 8.2. Diagram of some of the organisms which occupy the different levels of the
pyramid of biomass.

include the worms, insects, fungi, and bacteria that eventually decompose the
residual (waste) organic matter to water, carbon dioxide, and nitrogen.

Many species of bacteria, fungi, insects, and animals can and do attack
and live on plants, the producers, and animals at all levels of consumers.
Other species live symbiotically, together. So in any ecosystem of any size the
interactions of all biological systems may be so complex as to defy complete
definition; hundreds of species of organisms may live within a certain prescribed ecosystem. When we consider the entire earth as one system, the
complexities are so prodigious that it is most difficult to predict the effects on
the entire ecosystem of modifying the population of one or a few component
organisms, such as man and the organisms he nurtures for his food.

Recently there has been much public attention given to this situation.
It has been found that DDT—*d*ichloro*d*iphenyl *t*richloroethane, or 1,1,1-
trichloro-2,2-bis(*p*-chlorophenyl)ethane—a most effective insecticide of more
than three decades of use, decomposes slowly and tends to accumulate to

PRODUCERS	CONSUMERS		
	1st ORDER	2nd ORDER	3rd ORDER

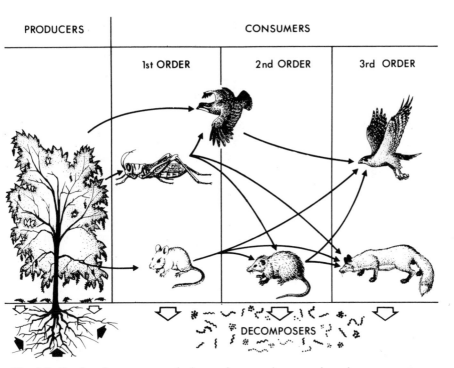

Fig. 8.3. Food webs among producing and consuming organisms in an ecosystem. (From *Bio-Learning Guide*, by C. B. Meleca *et al.*, 1971. Courtesy of Burgess Publishing Co.)

undesirable concentrations in some higher-level consumer organisms such as predaceous birds and fishes because of the length of their food chains. Mercury can similarly be concentrated sometimes to dangerous levels, as far as human food is concerned, particularly in tuna and swordfish, irrespective of whether the mercury reaches their habitats from natural sources or human pollution. However, most people probably do not realize that the human nutrient vitamin A is also concentrated from their food chains by such animals as the polar bear, swordfish, and sea bass to the extent that their livers may be toxic for human consumption because of excessive vitamin A content.

Anyone who enjoys excursions into relatively untouched wilderness areas and marvels at the interplay of biological organisms in areas uncorrupted by man is concerned about man's impact on these ecosystems and on the ecosystem of the entire earth. These concerns are popularly termed "ecology," although this branch of science is much more profound than is often publicly portrayed. The automobile and the almost insatiable demand for power by modern society surely have as much or even more effect than the use of fertilizers, insecticides, antibiotics, and growth regulators in food production. Freeways create barriers to migratory wildlife. Exhaust gases foul the air of our cities, causing public health problems, corroding buildings, and even affecting food production in adjacent areas. Demands for the comfort of heated and air-conditioned buildings, time-saving air travel, and many labor-saving devices and conveniences require more and more energy from fossil fuels and atomic sources. What are the effects of these desires of modern man on his chances for survival? We do know that the automobile kills more people than war, as well as an abundance of wildlife; that newer knowledge and technologies can profoundly affect birth and death rates; that millions of people are still starving; and so on and on. Man has evolved to this state of affairs through his mental abilities; he has evolved to the extent that he must survive by use of this intellect. It seems that man can survive only by understanding the system of which he is a part. Perhaps the only part of that ecosystem subject to a great degree of adaptation is man himself.

With this brief background, let us consider some major factors of the ecosystem of which we are a part in relation to our ability to provide food organisms for our survival.

SOME PROPERTIES OF ECOSYSTEMS DIRECTLY RELATING TO FOOD ORGANISMS

The Impact of Human Populations

For every action there is a reaction. So it is with any ecosystem, because for any change in one sector of the ecosystem there is a compensatory change somewhere else in the complete system. For example, in a relatively stable forest ecosystem, radical changes in the flora and fauna may result from a forest fire started by lightning or a new disease of certain plants or animals brought in by windborne bacteria or insects. If one studies the populations of different organisms over a period of time in an isolated ecosystem, fluctuations in populations will usually be found in relation to increases and decreases of food supplies for the different species of organisms. For example, climate during a particular time may favor the growth of a certain plant. It will then favor the insects, birds, and other animals which use that plant for food, and the proliferation of these organisms will favor other organisms along the food chain. Then later a shift in climate may favor different plants and thereby create a less favorable food supply for those organisms which fed on the first group of plants. The result of the change in climate is a shift in populations to another direction and so on.

Let us consider prehistoric man in relation to his ecosystem. He was a hunter and gatherer of food, and the limits of his numbers were determined by the numbers of organisms in his food supply. If he overharvested, he starved; if there was more food available than he needed, he had more children and more of them survived. Then prehistoric man learned to nurture and cultivate his food plants and feed and husband his animals so that their numbers increased. But by increasing the population of his food plants and animals, he was increasing food supplies for the insects, birds, and other animals that preyed on his food. So organisms which he could ignore when he was a simple gatherer now became his enemies and the whole ecosystem responded.

As early man cultivated his food organisms through generations, he protected and selected them for their yield of food and ease of cultivation. In time, his food organisms largely lost their ability to survive in the wild in numbers sufficient to support the number of humans needing food. After generations on generations of selecting seed and breeding stock on the basis of their food value and his ability to cultivate and protect these organisms to effect his own survival, modern man has developed an elaborate system of plant and animal breeding. The result is that today man himself and the highly bred organisms he uses for food would have great difficulty surviving in a wild, strictly competitive ecosystem. Pirie[1] tells about a highly productive strain of wheat that was planted in a field at the Rothamsted Experiment Station near Aberdeen, Scotland, and then ignored. In only 5 years it was almost impossible to find a single wheat plant, let alone harvest any for use. The highly productive dairy cow shown in Fig. 8.4 is so large of udder and so sluggish in movement that it is almost defenseless against predatory animals. In developing and cultivating high-producing food organisms in order to survive, man has accepted the defense of these organisms against predators, whether they be animals, insects, fungi, bacteria, or viruses, as well as the responsibility of supplying their food. The result is that today we have an estimated 4 billion people living in the ecosystem of the earth compared to only 250 million at the time of Christ. And still a large proportion of the world's population is ill fed (see Chapter 12).

As a simple gatherer of food, prehistoric man had a relatively insignificant niche in the world ecosystem. Through the millennia to the beginning of recorded history man's impact on the ecosystem of the earth was nil. Survival was his occupation. The estimated 250 million people in the world at year 1 of our calendar still had little effect on the world's ecosystem. The accumulated knowledge of agriculture and other technologies of that era were insufficient to cause much growth of human population, and by the year 1650 only about

[1] N. W. Pirie, *Food Resources, Conventional and Novel*, Penguin Books, 1969.

Fig. 8.4. Sir's Standard Bright Beauty, National Dairy Show Grand Champion. (Courtesy of American Jersey Cattle Club.)

500 million lived on the earth. The next doubling of the population, to 1 billion people, was achieved in only 180 years, by 1830. During this interval some ideas concerning the nature of contagious diseases were developed; yet the density of population and the resultant pollution caused high mortality rates in the cities. For the first time there seemed to be some concern for man's adverse effect on his own environment, and a few rudiments of public hygiene concerning water, air, and sewage were applied in some cities by the late 1700s—only two centuries ago. Still the numbers of people had little effect on the ecosystem as a whole.

Then the age of modern science began to flower. The technologies of agriculture and medicine began to develop rapidly but still with emphasis on survival as in ages past—to provide food and eradicate disease. By 1930 the world's population had doubled again to 2 billion, and it has doubled now to 4 billion. It is estimated that world population will be more than 6 billion by the year 2000, with the next doubling to 8 billion by the year 2010.

Wars and pestilence of the past had profound effects on rates of population increase. This has not been so true in the nineteenth and twentieth centuries, notwithstanding the human tragedies of this nature in the last 200 years. Now the numbers of people are putting increasing stresses on the world ecosystem. Furthermore, modern ways of life are complicating the ecosystem in ways unknown in the past. The nonconsumable, nonbiological trash that characterizes the residue of our modern culture is perhaps of even greater concern than numbers of people and their requirements for food. A lack of comprehension of the ultimate effects on the whole earth of modern agricultural and medical technologies interacting with equally sophisticated industrial technologies is causing great public anxiety. Now for the first time a dimension which has always been in the human equation is being recognized by large sections of society. We cannot be guided as in ages past solely by our instincts for survival and propagation of the race. We must learn the direction of the future by understanding the changes that our great numbers and the manner in which we live cause in the ecosystem which supports us all and in whose limited confines our species must survive. In such a situation, it is inevitable that many individuals begin to question the wisdom of modern practices which have so greatly increased the population of the world and allowed many to live longer. Let us now consider from an ecological point of view some of these practices on which food production depends.

Use of Fertilizers

Without fertilizers, the present level of food production throughout the world could not be maintained. In many areas of the world where there is insufficient food, fertilizers for crops are either unavailable or in very limited supply. The use of fertilizers to increase productivity of crops is an ancient practice; historically animal manures have been used to a great extent, but rotting plant material, animal offals and by-products, and some minerals have also long been effective in increasing crop yields. Every American schoolchild has heard the story of how the Indians taught early settlers in the western hemisphere to grow corn using fish as fertilizer. However, there is simply not enough organic material (biological waste) to meet current fertilizer needs, and mineral fertilizers manufactured by the chemical industry are now of prime importance in fulfilling fertilizer requirements. These developments have been made possible by increased knowledge of what nutrients crop plants need to thrive.

Green plants are primary producers which use the sun's light to convert water and mineral elements from the soil and carbon dioxide from the air to their own protoplasm. They do not require organic substances for growth. Plants differ a great deal with respect to the types of soil and the minerals needed for growth. When these requirements are known, fertilizers can be added to the soil to assure rapid and efficient growth. The modern technology of crop production results from the combined contributions of soil scientists, plant nutritionists and physiologists, agronomists, and, the chemical and machinery industries.

Fertilizers usually contain potassium, phosphate, and nitrogen in the form of nitrates or ammonia and also may contain sulfate, calcium, and trace elements such as boron, copper, and iron which might be required for a particular plant to thrive on a particular soil type. "Organic" fertilizers, thought by some to be superior, must also supply these same plant nutrients to be effective in promoting growth of crops. The organic matter of, for example, animal manure may modify the texture of the soil to facilitate aeration and movement of water and may support soil bacteria which grow until all organic material is completely used up. However, often these things may not be necessary or may be accomplished by the cultural practices employed by the farmer.

The question may still persist—why use fertilizers at all, because do not many plants thrive in their native habitat? Indeed they do—to some degree at least. But the native habitat might well have been a rather small, stable ecosystem where all organisms lived and died in the same locale and as minerals were taken from the soil by living things they were returned to soil by the dying and decomposing plants and animals. In the production of food to feed people or animals it is generally the case that those to be fed do not live in the area where the food is grown. When the food organisms are moved from where they are grown, with them are removed the plant nutrients which might have been returned to the soil if they had stayed and rotted. Continuous cropping and removal of crops from the soil where they are grown rapidly deplete plant nutrients from the soil. Without fertilizers to replace the nutrients carried away, future crops would not thrive or might not even grow at all. Also, many soils otherwise too poor to support cropping can be made extremely productive by proper fertilization, and so the use of fertilizers increases the amount of land that can be used for growing food as well as making a particular soil itself more productive. Without fertilizers, available food supplies would diminish rapidly and more people would starve and still more would move down to the semistarvation state of human misery.

Use of Insecticides, Fungicides, Bactericides, Herbicides, and Plant Growth Regulators

Since agriculture depends on the alteration of an ecosystem to favor the growth, development, and reproduction of particular food organisms, the result is an ecosystem which is unstable because increased numbers of food organisms will also increase the numbers of organisms feeding on the crop. Increased food supplies result in increased human population *and* increased populations of other organisms which can also feed on the same crop—bacteria, fungi, insects and other animals, and even other plants (see below). In order to be able to use a food crop for his own purposes, man must destroy or at least control organisms which compete for it. In primitive times, the farmer protected his crops by picking off and killing insects by hand, removing competing plants (weeds) by hand, and hunting and killing or chasing away animals. Manual control is tedious, not too effective, and in the present state of human cultural development impossible economically (see Chapter 9). Also, fungi, bacteria, and viruses cannot be manually controlled. Eventually, certain plant materials containing compounds such as rotenone and nicotine were found to have valuable insecticidal properties for food crops. Mineral preparations containing lead, arsenic, sulfur, and copper were also found to be effective in protecting certain crops. Many of these things are used today.

As we continue to learn more about nutrition, biochemistry, physiology, and genetics, new concepts of biological control are continuing to be developed. These concepts applied in medicine give us many effective chemotherapeutic agents. In agriculture, these same ideas lead to new types of insecticides, fungicides, bactericides, and herbicides—some of biological origin and some chemically developed. As in the case of fertilizers, mankind has come to the point that without proper use of these materials sufficient food supplies cannot be maintained.

The use of these substances to protect food organisms depends on their selective action and is subject to three general restrictions. First, neither the crop nor the people who cultivate or consume it should be adversely affected by the chemical. Second, the insecticide, fungicide, etc., should be effective against the target organism without undue harm to others of the ecosystem. This objective is very difficult to accomplish, for there are hundreds of thousands of species of insects, bacteria, and fungi but relatively very few are any threat to man and his food supply. In fact, many more species are quite beneficial. For example, the problem for a fruit grower is to protect the bees necessary to pollinate flowers so that fruit will be produced but to kill other insects that feed on developing fruit and leaves. Use of fungicides against plant diseases poses little threat to insects but might affect some soil fungi. Third, compounds used to protect food crops must not persist in the environment too long. In other words, their effect should be temporary and not cumulative. These conditions will protect other organisms of the ecosystem and preserve the effectiveness of the pesticide because organisms have more difficulty developing resistance to substances which quickly decompose to nontoxic substances.

Discussion of the mechanisms of action of pesticides is beyond our purpose here. Fundamentally the control of biological systems depends on understanding the dynamic chemical and physiological processes of all living things in order to protect desirable organisms and control or eliminate competitors. Both modern medicine and agriculture depend on this basic concept. Many highly skilled plant and animal physiologists, biochemists, organic chemists, pharmacologists, agronomists, microbiologists, entomologists, ecologists, and engineers are working in this highly developed scientific field. Chemical aids to food production are equally or more important than similar chemical aids to the field of medicine for it has long been known that a well-nourished person is a healthier person—more resistant to disease and to parasites. Yet there has been increasing public concern over the use of these substances. The nature of the anxiety can be better understood by discussing examples in terms of the three criteria for the use of these chemical aids mentioned above.

Insecticides

DDT was first widely used in World War II because of its outstanding performance in killing disease-carrying body lice of humans—a very dangerous health problem where people are living in very crowded conditions with limited sanitation facilities. DDT did not harm the persons using it even on prolonged body contact but it killed the target organisms. Soon it was found to be very effective against flies, mosquitoes, and pest insects on food crops, and it was promptly used to protect fighting men from insects and to increase food production so necessary in wartime. DDT is cheap to produce and so effective against mosquitos that the mosquito-borne disease malaria, a scourge of man throughout history, has been almost wiped out in large areas of the tropics and semitropics. Another mosquito-borne disease, equine encephalitis, also may be controlled in a large measure by DDT. This insecticide was used in increasing amounts in homes, farms, and other places where flies and mosquitoes were bothersome. Although apparently not harmful to man, DDT kills many desirable insects such as bees as well as insects deleterious to crop production, so its selectivity was not as limited as desired for some uses. Because DDT does not deteriorate quickly, some insect resistance did develop and its effectiveness decreased somewhat. However, the greatest concern was that, because DDT decomposes so slowly, it could accumulate in certain predatory species (upper-level consumers) of birds and fishes since it is fat soluble. Many people became concerned because there is some evidence that accumulation of DDT decreases the ability of the affected species to propagate. So the relative stability of DDT in certain parts of the ecosystem of the earth has led to severe restrictions on its use notwithstanding the pleas of many sectors of the agricultural economy. However, nothing as effective and as economical as DDT is yet available to control the malaria-carrying mosquito, and large amounts of DDT are still used for this purpose.

Government authorities and the farmer-producers of food know that the food required for our population cannot be produced without insecticides. So in place of DDT a class of insecticides called *organic phosphates* (malathion and parathion are examples) is being used more extensively. These materials can be very toxic or even fatal to users unless they are handled properly, and are not nearly as safe to users as DDT. They are very effective against target organisms but also act against other insects. Their major advantage is that in the presence of water these poisonous substances decompose in a matter of days to harmless substances. They do not persist, so the emergence of resistant strains of insects is difficult and there is no cumulative residue. Ecologically this is a great advantage, but the rapidity of decomposition requires that the insecticide be applied more often, adding considerably to the cost and to the danger to those who apply it on crops. It is also essential that no freshly treated food crop be consumed unless the insecticide is removed.

There are other useful insecticides and many potential ones are being investigated, although millions of dollars and several years of diligent effort are required to prove the usefulness of a new insecticide and get it approved by government authorities. But the perfect insecticide is not likely to be found soon. Biological control of pests by other insects or by introduction of diseases unique to the target insect is under study. Sterilizing the males by irradiation and then releasing them to mate has been found effective in controlling insect species in which the female mates with only one male. The eggs of a female that has mated with a sterile male are infertile and no offspring are produced.

Other approaches to insect control that are being investigated as a possible alternative to DDT—other chlorinated hydrocarbons, carbamates, organic phosphates, and so on—make use of the unique physiology of insects. The metamorphic changes in the development of insects, from egg to larva to pupa to adult, might be taken advantage of if a substance blocking any one change could be found. These metamorphic changes are caused by the rise and fall of a number of growth hormones in the developing insect. If, for example, larvae could be treated with an "insect youth hormone," they would never develop further. Also, most insects have a highly developed sense of olfaction (smell) and are attracted to each other by use of their antennae to detect certain substances produced by another member of the species. These substances are called *pheromones*. They are chemicals produced and released to the atmosphere or deposited along a trail to communicate to insects of the same species. Sex attractants fall into this class. Mating of insects can be interrupted by using a sex attractant to lure the insects to a trap or simply to confuse the insects. The difficulty is to get enough of the chemical to identify and imitate, although a few insect pheromones have been identified. The gypsy moth imported from Europe has few natural enemies in America and defoliates large areas of New England forest. It was held in check by DDT but is increasing again since the DDT ban. Only 10^{-12} gram of the sex attractant secreted by the abdominal tip of the female will attract and excite males from some distance away. Recently it has been shown that the active substance is 2-methyl-*cis*-7,8-epoxy-octadecane, but how effective this substance will be as an insecticide and what effect it may have in a larger ecological sense are still unknown.

Fungicides and Bactericides

Plants have infectious diseases, as do animals, wherein fungi, bacteria, or viruses may be the attacking organisms. Every home gardener or farmer has observed plants attacked by these organisms, which at times can completely destroy a crop. The great Irish famine of 1845–1847 which forced many people to emigrate from Ireland to the United States was due to potato blight caused by a fungus, *Phytophthera infestans*, that destroyed the potato crops on which the Irish peasants had become dependent. Fungi cause many plant diseases, while bacteria and viruses are somewhat less prominent as disease causers in growing plants. Fungi and bacteria destroy large amounts of harvested fruits, vegetables, and cereals if not stored properly. Fungicides and other antimicrobial agents are necessary to protect many crops.

Among plants, bacteria, and fungi, chemical protection is a common natural phenomenon. One organism often produces a chemical lethal to other organisms. Antibiotics, both antibacterial and antifungal, used in medicine and agriculture are produced commercially from fungi or bacteria. Penicillin, the first commercially successful antibiotic, is produced by various molds of the genus *Penicillium*; streptomycin is produced by the soil organism *Streptomyces griseus* and the tetracycline antibiotics by the soil organisms *S. aureofaciens* and *S. rimosus*.

Sometimes it is possible to breed suitable crop plants resistant to certain plant diseases in the same way that other desirable characteristics can be altered by breeding. However, such an approach to disease control is a long and sometimes frustrating process. In any event, the present high levels of food production cannot be maintained unless crops are protected from diseases by the use of antifungal and antibacterial agents, because, as noted above, cultivation of essentially a single type of plant in a large area creates an unstable ecosystem and makes the crop more susceptible to invading organisms.

Plant Growth Regulators and Herbicides

Every person who has ever had experience growing plants on a farm or even in the home knows that an unwanted plant (a weed) can kill or inhibit the growth of a desired plant if the two are growing in close proximity, because the weed uses soil nutrients, water, and sunlight needed by the desired plant. Removal of weeds by hand is tedious and uneconomic, and

HREE French sci-
sts, determined to
d the cause of
nkenness, decided
•ffer themselves as
•ratory subjects.
hey got drunk,
•e nights in a row.
first night they
nk gin and water,
second night scotch
water, and the
d night vodka and
•er. Their conclu-
• was inescapable.
he obvious cause of
nkenness, they
ded, was water, as
•as the only con-
•t factor.

•lso true for mechanized cultivation (weeding). For some
•thod is impossible. Knowledge of the comparative bio-
•iology of both desired and undesired plants offers new
•he age-old problem of weed control.
• of plant development are controlled by auxins or hor-
•y certain cells of the plant to control growth of other cells,
•, or to cause some other change in the normal life cycle.
•n by being present in certain physiologically prescribed
•o much or too little at the time when the auxin is needed
•effects. This is true for both plant hormones and animal
•ow). Hormones or other compounds possessing hormone-
•e used to control desired plants or weeds as needed to effect
•oduction. Indole acetic acid is an example of a naturally
•\ close chemical relative is 2,4-dichlorophenoxyacetic acid
•2,4-D. This compound has relatively little effect on grasses
and related plants but stimulates and causes disorganized growth in many broad-leafed plants. This eventually kills them. So 2,4-D is extensively used to kill weeds competing with such crops as wheat, corn, and grasses used as feed for livestock and to protect lawns.

Plant hormones are used to increase and improve food supplies in ways other than as herbicides for weeds. Spraying a minute amount of hormone such as naphthyloxyacetic acid on certain types of fruit trees can prevent premature abscission of the stem and excessive fruit dropping and spoilage. Another substance can kill a certain number of flowers on a tree so that a maximum yield of fruits is possible because too many fruits on a tree will result in fruits that are too small for most efficient use. Another use of plant hormones is to cause flower formation ahead of normal schedule. Pineapple plants in many locations may take approximately 2 years to flower in order to yield a fruit in the third year. However, with good cultivation a healthy plant can be produced in 1 year and then be made to flower months ahead of time by using a hormone, so mature fruit will be ready to harvest in a considerably shorter time. Thus the yield of pineapple from a given field in a given period is greatly increased, and the cost of this fruit to the consumer is decreased. Another important aspect of fruit production is maturation and ripening. Ethylene is a natural metabolite of plants involved in the ripening process of many fruits. Ethylene is also an easily produced gas, and it is used to control ripening of some fruits, particularly bananas. This delightful tropical fruit—picked when green and distinctly unpalatable—is widely used in temperate climates because its ripening can be controlled as it is transported long distances and prepared for market.

From these few examples it is clear that increased knowledge of the biochemistry and physiology of plants leads to more effective ways to protect food plants from undesirable weeds and at the same time to effect more efficient production and use. The net result is more and better food for more people.

Use of Nutrients, Antiparasitic, Antibacterial, and Antifungal Agents, and Hormones in the Production of Animals for Food

Increasing human population results in greater demands for meat, milk, and eggs, for these foods and products resulting from their processing are highly regarded in most cultures. They are high in nutritive value, particularly with respect to protein (Chapter 5). In the United States, more than half of the consumer's food dollar is spent for foods of animal origin. Intensive and efficient production of such foods in a limited ecosystem involves the same restraints discussed for plant crops.

Nutrients

Although there are some important differences, the nutrient requirements for the animals on which man depends for food in general closely parallel those of man. These have been discussed in detail in Chapter 5. Whereas fertilizers for plants need supply only inorganic nutrients to the soil in order for the plant to grow and synthesize the organic constituents of its protoplasm, animals, being heterotrophs, must be fed organic constituents plus mineral elements.

The nutritional requirements for the domestic animals on which we depend are known with greater exactitude than those for man. Furthermore, how to provide these nutrients efficiently and economically is more highly developed than the parallel technology of feeding people. There are a number of reasons for this situation. Almost since the beginning of the science of nutrition, agriculturalists have seen the compelling economic value of dependable nutritional information. The goal for the farmer is more product per unit of feed. For too many years there was truth in the facetious remark that there was always money for nutrition research if the product was valued in dollars per pound. The medical profession has traditionally been interested in nutritional science only incidentally. Then, too, domestic animals, being expendable, can be studied by precise experimental procedures not easily replicated in the case of man. However, increased interest in preventive medicine and social welfare is bringing greater emphasis to the nutritional needs of people and the food science and technology that can fulfill these needs.

The effectiveness of applying the science of nutrition to animal food production is dramatically shown in the cost of poultry and eggs. Only a few decades ago chicken and turkey were luxury foods and eggs were relatively expensive. Now, however, poultry products are among the more common and less expensive foods. The supply of milk and dairy products is being maintained in the United States, even though in the last 15 years the number of dairy cows has decreased from 19.8 million to 12.2 million, because the annual milk production per cow has increased from 6300 pounds in 1957 to 10,000 pounds in 1973 (from 2860 kilograms to 4550 kilograms).

To be sure, feeding animals properly is directly related to an adequate supply of plant material, but all plant materials do not have the same nutritive qualities and the nutritive requirements differ among animals. It is therefore necessary to use combinations of feedstuffs supplemented, if necessary, with minor nutrients such as minerals and vitamins in order to assure proper nutrition for animals bred for efficient production of meat, milk, and eggs. It is not possible here to discuss in detail the different nutritive requirements of the animals we use for food, but there are some fundamental differences which can help us appreciate in a general way how man has become so dependent on animals for food.

Animals used for food are of two types, monogastric and ruminant. They differ fundamentally in nutritive requirements. The pig, chicken, and turkey are monogastric and their nutrition is more comparable to that of man who himself is a monogastric animal. The nutritive requirements of the pig are quite similar to those of man. In a real sense nutritionally these domestic animals can be directly competitive with man for food supplies. They need carbohydrate as starches and sugar, fats, proteins, vitamins, and minerals of the same general quality as man. Perhaps the major difference is that these animals thrive on feedstuffs that can be produced cheaply in quantity and that man considers deficient for his own consumption, most likely unpalatable— field corn and certain other crops and by-products of the milling, meat, fish, dairy, fat and oil, sugar, and fermentation industries. Many of these feedstuffs might be made more acceptable for man by suitable processing, and indeed there is much activity in extending man's food supplies by doing this (see Chapter 12). By blending the feedstuffs and supplementing them, if necessary, with salt and other minerals and vitamins, highly efficient and palatable animal rations can be formulated. The animals thrive and convert otherwise unused biological materials to highly nutritious food for man.

Cattle, sheep, goats, buffalos, reindeer, llamas, etc., have since prehistoric times been domesticated to provide food and clothing for man. Some undomesticated or wild animals such as the moose and caribou have also been used for these purposes. The important role of these ruminants in the development of man to his present status is based on the fact that ruminants are capable of using for their feed plant materials generally not suitable for man and other monogastric animals. Thus ruminants convert unsuitable plants into highly nutritious food for man. Many diverse cultures on the earth have been primarily dependent on ruminants for this reason.

Whereas man, pig, chicken, and turkey must have starch and similarly digestible carbohydrate and fat for their energy supplies and high-quality protein and vitamins in their feed, ruminants do not. Although the ruminants can use these same materials, they are especially adapted to using for their energy need cellulose and similar carbohydrates which man is incapable of digesting and lower-quality protein or even simple nonprotein nitrogen compounds such as urea. Also, they need fewer vitamins—usually only A and E. So ruminants can use almost the complete plant whereas man and the pig may be able to use only the fruit or seed of the plant. Ruminants, of course, as they graze, prefer certain plants over others, and, given the chance, they will eat actively growing green grass rather than dry dead grass, although they can live on the latter. In fact, hay—dried forage—is a major feed for cattle. To understand why this is so it is necessary to appreciate a major physiological difference between monogastric animals and ruminants.

The gastrointestinal tracts of the two types of animals differ only in that ruminants have a more complex stomach—in fact, it might be said that they are animals with four stomachs (see Fig. 8.5). Forage is taken into the mouth, chewed, and then swallowed. The chewed material passes by means of the esophagus into a large paunch or rumen which acts as a large anaerobic fermentation vat. Here reside in symbiotic relationship a mixed population of microorganisms—mostly bacteria and protozoa—which can utilize the cellulose and other plant materials and even nonprotein inorganic nitrogen for their own growth and development. In this fermentation, cellulose is converted largely to acetic acid, which is absorbed directly into the bloodstream through the rumen walls and is utilized as the major energy source for the animal. In Chapter 3, it was pointed out that acetic acid occupies a focal

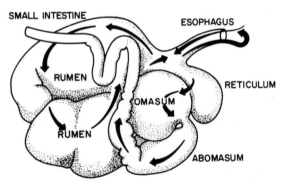

Fig. 8.5. Diagram of the four "stomachs" of the cow, a ruminant animal. Arrows indicate the flow of masticated and regurgitated feed.

point in the metabolism of carbohydrate, fat, and protein. The protein and inorganic nitrogen of the plant material are converted to bacterial and protozoal protein as these organisms multiply using energy derived from the rumen contents. To facilitate more rapid fermentation, rumen contents are regurgitated and chewed again and again. This is called chewing the cud. As fermentation proceeds, the gases carbon dioxide, methane, and perhaps even some hydrogen periodically escape by regurgitation and belching. As time goes on, the food particles and microorganisms pass to the reticulum and omasum, smaller chambers or stomachs, where fermentation continues. From here the mixture passes to the abomasum or true stomach. Beginning at the abomasum, digestion proceeds as in the monogastric animal. The hydrochloric acid kills the protozoa and bacteria, and digestive enzymes begin their functions, which continues down the small and large intestines. The amino acids, fatty acids, monosaccharides, etc., are absorbed by the small intestine. Besides proteins and amino acids, the microorganisms synthesize many of the vitamins, which are also absorbed. All of these materials absorbed from the intestine and from the rumen then are metabolized by the animal. Thus the ruminant thrives on feedstuffs which cannot be used by man because of their indigestibility and poor nutritive qualities and converts these materials into meat and milk, which are among the most nutritious of all foods.

In effect, the ruminant is a gatherer or grazer of forage (plants) and a walking fermentation vat. Furthermore, the nutritive requirements for the ruminant are in reality those of the bacteria and protozoa on which the animal depends rather than the nutrients usually considered essential for monogastric animals. The types of feedstuffs or plants that ruminants can use are quite variable. As the chemical composition of rations changes, there are compensatory changes in the relative numbers of different types of microorganisms in the rumen. Consequently, if rations are changed greatly there is a longer period of adjustment than in the case of monogastric animals.

It is of interest to point out here that ruminants accomplish normally the process being now investigated extensively to expand world supplies of high-quality protein—that is, the production of single cell protein or proteins from various species of microorganisms capable of growing on substances such as petroleum, industrial wastes, sewage, and wood which man cannot use. In this process, the bacteria or other organisms are harvested and their proteins refined for possible use in human food (see Chapter 12).

It is also worthy of note that A. I. Virtanen, an eminent Finnish biochemist and nutritionist, has recently shown that high-producing dairy cows can maintain production of milk on rations based on purified cellulose and nonprotein compounds of nitrogen as urea and ammonium salts. Such milk is equal in nutritive quality to milk produced by cows on normal forage-based rations, and is indistinguishable in flavor.

In view of ruminants' ability to live on sparse and rather poor vegetation and on plants unsuitable for human food, it is easy to understand why man early came to appreciate their unique value in providing food and even clothing and shelter. In the United States today, half of the meat consumed and, of course, all dairy products come from ruminants—mostly cattle.

Insecticides, Antibiotics, Vaccines, Antiparasitic Drugs, and Growth Regulators

As in the case of plants, intensive production of animals for food increases hazards of attack on these animals by other organisms. Usually large predatory animals and birds pose no problem where domestic animals are grown in protective confinement. Here even insects as such are of minor concern except as carriers of disease organisms. Parasites, bacteria, and viruses are generally of more concern than the fungi which pose the major disease threat to plants. Topical application of insecticides has long been used to control lice and ticks in the open range. When large numbers of animals are raised in small areas, any kind of disease poses a threat to the animal producer and may even wipe out his herds and flocks.

Control of Parasites. A healthy animal is a productive animal. Preventive medicine has traditionally been emphasized in the veterinary field. In order to be healthy, an animal must be well nourished, because then it can combat with its own body defenses invading organisms of many kinds. However, various types of internal parasites such as worms and protozoa are ever-present hazards for higher animals and can impair their efficiency of growth and reproduction. It is generally easier, safer, and certainly much more economical to prevent parasitic infestation than to cure it. For this reason, certain antiparasitic drugs are sometimes included in rations as prophylactic measures. Use of such substances of course must be governed by the three criteria already discussed with respect to insecticides for plants (see p. 199).

Antibiotics and Other Antimicrobial Compounds. Antibiotics and other antimicrobials are also used in animal feeds—particularly for monogastric animals. Such use can effectively improve feed efficiency from 5 to 10%. This translates into more, and thus less expensive, highly nutritious foods. Apparently these antibacterial agents serve as prophylactic agents to control subclinical infections and decrease numbers of potentially harmful bacteria that might cause disease outbreaks. The amounts of these substances added to rations is far below therapeutic needs for active disease states and even below the level that would cause major disturbances of the normal bacterial flora of the intestinal tract (see also Chapter 7).

The practice of including antibiotics in animal rations has been effective for a quarter of a century and how it came about is itself an interesting story—an example of scientific investigation in one area having unexpected practical value in another area. Nutritional scientists were hard at work trying to track down the nutrients needed by man and higher animals and discover their chemical nature. In these studies, research on the nutritional requirements of various microorganisms was particularly helpful, for some bacteria were found to require the same vitamins as higher animals. Also, early commercial production of some vitamins was possible only by fermentation of microorganisms from which the vitamins were extracted and purified. About this time, penicillin, streptomycin, and the tetracycline antibiotics became the wonder drugs for controlling some infectious diseases in man. These antibiotics also had to be produced by fermentation. Practical animal nutritionists and animal feeders had known for some time that the wastes or by-products from industrial fermentations, particularly from alcoholic beverage production, had value as animal feed. Quite naturally, fermentation by-products of other types were studied with respect to their nutritional value. Even after vitamin B_{12} was identified as apparently the last remaining nutrient, the residues from certain industrial fermentations seemed to have greater nutritive value than predictable from their contents of all known nutrients. Residues from commercial antibiotic production particularly showed this mysterious plus factor. In careful study to determine whether some unidentified nutrient still existed, it was found that the minute amounts of antibiotics remaining in the fermentation residues after extraction were what enhanced the feeding value of the residues. Further work clearly established that the addition of very small amounts of certain purified antibiotics to animal rations containing no fermentation residues or to highly purified rations improved the efficiency of feed utilization to the same degree. Therefore, antibiotics and some other antibacterial substances are now generally added to many animal rations, particularly for swine and poultry. After more than two decades of use in this way, the effectiveness of these additives to animal rations has not diminished. Furthermore, the usefulness of these therapeutic agents for clinical disease in humans appears unchanged, although there is some reason to believe from some types of laboratory studies that such loss of effectiveness might have been expected.

Vaccines. Domestic animals grown in large numbers are subject to viral as well as bacterial infections much like man himself is. A number of hazardous diseases can be controlled by immunization using vaccines. These include such highly infectious or fatal diseases as hog cholera, equine encephalitis, anthrax, brucellosis (undulant fever), and Newcastle's disease. Even so, some very costly animal diseases are only partially controlled by vaccines. Among those of worldwide importance are foot and mouth disease, rinderpest, and rabies.

Growth Regulators. As with plants, in order to understand the action of growth regulators it is necessary to appreciate something of the physiology of growth and reproduction in animals. Growth, development, maturation, and reproduction in animals (including man) are under the complex and synchronized control of hormones. These substances are produced by one part of the body—an endocrine or ductless gland such as the pituitary, thyroid, ovary, testis, or adrenal—and carried by the bloodstream to control the growth or function of other tissues of the body. Among the popularly recognized hormones are the sex hormones, the growth hormones, and insulin, which helps control carbohydrate metabolism (lack of it causes ordinary diabetes). There are many other hormones. There is even reciprocating action among hormones—as the amount of one decreases, another increases, etc. In order for a particular hormone to perform its proper function it must be present in an animal's bloodstream at the proper level—too little or too much can produce adverse results.

When we consume foods of animal origin, it is inevitable that we also consume the hormones in the food at that time. For example, in meat there will be male and female hormones, other steroid hormones from the adrenal gland, thryoxine, epinephrine, and various pituitary hormones in the amounts proper for the meat animal at the time of its slaughter and preparation for use. Consumption of these biologically potent molecules in this way is of no importance simply because the amount of meat consumed in a day would be 1% or less of our own body weight—and thus one-hundredth of that necessary to affect a human, more likely several orders of magnitude less. We are continually making our own similar hormones and the eating of so small an amount in our food is of no physiological significance. If it were, man would not have survived to become so dependent on animals for food.

In the production of foods of animal origin, efficiency with respect to feed consumed determines the ultimate cost of the food to the consumer. Only females (cows and hens) produce milk and eggs, but males are desirable for meat production because they grow faster and are more efficient converters of feed to meat than females of the same species. However, females produce meat with more fat, which to some people makes the meat more palatable. Castration of male animals to increase the fat content of their meat has been practiced since prehistoric times. Castrated males are also less aggressive and somewhat easier to manage. Even though castrated males do not grow as rapidly as normal males, they do grow somewhat faster than females. In view of these practical considerations and the increased knowledge of the physiology of growth, development, and reproduction and the involvement of hormones in these processes, it is understandable that animal scientists would want to put such knowledge to use for increased food production.

Two uses of a hormone as a growth regulator to increase meat supplies have been found practical to date. Both involve the use of *di*ethylstilbestrol (DES), a compound which can be easily manufactured and has female sex hormone activity. First, DES can be used in male poultry and cattle to modify some undesirable male characteristics in the direction of desirable female characteristics. Second, DES increases the growth efficiency of the male somewhat beyond normal, which means more meat per unit of feed consumed. Even in castrated male cattle, steers, growth efficiency is improved by adding small amounts of DES with no adverse effect on the quality of the meat. In fact, steers fed DES have somewhat leaner carcasses and less waste fat. Apparently at the proper level of feeding in the ration DES causes some changes in the hormonal patterns of the growing and developing animal to stimulate overall growth. The increased rate of growth, usually 5–10%, helps to increase meat supplies.

In order to give some perspective to this method of using a hormone to control some physiological developmental process, recall the use of the "pill" for birth control purposes. In the animal a hormone or hormone-like substance is fed to promote growth, in the human a hormone or hormone-like substance is fed to interfere with the development of a mature egg in the female, the fertilization of the mature human egg or the implantation of the fertilized egg in the uterus.

Conclusions

Up to this point in this chapter we have discussed some of the major principles and concepts relating to the cultivation of plants and the husbanding of animals by man to increase his food supplies. In doing so, man has increased his own numbers enormously as well as the numbers of the few favored species of organisms on which he depends. These accomplishments have been possible only by altering the ecosystems in which he and his food organisms live. In reaching the goals for food production, man has been obliged to accept responsibility for maintaining proper environmental conditions for the growth of these plants and animals and for protecting them as well as himself from competitive organisms. Now we shall discuss some of the specific biological characteristics of several important food plants and animals. From this, we can obtain still other perspectives on to how modern man feeds himself.

PLANTS

In food markets almost anyplace in the world will be found foods derived from a wide variety of plants. Markets in more economically advanced areas may offer a greater selection of such foods than markets in less economically

Table 8.2. Probable Area of Origin of Selected Plants Now Cultivated
for Food

Eastern and Southeastern Asia	Central Asia	Mediterranean Middle East	Central and South America
Rice	Barley	Wheat	Corn
Sugar cane	Rye	Millet	Peanut
Soybean	Apple	Date	Pineapple
Orange and other citrus fruit	Cherry	Fig	Avocado
	Plum	Melon	Cashew
Banana	Pear	Asparagus	Tomato
Peach	Olive	Carrot	Potato
Coconut	Lettuce	Cucumber	Kidney and lima bean
Persimmon	Pea	Cabbage	Cassava
	Spinach	Cauliflower	Sweet potato
	Mustard	Turnip	Cacao (chocolate)
	Onion	Coffee	
	Radish	(tropical Africa)	

Adapted in part from R. B. Duckworth, *Fruit and Vegetables*, Pergamon Press, Long Island City, N.Y.; N. W. Pirie, *Food Resources, Conventional and Novel*, Penguin Books, Baltimore; and William H. Chandler, *Evergreen Orchards*, 2nd ed., Lea and Febiger, Philadelphia.

developed areas. Yet for any market if we trace the historical origins of the plants yielding the foods offered for sale, we will find that many of them were originally discovered in areas of the world quite remote from the regions where they are grown today. For example, even though for all practical purposes Europeans did not discover the western hemisphere until the close of the fifteenth century, potatoes, which are indigenous to and were first cultivated in the high plateau of South America, are grown and used much more extensively in northern European countries than elsewhere. Peanuts and cassava also originated in the western hemisphere but are grown much more extensively in many areas in Africa. Whereas rice is the major cereal crop in the highly populated areas surrounding its place of origin and likewise corn is a major crop in the western hemisphere where it was found and first used, wheat, which had its origins in the Middle East, is grown in many regions throughout the world and contributes more to the world's supply of food than any other plant. Although bananas and sugar cane came originally from Southeast Asia, they are major economic crops in tropical and subtropical America. In Table 8.2 are shown the regions of origin of many of the familiar plants which contribute in a significant way to the world's food supplies at the present time. Although the lists of plants are by no means complete, they are sufficient to illustrate that there is great diversity in types of plants from a botanical point of view. Even so, it is also clear that of the many thousands of species of plants known only a few are used for food.

To be useful as food, some part of the plant—root, stem, leaf, flower, fruit, or seed—must have some value as food without undue toxic side-effects. But though this is the first requisite, it is by no means the only requirement. The plant must respond to cultivation, give high yields, and be adaptable to large areas or regions throughout the world. This means that the plant must be able to thrive in some range of soil-types and climatic (light, temperature, rainfall) conditions. In addition, the physiology of the plant itself and particularly of that portion used for food must be such that the food is usable over a period of time. This is because most plants are seasonal in their yield of food but man requires food each day. Let us illustrate these points by two common examples.

The tomato is indigenous to certain areas of tropical America and is a warm-weather crop requiring a reasonable amount of water. Because the plant matures and begins to produce fruit in a short time, it can also be grown in many areas in the temperate zones of the world where there are a few months of rather warm "tropical" weather. Tomato plants produce an abundant crop and, in some localities, can be planted in sequence to prolong the time of availability. The tomato fruit itself is a metabolizing tissue which, like many similar fruits, is usable for only a few days without spoilage resulting from normal degeneration and invasion of fungi, bacteria, and insects. The rate of metabolism (deterioration) can be slowed somewhat by refrigeration to permit transport and use for a longer period of time. The tomato is easily preserved by heat sterilization (canning), further extending its use. Corn, like tomatoes, is a warm-weather crop requiring reasonable amounts of water and only 3–4 months to mature. As it matures, the grain (seed) becomes dry. The metabolic rate of the dry seed, which is a living system, drops to a low level. If harvested at this stage and stored dry and away from insect and rodent pests, corn may be kept for long periods of time. Thus corn can be effectively produced in some semitropical and tropical areas where the growing period coincides with the rainy season and the seed maturation occurs at the onset of the dry season. In higher latitudes, maturation of the seed coincides with the onset of dry, cool autumn weather. Cold weather of winter can facilitate extended storage for later use. For both corn and tomatoes, it is necessary to harvest and protect some of the dry seed from one growing season in order to start the crop in the next growing season.

These two simple examples show how man, as he learned to cultivate his food plants, could recognize other locations where such plants could be grown. At the same time he began to understand that plants from other areas might be grown in his own fields. Accounts of the great explorers of history such as Marco Polo, Columbus, Magellan, and Cook reveal that they brought possible new food plants back to their patrons.

Different varieties of single species of food plants were recognized for their ease of cultivation, resistance to disease, palatability, or other characteristics. As time went on, it was recognized that some varieties of a particular food plant might thrive better in certain locations than others. Finally, plant breeding techniques were developed, and with the advent of the modern scientific era the disciplines of botany, plant physiology, and genetics have led to such a sophisticated technology of plant breeding that food plants can be bred and propagated to yield varieties adapted to specific soil types, climatic conditions, length of growing season, disease and insect resistance, ease of mechanical cultivation and harvesting, nutritional value, color, texture, flavor, and storage and processing characteristics as required for milling, canning, dehydration, etc. The overall result of such developments is that yields of food are maximized with respect to production location and ultimate use. Perusal of a seed or nursery catalogue will reveal, for example, that some varieties of peas, peaches, etc., are better than others for specific purposes. In the relatively small state of Ohio, 245 varieties of corn and 33 varieties of soybeans are offered to farmers on the basis of days of growing season to maturity (95–125 days for corn, 110–149 days for soybeans), response to soil types, cultural practices, regional climatic conditions, quality attributes of grain, disease and insect resistance. Just choosing the particular variety of seed to plant requires the farmer to have a considerable degree of technical sophistication. Yet it is just such developments of modern agricultural science which can improve our food supply and offer some glimmer of hope for the still hungry millions of the earth. The "green revolution" of the last decade which greatly improved food resources in many areas of the world was not a revolution at all but rather the result of many dedicated people applying their technical skills to the problem of producing more food. The development of improved varieties of both rice and wheat adaptable to food-deficient regions was a key contribution to the "green revolution." For his work in this field, Norman Borlaug was awarded a Nobel Prize; however, he has warned that these advances offer only a brief respite from the urgency of producing enough food for the rapidly increasing population of the earth.

The dynamic, changing character of biological systems—man and the organisms he uses for food—has been emphasized repeatedly. Time is therefore a fundamental parameter in the existence of all organisms. A puzzling basic problem for biologists is the nature of the biological clocks which seem to control genesis, growth, development, and maturation of organisms. Time factors are of particular significance in our use of plants for food. There is, as with any organism, the time period of the life cycle itself. Also, green plants must have proper light, water, and temperature for growth, and these are all related to the rotation of the earth around its own axis and around the sun (the primary energy source). Since these phenomena are responsible for the daily and seasonal variations of these three environmental factors, there are obviously time effects related to seasons of the year. Plant species evolved in concert with the seasonal variations in environment. How these time-related factors control the biochemical and physiological events of plant growth is not well understood. Nevertheless, we can observe some practical aspects of the interactions of time, temperature, water, and light.

Most plants require alternating periods of light and relative darkness— day and night. Some will thrive only when days and nights are of approximate equal length such as in the tropics; others may thrive on long days and short nights as in the more northerly or southerly regions of the world. Some plants will grow only vegetatively when days are short and will not flower to form a seed or fruit until days become longer as in the temperate zones, but still others display just the opposite effects of photoperiodicity. Some plants are induced to flower when temperatures fluctuate or reach a certain narrow range for specific lengths of time. Lack of water, changing length of day, or low temperatures can induce a period of dormancy (almost suspended animation) in some plants which then can resume normal growth when conditions become more favorable for a period of time. Yet similar changes will kill other plants. From these brief considerations, it is easy to see that seasonality is an important characteristic of most plants.

Seasonality means that many food plants produce only one crop per year, particularly in temperate climates even for plants of short life cycles such as corn and wheat. However, the winter weather that limits food production is not without some compensations. In warm climates, plentiful harvest may be illusory because of difficulty in storing and preserving food from one crop to the next. Much food spoils and is lost to fungi, bacteria, and insects. The cold of winter can be utilized to retard spoilage of stored food. Furthermore, winters have a positive effect on production for next season in that cold weather kills off many insect pests and retards the metabolism of decomposer organisms of the soil—particularly the bacteria. Residual organic matter from the roots and other parts of the plant left to rot on the ground slowly deteriorates to humus, which helps to maintain proper mineral and water-holding properties and physical characteristics of the soil for the next crop. Where the weather is warmer, soil bacteria will continue active decomposition of the humus to, eventually, carbon dioxide and water. The result is that minerals the plants need from such soil are more easily leached out by rain and the loss of organic matter can cause compaction of soil to the extent that it will become much less productive or even unfit for agricultural uses.

With this background of information, let us now look at some of the characteristics of a few individual food plants as a step toward appreciating what is required in producing a particular food crop and also what makes certain plants so valuable as food.

Rice

Rice (*Oryza sativa*) requires 4–5 months of sunny, warm weather and plentiful supplies of water from either rain or irrigation, which reflects its origin as a swamp plant. In many areas of the world, seedling plants are transplanted into a flooded paddy where during 60–90 days of the growth period the floodwater serves to control weeds as well as furnish moisture. However, some varieties of rice have been adapted to field culture and thrive in many areas where soil, temperature, sunny skies, and water are available. In tropical areas where there is irrigation water, two crops of rice can be grown per year. In terms of production of food calories available for human use, careful cultivation of rice can yield about as much food value per unit of land area as any other crop. Like other major cereals, properly dried and stored rice can be used over long periods of time (see below). Rice varieties differ in food quality as well as in ability to grow under different agronomic situations. Generally the polished whole grain is eaten after boiling in water and is not milled into flour as is the case with wheat. Since rice is the major food for so many people in the world, it is understandable that certain varieties of rice are preferred to others on the basis of eating quality alone.

Wheat

More than 30,000 species of the wheat genus *Triticum* are known. These include emmer, which was used as food in prehistoric times and was known to have been cultivated in Iraq about 7000 years ago. Today most cultivated wheats are varieties of *T. aestivum* (also referred to as *T. vulgare* and *T. sativum*) and *T. durum*. There are a great many varieties of wheat adapted to different soils and climatic regions. Wheats generally require a growing season with an average mean temperature above 56°F (13.3°C), and from planting to harvest requires $2\frac{1}{2}$–5 months depending on variety, moisture available, and temperature. Wheat may be grown in tropical, semitropical, temperate, and even cool climates and thrives best when moisture is available during early growth with less during maturation. Another very important aspect of wheat is that some varieties can tolerate mild winters. These so-called winter wheats are planted in the fall so that the seeds germinate and the young plantlets get started before the onset of freezing temperatures. If winters are not too cold and, particularly, if the small plants are protected from extreme cold by a snow cover, the young wheat plants survive and begin growing again immediately with the onset of spring. In areas where there is an early spring, the wheat can be harvested and another crop such as fast-maturing corn or soybeans grown. In areas of severe winters and short growing season, wheat is planted in the spring and grows to maturity without interruption. Such varieties of wheat, known as spring wheats, are grown extensively in the northern plains states, Canada, and the USSR.

Unlike rice, wheat is milled and the resulting meal or flour is used for mixing with other ingredients to make foods of many kinds. Wheats are commonly spoken of as "durum," "hard," and "soft." These terms reflect milling characteristics—whether the milled product is hard and granular or soft and fluffy. Hard wheats usually have higher protein content than soft wheats. Wheat varieties differ greatly as far as food quality is concerned. Durum wheats are used to make macaroni, spaghetti, etc., but flours made from durum wheats are generally unsatisfactory for breads, cakes, and cookies. Hard wheat flours are superior for bread but inferior for cakes, cookies, and crackers. Just the opposite is true for soft wheat flours. These differences are related not only to the quantity of proteins present in the wheat but also to the chemical makeup of the individual proteins of the wheat. Interestingly, the quantity and quality of proteins in wheat varieties are directly related both to their genetic makeup and to the climatic conditions in which they are grown.

Growing wheat, like most other crops, is susceptible to many plant diseases and to attack by insects. The mature grain must be protected from insects, birds, and animals. Rats and mice particularly cause great losses in stored grain. With proper storage, wheat can be kept in satisfactory condition for use between harvests or even longer. But what is proper storage? Since wheat is so extensively used for human food, storage of wheat is itself a rather sophisticated technology. Furthermore, storage of other seeds such as rice, corn, beans, and peanuts used for food can present similar problems. Let us look into this matter more deeply.

Wheat and other viable seeds are living, respiring systems. They consume oxygen and give off carbon dioxide, water, and heat. At 14% moisture and 68°F (20°C), which are common baseline or reference storage conditions, respiration of wheat and other seeds is usually at a quite low rate. Of course, at temperatures below and above 68°F (20°C) respiration and heat production decrease and increase, respectively, in an exponential fashion as discussed in detail in Chapter 4, pp. 35–40. Even if the rate of respiration and simultaneous heat production is low, the temperature of grain stored in large quantities will increase because grain is an extremely poor conductor of heat. As temperatures increase, respiration increases in an exponential manner as does water and carbon dioxide production from the oxidation of sugars and starches. So in order to counteract this effect the grain must be "turned" by moving it from bin to bin. In such a process, water is evaporated from the grain and the grain is cooled by air movement. Furthermore, by turning, the seeds are not smothered by excess carbon dioxide. Wheat, for example, can be stored in sealed containers; however, when all oxygen is used up respiration will cease but a number of biochemical reactions will continue in the killed seeds and the wheat will deteriorate and become unfit for use.

As moisture content of seed increases respiration increases. Furthermore, if not sufficiently dry, stored grain becomes an ideal medium for the growth of fungi. For example, fungi will grow on wheat containing 15.6–30.0% moisture. So if harvested grain contains too much moisture, it must be dried below this level, usually to 14% or less as a practical margin of safety. If the drying is to be done by hot air, care must be taken not to overheat and kill the seed. Beyond 30% moisture, bacteria will begin to grow on the grain. Such grain soon becomes unfit for human food. The growth of fungi including molds and then bacteria on grain as moisture increases from below levels which support microorganisms to those which do is a practical reflection of the fundamental environmental parameter of osmotic pressure or water activity in the support of living systems (Chapter 4, pp. 52–59). If water (moisture) content of stored grain is sufficient to support microorganisms, they respire, consume organic matter, and produce more moisture and more heat. If not checked, the temperature of the grain may relatively quickly reach 140–158°F (60–70°C) where the fungi and bacteria may be killed but other oxidative chemical reactions continue, eventually to cause fire. So temperature control and moisture control are extremely important in storage of grain. To put this in perspective, the heat necessary to raise the temperature of stored wheat from 68°F to 140°F (20–60°C) represents the metabolism of only 0.5% of the weight of the grain irrespective of whether this organic matter is consumed by respiring wheat, fungi, bacteria, or insects. In addition to all the other problems of storage of seeds (grain), metabolic products of invading organisms may render the grain unfit for human or animal consumption (see below under Peanut).

Corn

Corn (*Zea mays*) sometimes called maize, has already been described in some detail. Compared to rice and wheat, corn is not used as extensively for human food, but great quantities are grown as feed for animals and for this reason corn must be ranked among the most important food crops. Corn requires a somewhat warmer climate than wheat and therefore it is not cultivated extensively in European countries. Like wheat, corn is generally ground and the resulting flour or meal is used in preparing various items of food. The varieties of corn known as popcorn are consumed as whole grain after popping.

Sweet corn is used extensively as a vegetable, particularly in the United States and other corn-producing countries. It must be picked at the so-called milk stage of maturation, when it is soft and sweet (because sugars have not all been converted to starch) and has not acquired the stronger and characteristic, somewhat objectionable flavor of mature, dry grain. Sweet corn can be preserved by canning or freezing. These processes stop the dynamic metabolic activity of the immature corn so that properly canned or frozen sweet corn maintains its desirable quality and can be used over long periods of time, in contrast to the 2 or 3 days that the sweet corn is of satisfactory eating quality on the rapidly maturing plant.

Deciduous Fruits

Deciduous fruits include apples (*Malus sylvestris*), pears (*Pyrus communis*, plums (*Prunus domestica*), peaches (*Prunus persica*), and cherries (*Prunus avium, P. cerasus, P. glandulosa, P. laurocerasus*). Usually the portion eaten is the pulp surrounding the seeds of the maturing fruit from perennial plants which shed their leaves and bear fruits annually. Although crops are generally available annually, usually several years, often 5 or more, are required from the planting of the seed until fruit is produced in sufficient amount to be economic. Today most of the trees are not simply grown from seed. Rather, they are roots of one variety and aboveground structures of another variety fused by grafting. In this way, the desirable characteristics of root systems of one variety can be combined with desired fruit characteristics of another variety. Trees are susceptible to diseases and insect pests which can be controlled by suitable insecticides, fungicides, and bactericides, but climate, which is not controllable, often is a hazard. Trees themselves may be killed or only the annual crop may be lost. For most fruits, a full year is required to reach maturity. Buds which will become leaves and flowers in the following year are formed during the growing season and become dormant as the season comes to an end with the shedding of leaves and maturing fruit. The buds remain dormant until spring, when flowers and leaves begin to grow. Fruits begin to form on the flowers which have been successfully pollinated by bees, other insects, or wind and grow until maturity. The time required from flowering to maturation of the fruit varies with species and with variety within species. Cherries generally require less time than peaches and peaches less time than apples and pears.

All of these fruits are usable for very limited times at ambient temperatures. They are, of course, active metabolizing systems and will deteriorate rapidly on maturation. Their metabolic rates and respiration decrease significantly if the fruits are refrigerated under controlled atmospheric conditions. They can then be used over somewhat longer periods of time. Some varieties of apples can be stored for up to almost a year, while peaches under the best conditions can be kept only a few weeks. Otherwise, canning, freezing, or drying is necessary to extend usefulness of these fruits.

Oranges

Oranges (*Citrus sinensis*) are perhaps the most familiar fruit of the citrus group. As might be expected, there are major physiological differences between the evergreen orange tree and the deciduous apple tree even though both annually yield fruits of many excellent qualities. As with apples, one major advantage of a number of citrus species and varieties is that their usefulness may be extended over a fairly long period of time by storage under proper conditions of temperature and controlled atmosphere.

One and one-half to two years is required for seedling trees to reach the grafting stage and then another 2 years in the nursery is required before the young trees may be transplanted to the orchard. After 2–3 years in the orchard, trees begin to bear and reach maximum productivity in 12–30 years depending on the climate. Trees generally produce satisfactorily for another 30 years.

Orange trees grow best in sunny, warm climates where there is sufficient water. They can be cultivated well in arid regions where there is water for irrigation. The trees are tolerant to periods of cool temperatures and some varieties may even withstand frost and a few degrees of subfreezing temperature for short periods. When temperatures drop to 55–60°F (13–15.5°C) or below, the metabolic activity of the trees drops markedly and they appear to stop growing. Some plant physiologists say that it is this shift to very low metabolic activity that makes oranges and many other citrus trees so tolerant of lower temperatures. When ambient temperatures go above this critical level, growth resumes. In warm, tropical climates the time from bloom to maturation may be 9–10 months, whereas in subtropical and cooler climates 15–16 months may be required.

There are two important physiological characteristics of the trees that contribute to more efficient use of this fruit. First, as with many evergreen seasonal warm-weather plants, the flowering period lasts 4–6 weeks, whereas for deciduous trees springtime flowering is complete within a week or so. Therefore, ripening of citrus fruits takes place over a longer period of time, making fruit available for harvest and consumption over a period of 1–2 months. Second, whereas in deciduous trees ripe fruit falls off unless picked first and will be lost or severely damaged by falling to the ground, in orange and other citrus trees there is no similar abscission and the fruit remains on the tree. If not picked off, the fruit eventually dries up long after its prime. This also increases the harvest period for citrus fruit.

Because of these advantages, oranges and other citrus fruits have been cultivated since earliest times. Yet in modern times, particularly in highly developed countries where labor costs are high, it is quite expensive to select prime fruits for harvesting over a period of time, although the fact that they remain on the tree does counterbalance this problem somewhat.

Fruits in general tend to be expensive compared to cereals and many vegetables. We have already noted that a number of nonproductive years are needed before a profitable harvest can be obtained and that fruit crops are perhaps a bit more vulnerable to insects, diseases, and adverse climatic conditions because of the long periods of time from the inception of a bud to the harvesting of fruit. A third major factor is that in spite of mechanization of many cultural practices such as fertilization, spraying with insecticides, and irrigation, two essential operations necessary for efficient fruit production require hand labor that is sometimes tedious. These are (1) pruning to remove unproductive branches and at the same time to shape the tree for most efficient photosynthesis and (2) harvesting, Harvesting requires many hands for short periods of time. The need for such hand labor is, in a real measure, the root of a major socioeconomic problem—the migrant worker in more developed countries and the poor agricultural worker in many underdeveloped nations.

Coconut

Coconut (*Cocus nucifera*) is produced by a tropical palm. It is used in the United States only to a limited degree, usually in confections in a dried form and as a source of edible oil or oil for nonfood uses. However, coconut is used much more extensively as a food in certain of the tropical areas where it grows. The production of coconut is not too seasonal, because new crops of coconuts reach maturity about once per month. Coconut palms require temperatures in the range of at least 75–86°F (24–30°C), with an annual rainfall of at least 60–80 inches (152–302 centimeters). These palms cannot live at temperatures as low as 68°F (20°C) or when there is less than 40 inches (100 cm) of rain per year. The trees begin bearing after 5–6 years. As a new leaf forms, an inflorescence of male and female flowers occurs and fertilization is possible only for 1–2 days. The fertilized flowers which set will produce mature fruit in 12–14 months. There is no dormancy and flowers appear almost each month with each new leaf. Trees bear for 70–80 years. With continuous production throughout the year, it is understandable why coconuts became so important as a tropical food crop.

Banana

Banana (*Musa sapientium*) is a tropical food popular the world over. There are a great many varieties, perhaps as many as 300, that are eaten without cooking. Closely related species called plaintain (*M. paradisiaca*) are preferred for cooking. The banana tree, which is not a tree at all but more like a perennial herb, may require as much growing space as a small tree because its height ranges from 10 to 33 feet (3–10 meters). Bananas do not grow well at temperatures below 60°F (16°C) or above 95°F (35°C), and better production is obtained when temperatures remain at 75°F (24°C) with plentiful rainfall or irrigation water. From $9\frac{1}{2}$ to 30 months is required from rhizome or sucker to harvest. Longer periods are required in milder climates. For commercial production warmer climates are preferred where only 10–14 months is necessary. Each banana plant produces only one stalk of several "hands" totaling 50–100 pounds (25–46 kilograms) of fruit. It will also produce several suckers or rhizomes that may eventually develop into new plants which produce bananas.

Whether the bananas are to be consumed in the tropics or transported to markets a few thousand miles away, the stalk is harvested green and held in storage or ripening rooms until ready for market. It is this one factor which makes it possible for the banana to be such a common fruit in temperate climates. During ripening, the starches change to sugar and the characteristic aroma and flavor develop. The time required for ripening depends on the degree of development of the fruit when the stalk is harvested, less mature fruit requiring a longer ripening time. Freshly cut stalks of bananas may be stored up to 20 days at 55–58°F (13–14.5°C) and then fully ripened for market by holding at about 68°F (20°C) for several days. The temperature and the composition of the atmosphere in the holding and ripening chambers also affect the time for ripening. Cut stalks of fruit continue to respire and give off carbon dioxide and ethylene (a natural ripening stimulant or hormone). So if the maximum holding period is desired as in shipping, it is not only necessary to keep ship holds at the proper temperature but also to remove ethylene by ventilation. Although bananas may be held (and transported) for some days at 55–58°F (13–14.5°C) without injury, the metabolism of bananas is such that at temperatures below 55°F (13.0°C) biochemical changes result in browning, softening, and undesirable flavor development, a phenomenon sometimes called "chill damage." Most of the cost of bananas in markets in temperate climates is cost of transportation, ripening, and distribution rather than production (see Chapter 9).

Pineapple

Pineapple (*Ananas comosus*) is consumed throughout the world in fresh form or preserved (canned). Unlike the banana, pineapple reaches its peak of quality when it ripens on the plant. Ripe fruit will keep only a very short time, and not more than 24 hours must elapse from harvest in the field to finished canned pineapple; if kept for 3 days, the fruit is unsuitable. When sold fresh, the fruits are usually cut from the plant some days before full ripeness.

Pineapples are unusually resistant to dry weather but not to low temperatures. As with many tropical plants, chill damage can result from extended exposure to temperatures below 50°F (10°C), although short periods of lower temperatures may be tolerated. The leaves of the pineapple plant are particularly adapted to conserve water. About 6–7% of the total water absorbed by the plant remains in the plant and the remainder is transpired. In most other plants, over 99.5% of water absorbed is lost by evaporation (transpiration). Only 30 lb of water needs to be absorbed to produce 1 lb of dry (organic) matter in the pineapple plant, whereas most other plants absorb in excess of 300 lb of water in order to produce 1 pound of organic matter.

The pineapple plant is a perennial and is known to produce fruit for up to 25–30 years. However, after the second or third year the fruits are usually too small for efficient use. The plants are not developed from seed but from suckers or shoots from the parent plant or even from the crown of the harvested fruit. Depending on what is planted and climatic conditions, 15–32 months or more is required for the first fruit to ripen. Then another crop is harvested the next year. After another year and a third crop, the plants are usually dug up and new plants started. It is important that a plant be vigorous and of such size that it can produce well before it forms a flower. If a plant flowers too soon, only small fruit are produced. Flowering may be induced by cool temperatures—60–62°F (16–17°C). If plants are grown in climates which never get that cool, flowering may be delayed for many months and so also the harvestable fruit. Fortunately, the plant is easily induced to flower by application of naphthyloxyacetic acid or other similarly acting plant hormonal substances. Such control of flowering also controls period of ripening, and in this way efficiency of production is increased greatly. Synchronization of harvest decreases labor costs and permits more efficient utilization for marketing as fresh fruit and especially for canning.

Cassava

Cassava (*Manihot esculenta, M. utilissima*), also called manioc or mandioca, is of only minor importance in the United States as the processed starchy product tapioca, but in the tropics it is one of the most important food plants. The shrubby plant, native to the lowland areas of Brazil, grows to about 9 feet tall and matures in 8–16 months. The major part of the plant used for food is the root, although the leaves are used to some extent as a cooked vegetable. Ten tons or more of edible roots is usually produced per acre. They contain up to 30% edible organic matter—mostly carbohydrate and very little protein (see Chapter 5, pp. 90–92), so except for calories they are not particularly nutritious. But since the yield is so high and since the roots are usable when left in the ground over a period of time (although they may become woody if left too long), cassava has almost become the "staff of life" in tropical areas where food is often chronically scarce.

The many varieties of cassava are grouped into sweet and bitter. This reflects the amount of cyanide-containing glycoside present—the bitter varieties usually contain 0.02–0.03% available HCN and the sweet only about 0.005–0.01%. Although cyanides are considered quite toxic to man, there is relatively little danger from consuming this amount of potential HCN. Fresh cassava root may be eaten boiled or fried. More often, the roots are processed by fermentation, grinding, pressing out liquid from solid, using the liquid to form a beverage or soup, and drying the solid to give a farina-type product or flour which keeps almost indefinitely if protected from moisture, insects, and rodents. Such processing, often done in a very primitive way, removes potentially toxic HCN. There are a great many different foods made from cassava which are characteristic of different cultural groups.

Potato

Potato (*Solanum tuberosum*) is perhaps the most important vegetable in the world. It is a cool-weather crop native to the Andean highlands around Lake Titicaca in South America and so has been adapted to cool temperate areas of the world. Although a great many potatoes are consumed in the United States, more are used in northern and eastern Europe and the USSR than anywhere else in the world. Some varieties are used almost exclusively as animal feed in these countries. Only about 100 days is required to produce a crop from whole or cut "seed" tubers containing one or more "eyes." Moderate rainfall and ambient temperatures of 59–75°F (15–24°C) are most favorable. Although, like cassava, potatoes contain besides water mostly carbohydrate, they are somewhat more nutritious in terms of available protein. They are also a good source of ascorbic acid, vitamin C. Tubers are usually harvested in the autumn after the vines die and before the ground freezes. They become dormant and can be stored for some months at temperatures above freezing and, with care, can be kept from the end of the growing season until a new crop becomes available. The importance of the potato as food has led to the development of a great many varieties adaptable to the soils and climates of different regions. The potato, sometimes colloquially called the Irish potato, should not be confused with the sweet potato (*Ipomoea batatas*), which, like cassava, is not a tuber but a swollen root, as are the yams of the *Dioscorea* species. Both sweet potatoes and yams are of tropical origin, are warm-weather crops, and are consumed in great quantities in tropical Africa.

Sugar Cane

Sugar cane (*Saccharum officinarum*) is one of the world's major crops. Originating in India, it was first cultivated for its sweet juice. It was not until about A.D. 500 that crude sugar was first crystallized from boiled sap. By A.D. 900, cane culture was stimulated by the commercial development in Egypt of sugar refining. Cane culture reached Hawaii about A.D. 1000 from the more western Pacific islands. Columbus tried to establish cane cultivation in Hispanola. This failed but later much of tropical America was found adaptable to the production of cane. The demand for sugar in world trade stimulated cane culture in the western hemisphere. Because sugar cane production required much hand labor, there developed in the new world a system of indentured labor and slavery to provide necessary manpower. To this date, the ill effects of such a system are still manifest in the social structure of many countries.

Sugar cane, a perennial plant, is asexually propagated by planting stem cuttings or sections of cane with viable buds which will become new plants. The first crop is usually harvested in 16–24 months of wet, warm weather. Rain in excess of 50 inches (125 cm) per year with soil temperatures between 70 and 95°F (21–35°C) is best for growing cane. Growth stops when temperatures fall to 54°F (12°C) or below or when rainfall or irrigation water is not sufficient. Some varieties of the closely related species *S. baberi* can be used for sugar production in some wet subtropical and temperate regions with long growing seasons. Cane is cut just prior to flower formation at the apex because the sugar content of the plant, about 10–11%, drops when flowers mature. Another crop will follow from the stubble in 12–13 months and still a third after another 12–13 months. But yields go down each year and after the third year's crop all old roots are plowed up and a new planting is made.

Cane must be milled as soon as possible, preferably within hours, after cutting because delays will decrease yields of sugar. The first step in milling and refining of sugar is pressing and recovering the cane sap. After heating and clarification or filtering, the sap is evaporated and sugar is crystallized from the resulting syrup. The technology of sugar refining is highly developed and mechanized. In order to produce the 102 lb (46.4 kg) of sugar the average American consumes each year, 1020 lb (464 kg) of cane must be produced.

Putting this another way, each of us requires 2.8 lb (1.3 kg) sugar cane each day—a formidable amount of plant material. In a small way the sugarbeet, *Beta saccharifera*, a cooler-weather, shorter-term crop, is now used to augment cane sugar supplies. Whether from cane or beet, refined sugar is sucrose—a pure chemical compound (see Chapter 6, pp. 124, 126) and a major item of food in much of the world today.

Beans

Beans (*Phaseolus vulgaris*) are grown and widely used the world over. Originating in Central America, beans of many varieties were an important food for most natives of the western hemisphere when Columbus discovered America. Since that time, the unique values of beans have also made them a popular food in the eastern hemisphere.

Beans are a warm-weather annual crop grown from seed. Bean plants are sensitive to low temperatures and are killed by frost. They grow well on a variety of soil types, require moderate rainfall, and mature in 90–120 days depending on temperature, variety, and rainfall. This means that in warm climates more than one crop a year is common. They belong to the large family of Leguminoseae or leguminous plants which are unique in being able to use nitrogen from the atmosphere to make their protein. This is because nitrogen-fixing bacteria live symbiotically on the roots, converting nitrogen to a form that the plant can utilize for synthesizing protein and other nitrogenous compounds. Bean plants flower over a period of time so that maturation of the seed takes place over an equal period of time. However, during maturation the pods holding the beans will usually change color and lose moisture and the beans will dry in a manner like wheat and corn (see above). Properly dried beans may be stored for long periods of time, so the advantage of beans as a food is obvious. The percentage of protein in beans is quite high compared to that in other vegetables and cereals, and this is of great nutritional importance (see Chapter 5), particularly since beans are relatively low in cost.

There are a great many varieties of beans—e.g., navy, kidney, pinto, black, and red. Consumers generally prefer some varieties over others. Although most beans are consumed mature, generally soaked and cooked in water, immature beans, called green or wax beans, are also consumed as a vegetable. Lima beans, a common food in the United States, belong to a closely related species, *Phaseolus limensis*, and mung beans (*P. aureus*) are popular in many Oriental countries as a vegetable—bean sprouts.

The term "bean" is often used in connection with plant species of closely related genera, e.g., black-eyed peas or cowpeas (*Vigna sinensis*) and soybeans (*Glycine max*).

Soybean

Soybean (*Glycine max*) is a warm-weather crop and requires about the same growing conditions as corn and, depending on variety and climate, 110–150 days from planting of the seed to maturity. Soybeans, unlike common beans (*P. vulgaris*), are not particularly palatable. However, especially in the Orient, soybeans are the raw material from which other foods are prepared by fermentation, extraction, etc., applied even at the home or family level. Whereas 50 years ago they were not grown in any significant amount in the United States, today soybeans are a major crop and the increase in production has been more than fourfold since 1949. This has occurred because soybeans contain both oil and protein in large amounts and both of these major constituents can be processed and refined so that they are excellent for use in human foods. Still, as of now, most soybean protein is used as animal feed in the form of seed meal remaining after extraction of the oil. However, purified soybean protein is being used in increasing amounts in processed human food (see Chapter 12).

Peanut

The peanut (*Arachis hypogaea*) is another leguminous New World plant that has been adapted to many other parts of the world, particularly Africa. Unlike beans, which can thrive on a wide variety of soils, peanuts require what soil experts call a light or friable, usually sandy, soil. Peanuts are very sensitive to cool weather and will not survive any frost. They are a warm-weather crop requiring good rainfall throughout the growing season, which is from 120 to 150 days depending on temperature and variety. Although some peanuts are consumed as food with little processing (only cooking, roasting, and grinding to peanut butter) and some are used directly for animal feeding, most of the peanut crop is processed like soybeans into oil and meal for animal feeds.

In many tropical areas, two crops of peanuts can be produced in a year. The plant itself is quite unusual in that the flowers after pollination are thrust to the ground by pedicles where the seed in its shell develops below ground. This characteristic plus the fact that warm, moist weather is required for growth is the source of a major spoilage problem. Mature nuts must be quickly harvested and dried. If allowed to remain in the moist soil too long or if the moisture content of the harvested nuts is too high, they are an excellent medium for growth of the ubiquitous mold *Apergillus flavus*. The mold produces very toxic and carcinogenic metabolites called aflatoxins. Moldy peanuts are therefore to be avoided as food or feed. Even meal resulting from processed moldy peanuts may be toxic to animals. Aflatoxins were discovered as the cause of the "X" disease in turkeys fed meal from moldy peanuts. This example illustrates not only the importance of proper handling of harvested food crops but also the problem of growing seed crops in moist tropical areas—saving the crop from spoilage after it is produced.

ANIMALS

Animals have been used for food from the time man learned to hunt and fish. Prehistoric man survived as a gatherer of edible plants and as a hunter of animals. Today almost all plants used for food are cultivated; only minute quantities of foods of plant origin are gathered from wilderness areas. This is not true in the case of foods of animal origin, because fish and other aquatic animals which supply food in commercial quantities are harvested from their natural habitats. The Eskimos, whose culture has survived to this day, are hunters and traditionally consume mostly sea mammals—whales, walruses, seals, etc.—and to a much lesser extent terrestrial animals and fish. As consumers of animals only, Eskimos are truly second-, third-, or higher-order consumers in an ecological sense. There are also some cultures which are strictly vegetarian in food habit that would be ecologically classified as strictly first-order consumers. However, most present-day cultures rely on both plants and animals for food. Almost the complete spectrum of biological organisms has been used for human food at one time or another. As present-day civilizations developed, man learned not only to cultivate plants but also to husband animals which could be used for food, for clothing, and for doing work.

Of the many species of animals including birds and fishes that were originally hunted and found to be suitable for food, only a relatively few have been domesticated. These include cattle, sheep, goats, horses, pigs, water buffalos, camels, yaks, llamas, rabbits, dogs, chickens, ducks, geese, and a few others. Although husbanding of fish and some other aquatic animals such as oysters is practiced in a small way, almost all aquatic animals used for food are hunted. The increasing demand for seafood caused by increased human population has brought new technologies of fishing. This is of grave concern because populations of fish and other marine animals, many of which are second- or even higher-order consumers, are vulnerable to overharvesting. Some species of marine animals, e.g., whales, are decreasing almost to the point of extinction. Certainly the waters of the earth are not unlimited in their ability to provide food. We have much to learn concerning how best to use our oceans, lakes, and rivers for the production of food.

Man, himself an animal, looks upon the animals he uses for food in diverse ways. He may be indifferent, he may consider them simply as wealth, or he may show great affection toward them. A sportsman may reap pleasure from hunting or fishing or using animals for exhibition. But in reality man has survived by using animals directly as food (meat and fat) and as clothing (pelts, hides, and feathers) or as producers of food (milk and eggs) and clothing (wool, silk, and similar fibers). In large measure man has depended on a relatively few species of animals which, since prehistoric times, he has been able to domesticate and raise in captivity. Some of these species are closely related and will be described in more detail later.

At this point, let us consider some general characteristics of animals which make them so valuable as producers of food. Of course, animals must be fed in order to grow, develop, and reproduce. In domesticating animals, man assumes responsibility for providing feed and protection from predators. The most useful animals and least costly to maintain are the primary consumers, or animals that can eat plants, the primary producers. Carnivores or secondary consumers would be much too expensive to husband for food. You will recall that the energy requirements of primary consumers in terms of biomass of primary producers (plants) are very large (see Figs. 8.1 and 8.2). It is this type of argument that many concerned persons use in saying that future generations of man must look to plants for food rather than to animals. However, since many of the terrestrial animals used as human food are capable of thriving on plant materials unsuitable for human consumption (see pp. 204–208), they may be considered gatherers that convert otherwise unusable biological material into some of the most nutritious of all foods (see Chapter 5).

There is much evidence that neolithic or stone age man learned to domesticate animals and the ones he selected are the same species which we use today. In their migrations the progenitors of modern man took their animals with them and now the more important species of domestic animals are found throughout the civilized world even though most of these species apparently originated in the eastern hemisphere. Some breeding lines of livestock are more adaptable than others. For example, horses are husbanded in very warm (Africa) and in very cold (Siberia) climates. Animals not only accommodate to differences in climate but also adapt to different plants for food. This is not to say that cattle, for example, will eat any plant; they do have preferences and even avoid certain plants in any locale. However, if plants grow at all, usually some local species can be found that is acceptable as feed. For example, in New Zealand cattle may have green grass almost all year, in our far west they may live on scrubby vegetation characteristic of the area, in the arid regions of northeast Brazil they may live on native cactus, in the Amazon area on native plants of the rainforest, and so on in other parts of the world. In Poland, Russia, and northern Europe pigs may be grown largely on potatoes, whereas in the United States corn is the basis for many pig rations. The animals man learned to use as food as well as man himself have adapted to diverse regions of the world.

Humans have survived by learning to use the plants which can thrive locally for their own food and for livestock feed. Fundamentally this adaptability characteristic of many animals is a reflection of the fact that, given a supply of food, water, and air and some protection from extremes of hot and cold, warm-blooded animals have mechanisms for controlling rather precisely their internal environment so that all cells of the body function properly (see Chapter 4). On the other hand, plants are less able to control their own environments and so are somewhat more fastidious as to where they can thrive. The result is that the natural floras of various areas of the world are characteristically different.

Another important biological factor that has made animals so important for the growth of civilization is that they provide a more continuous food supply. It has already been noted that the seasonal or annual nature of most plants limits their usefulness as food sources because man must have food daily. Thus foods of plant origin must somehow be stored or preserved from one harvest period to the next. This is not necessary with animals, for the live animal is its own storage facility. If the animal is to be used for meat, it is necessary only that it be alive until needed. Of course, freshly slaughtered meat is perishable—generally more so than fresh fruits or vegetables—but size and numbers of animals to be slaughtered can be coordinated with the population to be fed over the period of usefulness of the meat. If an animal is to be used as a producer of milk or eggs in excess of that needed to maintain herds or flocks, the availability of highly nutritious foods is greater. The lactation or milk-producing period may continue many months after the birth of a new generation of animals, and the egg-producing period is also in terms of months in the case of some birds. Of course, when animals are no longer efficient producers of milk and eggs, they can be used as meat.

Even though a great advantage of animals is that the food they provide is available over longer periods of time, there is some seasonality associated with the reproductive performance of some domestic animals. However, centuries of selection and breeding along with certain management practices have, for the most part, reduced this factor of seasonality and, perhaps even more importantly, have led to the development of breeds of animals which are efficient producers of milk and eggs beyond requirements for maintaining stocks. In recent years, the sciences of nutrition and physiology have led to more efficient feed usage and reproductive performance so that meat, milk, and egg production is less periodic and more efficient. Consequently, these foods are available in relatively constant amounts throughout the year in more highly developed countries of the world.

However, advantages of foods of animal origin are not without their price, as anyone who buys food knows. Even though an animal may live on feed unsuitable for humans, animals, like man, must have feed and water available daily. Animals must eat feed of biological origin—not just the minerals, sunlight, water, and carbon dioxide from the atmosphere that plants require—and so the cost of producing animals is higher than for plants.

Animals are generally more expensive to reproduce than plants. Many plants, as noted previously, are vegetatively propagated and, if sexually reproduced, male and female organs are often on the same plant. Also, when plants are grown from seeds, the amount of seed required is only a small percentage of the expected yield. In the case of animals, male and female animals must both be properly fed and maintained and so the cost of a new generation of animals is relatively high. Table 8.3 summarizes the reproduction characteristics of man compared to those of the most important animals used for food production. For example, a cow usually produces only one calf per year. The cost of that calf is not only the cost of feeding and maintaining the cow for a year but also similar costs for the bull for the same period. Of course, a bull may serve more than one cow, which cuts down service costs per calf, but it remains a significant cost factor. Similar reproductive costs pertain to other animals. Current knowledge about reproductive physiology has led to the practical application of artificial insemination in cattle. This development has cut bull service costs, particularly for dairy cattle. An equally or even more important factor is that artificial insemination permits more efficient use of germ plasm to improve milk and meat production (see below). Essentially the technique is that bull semen is collected as the bull serves an artificial vagina rather than a cow in heat. The semen is diluted and preserved by refrigeration or freezing. The semen from a single ejaculate usually contains 4 billion or more live sperm and after dilution can be used to artificially inseminate 100–200 cows. Therefore, a single bull can sire thousands of offspring. Although artificial insemination has changed dairy cattle production notably, as yet this technique has not been made practical for pigs and some other species because of physiological differences in the reproduction of these species.

The production cost of an animal for the next generation can be cut materially if the periods of gestation and of weaning are shorter, permitting greater rates of reproduction. Then, too, if the animal produces a number of offspring in a single gestation, the cost of a new animal is reduced further. These characteristics are found in the pig. In a sense, the chicken fits the same pattern, for a hen may produce 200–300 eggs per year.

If an animal is to produce human food, the cost of that food is going to be determined largely by the cost of the feed consumed by that animal as it produces meat, milk, or eggs. But all the feed consumed is not converted

Table 8.3. Reproduction Characteristics of Animals (Average or Range of Normal Values)

	Cattle	Swine	Sheep	Horses	Chickens	Humans
Age at puberty (sexual maturity),[a] months	6–9	5–7	4–10	12–30	4.5–6	102–204
Number of breeding females served by one breeding male	15–60	20–30	15–50	10–60	8–20	1
Male common name	Bull	Boar	Ram	Stallion	Cock	Man
Millions of sperm per milliliter of semen	800	300	2000	200	4000	150
Semen volume per ejaculate, milliliters	5	250	1	75	1.5	4
Site of semen deposit in female	Anterior vagina	Uterus	Anterior vagina	Uterus	Everted vagina	Anterior vagina
Female common name	Cow	Sow	Ewe	Mare	Hen	Woman
Frequency of ovulation, days	20–21	21	16–17	22	1	28
Duration of estrus or period when female will accept male ("heat"), hours	14–18	48–72	24–48	120–168	Frequently during egg-laying period	No fixed time
Time of ovulation	14 hr after end of estrus	Near end of estrus	Near end of estrus	24–48 hr before end of estrus	24 hr before egg is laid	14 days after onset of menstruation
Gestation period, days	283	114	146	336	20–22[b]	280
Number of offspring per gestation	1	8–14	1–2	1	1	1

[a] Males reach puberty later than females.
[b] In birds, which develop in eggs incubated outside the mother's body, the time for hatching varies according to the temperature of incubation.

directly to these products. The feed must supply the maintenance needs of the animal before any can be used for growth or for milk or egg production. It was pointed out in Chapter 5 that the food energy required for maintenance is largely a function of surface area rather than weight of an animal. Thus on the basis of body weight (meat) more feed is required to maintain a small animal. Feed economy favors the large animal as far as human food production is concerned provided that the animal grows rapidly if used for meat or produces a lot of milk and eggs.

Assuming an adequate, nutritious feed supply and good health, whether animals are poor or good producers is largely genetically controlled. That is, if a cow is a good producer of milk her daughters are also likely to be good producers, particularly if their sire is one that has produced other daughters that are good producers. The same is true for meat and egg production. So farmer-producers of animals are vitally concerned about their breeding stock and the techniques of selective breeding. Because males are used to breed numerous females, animal producers show much greater concern for the breeding quality of males even though the genetic makeup of the offspring is equally contributed by both parents. A single male having so many offspring transfers more of its germ plasm (genetically controlled traits) to the next generation than does a single female.

If two cattle are to be raised or grown to a certain weight for meat production and one on adequate feeding requires 18 months to reach the desired size and the other on marginal feeding requires 54 months, they both may eventually yield the same amount of meat. However, the cost of producing the meat (the amount of feed consumed) will be much higher for the slower-growing animal because it has to be maintained $4\frac{1}{2}$ years instead of $1\frac{1}{2}$ years. This is precisely the situation in many underdeveloped areas of the world in comparison with the more highly developed areas. From this illustration, it can be appreciated that a well-fed animal is a much more efficient producer of human food than a poorly fed one (see pp. 204–208).

Figure 8.6 gives the general growth curve relating body weight to age for any species of animal beginning from fertilization of the ovum by a sperm.[2] The most rapid period of growth coincides with sexual maturation and development of reproductive capabilities. After this point is reached, the growth rate slows rapidly until finally a plateau is reached. The increase in body weight results from new protoplasmic material, and the nutrients and energy for such growth are available from the feed consumed only after maintenance needs are met. Growth itself is programmed physiologically and is under hormonal control. It is important that adequate nutrients be available to meet all growth demands as they occur or some growth potential may be

[2] This same type of curve also depicts the growth of plants or the population growth of a species in an ecosystem.

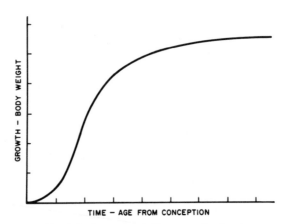

Fig. 8.6. General growth curve of animals.

lost and the animal may be permanently stunted. (This is also true of humans.) If feed in excess of growth demands is consumed, the animal will likely store the excess food energy as fat. But since the energy stored in a unit weight of fatty tissue (80% fat, 4% protein) will be almost seven times that stored in an equal weight of skeletal muscle (3% fat, 20% protein), the feed energy required to produce growth of muscle tissue (meat) is considerably less than that required to produce the fat deposited in or on the meat.

In the case of milk or egg production, the same situation holds. The maintenance feed cost of the cow or hen is essentially the same whether she is a poor or an excellent producer of milk or eggs. Because of this, the overall costs of a quart of milk or of a dozen eggs from poor producers are higher. We can illustrate this principle with milk. If a cow gives little milk, almost two-thirds of the feed consumed may be required just to keep the cow alive. To just keep a high-producing cow of the same size alive, the same quantity of feed will be necessary. However, the high producer will consume more feed and produce more milk so that perhaps only a third of the feed eaten is for maintenance and two-thirds is used to produce milk. Therefore, overall feed costs to produce a unit of milk will be greatly decreased in the case of the high-producing cow. Figure 8.7 shows that the feed cost per unit of milk produced by a cow giving 15,000 lb (6820 kg) of milk per year is less than half that for a cow whose annual production is only 3500 lb (1360 kg).

High-producing varieties of food plants are as essential as high-producing breeds of livestock in providing food to people at low cost. Although in animals it may be somewhat easier to visualize maintenance requirements apart from feed requirements to produce meat, milk, and eggs, there is a counterpart cost in plant production. A low-yielding variety of a plant essentially occupies the same amount of land and uses the same amount of

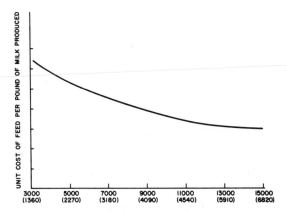

Fig. 8.7. The cost of producing a pound (or kilogram) of milk decreases as the productivity of the cow increases. The abscissa represents the pounds (kilograms) produced in a year. (Adapted from H. H. Cole, *Livestock Production*, 1962, W. H. Freeman Co.)

sunlight and rainfall as a high-producing plant. So, apart from the costs of fertilizer, etc., land costs for plant crops are somewhat analogous to feed costs to maintain animals. Costs of fertilizer, insecticides, etc., are somewhat similar to feed costs strictly for meat, milk, egg, and wool production, and seed costs correspond to costs directly related to getting a new generation of animals.

With the foregoing general considerations concerning animals in mind, let us discuss specifically the most important species of animals on which man depends.

Cattle

As producers of meat and milk, cattle (*Bos taurus*) represent the most important animals for food production. This adaptable species probably descended from *Bos primigenius*, which inhabited northern Africa, Europe, and eastern Asia and was domesticated about 8000–10,000 years ago in the developing civilizations of the Middle East. Breed lines adaptable to widely divergent climatic conditions and areas of the world have been developed as well as breed lines directed toward milk or meat production. Cattle range in weight from about 495 lb (225 kg) to over 2530 lb (1150 kg).

Cattle are ruminants (see pp. 205–208) and can use cellulose, the most abundant carbohydrate and the skeletal substance of plants, as their main source of energy. They can also use proteins of limited or poor nutritive value to man and other simpler nitrogenous compounds such as urea, and they require little in the way of vitamins (except A and perhaps E). Therefore, cattle, when assured a supply of drinking water, can live on a wide variety of plants and plant materials in diverse forms such as actively growing plants, dormant and desiccated plants common in winter and dry seasons, and preserved plant materials. Plant materials for cattle feed are usually preserved by drying (hay) or ensiling (silage). Hay is made by cutting actively growing plants and drying them, usually in the sun. Silage is made by harvesting and chopping plants still in the growing stage; the chopped material is placed in a silo or trench protected from air. The sugars in the material are anaerobically fermented by bacteria to lactic acid, acetic acid, and sometimes other acids, to the point that the pH drops below that at which microorganisms can survive and the fermented plant material is preserved (see pp. 40–52 and 64). Also, cattle, like most other ruminants, have no difficulty utilizing plant materials in addition to cellulose that contain mostly starch and simpler carbohydrates as sugars, and so they can use grain, such as corn, or molasses, a by-product of sugar production. Animals which can use such diverse plant materials for feed quite understandably can adapt to different regions of the world and to different management practices—some directed toward meat production and some toward milk production.

There is much land in the world that is unsuitable or uneconomic for intensive agriculture because of insufficient rainfall or other climatic conditions or because of unfavorable soil characteristics but that still can support vegetation and the grazing of cattle. In these places, cattle roam about eating the vegetation, which in arid regions may be sparse. If the numbers of cattle are controlled properly, the calves that the cows produce each year can be left to graze and grow, or they may be taken after weaning and placed on confined feeding regimes to hasten growth. When they reach market size, these cattle are used as meat. Cattle adapted to such management usually are rather poor milk producers. Range cows which give more milk than their calves can use may have serious difficulties, so it is desirable that these cows give only enough milk to support a calf until it can forage for itself.

Cattle used primarily for milk production are bred for their ability to produce milk in excess of that needed to feed the newborn calf, and the sooner the calf can be weaned the more milk is left for human consumption. Dairy cattle must consume large quantities of feed for efficient milk production, and so dairy cattle are usually found where feed crops are produced in higher yields than on the open range or where feed can be efficiently transported to the herd. If the calves produced by dairy cows are not needed for herd maintenance or development, they can be used for veal, which is meat from young calves, or for beef if grown to maturity. As milking cows pass the age of peak production, they can also be used for meat. Beef and veal as by-products of the dairy industry account for about 40% of the beef consumed in the United States. In many countries, almost all of the beef is of dairy origin. The amount of care and labor required per animal is much greater for growing dairy cattle than for growing beef cattle because lactating cows producing milk for human consumption require daily attention.

On a per cow basis, the amount of feed required for efficient milk production is much greater than that necessary for a beef cow which produces only one calf per year. The rumens and digestive tracts of dairy breeds are larger and may be even more efficient for conversion of cellulosic feedstuffs than in the beef breeds. These anatomical differences of course also affect the overall shape of the animal and the percent of skeletal muscle (meat) in the live animal. In some parts of the world market grades of beef are established on yield and shape of carcasses of so-called beef breeds, and meat from dairy breeds is discounted. However, the breed lines are physiologically very similar, and, if fed and managed in the same manner, dairy and beef breeds give meat that is indistinguishable in eating quality.

Because breeds generally have smaller rumens, they tend to consume less feed per day. To overcome this limitation, beef cattle in the United States are fed large quantities of grain and soy or other proteins which require relatively little in the way of rumen fermentation in order to be utilized, so the animals tend to grow faster and deposit more fat. Such feeding regimes, which are somewhat low in overall efficiency, put cattle in more direct competition with man, who can use these same feedstuffs as food sources. These feeding practices do not utilize cattle's unique ability to produce food for man from sources that are otherwise useless to him.

Traditionally dairying has been a family enterprise because of the daily attention required by dairy cattle. The success of a family farm dairy herd depends on effective and efficient herd management—continuous removal of poor or nonproducing animals, acquisition of good breeding stock, good feeding practices and health maintenance, and so on. The importance of good management is obvious when we consider that some cows have an annual production in excess of 36,000 lb (16,400 kg) of milk, of which 1830 lb (830 kg) is butterfat, and many cows reach 25,100 lb (11,400 kg) of milk annually, and yet the average cow in the United States produces considerably less—9600 lb (4360 kg). Today with modern equipment and hard work a single dairy farmer may have a herd of 40–60 milking cows which must be milked twice daily. Institutionalized or corporate dairy operations may have many more animals.

Heifers, virgin females, are usually not bred until about 16–18 months old rather than at puberty to permit more growth before the first pregnancy. The first calf is born when the cow is about 25–27 months of age. It is very important that the newborn calf consume the first milk or colostrum. Colostrum, which is produced for a few days, differs markedly in composition from mature milk in that its high protein content consists of immunoproteins (antibodies) that are absorbed directly into the bloodstream of the newborn calf to protect it from disease and invading microorganisms. At birth, a calf's rumen is not yet functioning and the animal physiologically is more like a monogastric animal. In these early days, a calf may consume 7–10% of its weight in milk, or in comparable feed if the calf is to be taken from its mother and the milk marketed. In a few months, the rumen begins to function and the calf begins to eat cellulosic materials (roughage); when the rumen reaches maturity, the calf consumes usual cattle rations. Although lactation may continue well beyond a year, cows are generally bred about 2 months after parturition. Lactation is terminated simply by not taking milk from the cow for about 2 months prior to the next calving. Such "dry" periods permit better fetal development, body maintenance, and nutrient storage, which produce a healthier calf and more milk during the ensuing lactation.

Although today in the United States cattle are grown almost entirely for meat and milk, this was not so in earlier times nor is it so today in many other cultures. Since cattle were first domesticated, they have also served as beasts of burden—doing work such as plowing the land, pulling wagons and carts, and furnishing power for grinding grain or pumping water for irrigation.

Sheep

Sheep (*Ovis aries*) and goats are closely related species of ruminant animals. They are indigenous to mountainous regions, and the domesticated sheep of today are thought to have descended from sheep native to the highlands of Central Asia. Mature sheep may exceed 220 lb (100 kg) but are usually smaller. They are used primarily for meat and wool production. Meat from mature sheep (mutton) and goats is somewhat stronger in flavor than that of lambs, and fat lambs have been prized for their meat since ancient times. As with cattle, breed lines of sheep have been developed for hardiness to variations in climate and terrain, and for specific market qualities of wool or meat. Like cattle, they can forage on sparse vegetation and thus use land for range which is unsuitable for intense cultivation, or they may be intensively fed. Being small, sheep are easier prey than cattle and thus require more attention and protection. In New Zealand, where there are no natural predators, sheep and lambs thrive and their production is a major part of the economy.

Sheep tend to be more seasonal in their reproductive habits than other important food animals. Onset of puberty and ovulation in mature ewes tends to coincide with decreasing daylight of late summer and autumn. This results in a lamb crop in early spring, which can then grow to maturity in the summer before the rigors of winter weather set in. This is something of a disadvantage, as is the fact that most ewes produce only one lamb per gestation. Two avenues are currently being explored to increase the productivity of breeding ewes. First, since the gestation period is only 146 days, an ewe should be able to complete two gestations per year if she could be made to ovulate in the spring as well as autumn. Since ovulation is under hormonal control, induction of ovulation twice a year is a possibility. Second, although most ewes produce only one lamb per gestation, twins are rather common and triplets occur occasionally, and a healthy ewe can support these extra offspring. Multiple ovulation and multiple births are inherited traits as well as being under hormonal control, and so sheep geneticists are developing breed lines in which twins or even triplets are common.

Swine

Wild boars are indigenous to the Old World. Present-day domestic swine (*Sus scrofa*), of which there are almost 300 different breed lines, are thought to have developed from breeding of the larger European wild boar with the smaller pigs of Eastern Asia. Neolithic man domesticated pigs, and pork is used throughout the world today except in a few cultures, notably the Jewish and Moslem, which ban pork for religious reasons.

Swine are omnivorous monogastric animals and are unable to digest and use cellulose as ruminants do. Consequently, swine require carbohydrate in the form of simplet sugars and starches for their major source of energy. In many respects, their nutritional requirements are similar to those of man except that pigs do not require ascorbic acid (vitamin C) in their feed because they can synthesize their own. In some respects, this similarity to man in nutritive requirements is a disadvantage because it means that men and pigs compete for the same food. However, swine will eat many animal and plant tissues man has difficulty eating—dry corn, low-quality grain, meat, dairy, and fish by-products, seed meals, mill by-products, garbage, etc. Because of their eating habits, pigs usually require more attention than cattle and sheep raised for meat. Hence pigs are usually raised in confinement in close proximity to feed supplies produced by intensive cultivation—e.g., corn or other grain, soybeans, or potatoes. Given adequate feed and water, swine can adjust to different climates, although they usually must be protected from extremely hot or cold weather.

Pigs for market are usually sold at about 200–220 lb (90–100 kg) body weight. This weight is reached shortly after the age of puberty, when growth rates decrease (see Fig. 8.6). If pigs eat more than is necessary for growth, as is usually the case if fed *ad libitum*, excess fat may be deposited in the body, usually in subcutaneous (below the skin) regions. Breeding sows and boars will continue to grow to over 605 lb (275 kg).

The great advantage of pigs as producers of food is their fecundity and rapid growth. With a gestation period of only 114 days, a sow can produce two crops of pigs per year with little difficulty. Furthermore, if well nourished and healthy, she may produce 10–14 pigs per litter. On good feed, a pig may reach market weight in 5–6 months. So with proper management a good brood sow can produce 20 pigs per year, which at 220 lb (100 kg) market weight is 4400 lb (2000 kg) liveweight per year, whereas a brood cow at best can only produce about 704–802 lb (320–365 kg) liveweight of offspring in the same period of time.

Overfed pigs deposit much fat in their bodies. In highly developed and rather sedentary societies, too much fat in the human diet is not desirable and the emphasis is on lean meat and lean pigs (hogs). However, fat pigs were desirable in the United States in the not too distant past. Before the advent of the vegetable oil refining industry early in this century, edible fats were at a premium on food markets and swine were produced for their fat (lard) as much as for their lean meat. People in some parts of the world still prefer fat pigs. Indeed, swine breeds are usually classified as "lard" type and "meat" type. Meat-type breeds have a higher ratio of lean to fat than lard types. Also, if lean pigs are desired, feed intake is limited so that the feed supplies all nutrients needed for growth without excessive fat deposition. Such pigs gain weight at a slower rate, but the ratio of lean to fat in the finished carcass is higher.

It was noted on p. 210 that males of a species are more efficient converters of feed to meat than females but that for management reasons castration of males to be used for meat is common. Male pigs to be used as meat are castrated for still another reason. In swine many mature males acquire an objectionable so-called sex odor which permeates the meat. It is described as "perspiration" or "body odor." The causative factor is the steroid compound 5-α-androst-16-en-3-one, a normal metabolite of male sex hormones. Whereas 96% of women can smell this compound only 46% of men can do so. The presence of this substance does not affect the wholesomeness of the meat.

Horses

The use of horse (*Equus caballus*) meat as food is perhaps incidental to other values of this intelligent and often beloved animal. Horses have been traditionally used as beasts of burden more than as food. They have served as draft animals for pulling plows, wagons, stagecoaches, and carts and also have been used to carry people and cargo on their backs. Between 1920 and 1960, when the tractor and truck were replacing the horse in U.S. agriculture, the population of horses dropped from 26 million to 3 million, which indicates their former importance in farm work. Warriors and sportsmen for many centuries have used horses in war and in racing because of their strength and speed. There are many breeds of horses which excel specifically for draft, riding, racing, show, etc. Also, some breed lines have been adapted to very cold or very warm climates. After their usefulness for such duties is ended, horses can be used as meat. In the United States, horse meat is used for food by very few people, but in cultures which depend on the horse as a beast of burden, horse meat is used quite extensively and mare's milk may also be used.

Horses range in size from a few hundred pounds to over a ton. They are herbivorous but require somewhat higher-quality rations (usually more grain) than ruminants because they are less efficient in using cellulosic material. Rather than a rumen preceding the true stomach, the horse has a large cecum in the lower digestive tract where intestinal bacteria facilitate digestion of plant material. The precise nutritional requirements of horses are less well known than those of other farm animals, perhaps because the drop in the economic value of horses resulting from the mechanization of agriculture occurred just as the science of nutrition was developing.

Horses have a longer period of gestation and grow slower than other animals used for food production. Consequently, use of horses solely for meat production is less efficient. With a gestation period of almost a year (336 days), it is difficult to get a foal per year per mare, particularly since horses are quite seasonal in breeding habits. Ovulation appears to be induced by the increasing day length of early spring.

Chickens

Chickens (*Gallus domesticus*) are the most important birds used to produce human food. Their progenitors were wild red fowl indigenous to the jungles of Southeast Asia. Other domesticated birds used for food include turkeys (*Meleagris gallopavo*), native to the western world and originally domesticated by the American Indians of Central America long before the Europeans came to America, ducks (*Anas platyrhynchos, A. boschas*), geese (*Anser anser*), guinea fowl (*Numida meleagris*), and some other species of minor significance. None of these other birds matches the chicken for efficiency in producing meat and eggs, although turkeys are excellent for meat production.

Research on the nutrition, physiology, and genetics of chickens has been unusually rewarding. A generation ago, chickens were rather expensive and the eating of chickens was reserved for festive occasions; now chickens are our cheapest meat. During the same period, the real cost of eggs has been almost halved. In terms of converting feed to high-quality meat, chickens are among the most efficient of all domestic animals. On the basis of converting the protein in feed to human food protein—egg protein, which has the highest nutritive quality of any common food—chickens are indeed the most efficient. However, when it is considered that ruminants can use nonprotein nitrogen, the dairy cow is most efficient. On the basis of conversion of total animal feed energy to human food energy, only the pig and a high producing milk cow are better than the laying hen. Chickens are monogastric animals and their nutritive requirements put them in competition with man. However, like pigs, chickens will use feedstuffs humans have great difficulty eating. Many of the ingredients of poultry rations are similar to those used in feeding swine.

Chickens, whose normal life span is relatively short—3–4 years—adapt quite readily to management practices designed to maximize meat and egg production, so that now rearing chickens in great numbers in confinement is common practice. Some breeds of chickens are better for meat production and others for egg production.

Chickens grow rapidly. Broiling or frying chickens reach market size of 3–4.5 lb (1.4–2.1 kg) in 8–12 weeks. This age coincides with the end of the maximum growth rate (Fig. 8.6). If larger birds (roasters) are desired, the chickens are fed longer, but the efficiency of feed conversion decreases and so larger chickens tend to be somewhat more expensive to produce on a weight basis. Hens and cocks (roosters) from breeding stock and hens which have outlived their periods of efficient egg production are also a source of meat.

Hens on adequate rations begin laying eggs at about 8 months of age. Cocks are not necessary for egg production, and market eggs are generally nonfertile. If young hens and cocks are left to their own devices, egg production is controlled by a number of factors. First, it is affected by hours of daylight. Next, hens normally lay eggs in clutches, i.e., an egg per day for one or more days in succession, then a rest period, and then another clutch. When the number of eggs in the nest reaches a certain number, the hen becomes "broody" and sits on the eggs to incubate them to hatch a new generation. Hens lay no eggs during brooding periods. Such a pattern of egg laying and broodiness continues until the hen molts at age 18–20 months—replaces old feathers with new—during which time no eggs are laid.

When hens are raised in confinement in large numbers, egg production is increased by controlling hours of light and darkness artificially. By removal of the eggs as they are laid, periods of broodiness are decreased in frequency and length so that egg production continues in the normal clutch rhythm of the particular hen. Since the artificial incubator became practical some years ago, hens have been relieved of the chore of incubating their own eggs.

In order to pay for her keep at the present time in the American market scheme, a hen must produce more than 200 eggs per year after the first egg is laid. This means that hens to be economic producers must have a clutch rhythm of at least two eggs before rest and have few broody periods. Generally, the bigger the clutch, the more efficient the hen for egg production. Some hens can actually produce an egg a day continuously for a year before molting. However, the hens of good commercial flocks are expected to produce 240–260 eggs per year on the average. Another characteristic of efficient hens is a short molt period. Some breeds under certain management conditions will require 6 months to molt, but the most efficient egg producers require only $1\frac{1}{2}$–2 months. In many commercial operations, hens are sold for meat when they reach the first molting period at age 18–20 months, but in some flocks hens are kept another year. After that, at age 30–32 months, egg productivity drops and becomes uneconomic. If a good hen produces 500 eggs in her lifetime of 32 months or 960 days, it is clear that two good hens are necessary to provide the one egg per day which is the average per capita consumption of eggs in the United States (see Table 8.1).

The chicken has extraordinary biological potential and adaptability for producing human food. The chicken as fed, grown, and bred today is perhaps the best example of how man, through the application of modern science, has been able to harness biological potential for his own benefit. Similar understanding of other species of both plants and animals which can be used as food offers the greatest hope for increasing food supplies for the hungry people of the world.

FOOD SPOILAGE AND THE REQUIREMENTS FOR PRESERVATION AND DISTRIBUTION

Food processing in the forms of peeling, milling, mixing, slaughter of animals, cooking, baking, etc., and food preservation in the forms of drying, canning, cooking, freezing, etc., are so much a part of American culture and that of other economically developed areas of the world that few people ask why foods are processed and preserved in a particular manner even though without these devices our society would be doomed. Primitive cultures developed systems of food processing and preservation that, although they may not have been as sophisticated as our own, were equally important. Many of these very old techniques are still practiced. In previous chapters we have alluded to certain aspects of the need for food processing and preservation, and now we shall look into these matters more systematically.

Important factors to keep in mind are (1) that man is a daily consumer of biological systems, (2) that living things are dynamic, (3) that the maturation of food organisms is periodic and rarely coincides with man's daily food needs, (4) that man usually consumes only certain tissues or parts of tissues of any living organism since he needs only certain chemical compounds or nutrients, and (5) that tissues used for food deteriorate or spoil in time and hence their period of usefulness is limited. Table 9.1 shows, for a number of common foods, how many people can be fed from a single organism over its period of usefulness provided that there is no wastage and processing or preservation except as noted.

Table 9.1. Numbers of People That Can Be Fed by a Single Food Organism
if No Food Is Wasted

	Daily consumption per capita	With refrigeration in U.S.A.	Without refrigeration	
			Cool climates	Warm climates
Period of usefulness		14 *days*	4 *days*	1 *day*
Beef, 1000 lb (450 kg)/steer[a]	0.50 lb, 0.23 kg	85	300	1200
Pork, 200 lb (90 kg)/pig[a]	0.50 lb, 0.23 kg	14	50	200
Chicken, 3.3 lb (1.5 kg)/bird[a]	0.50 lb, 0.23 kg	0.28	1	4
Milk, 26.4 lb (12 kg)/day	1 qt, 1.0 kg, 1 liter	0.86	3	12
Period of usefulness		3 *days*	1 *day*	0.5 *day*
Milk, not pasteurized		4	12	24
Period of usefulness		150 *days*	40 *days*	15 *days*
Apples, 1000 lb (450 kg)/tree	0.66 lb, 0.30 kg	10	37	100
Period of usefulness		20 *days*	5 *days*	2 *days*
Peaches, 600 lb (270 kg)/tree	0.66 lb, 0.30 kg	46	184	460

[a] Animal weights are before slaughter and not weights of usable meat.

These data representing only a few foods are sufficient to illustrate a number of the fundamental problems of providing nutritious food to people on a continuous basis:

1. Because we must have food daily and because our food comes from many different kinds of organisms which reach their period of usefulness at different times, it is almost impossible to provide necessary food throughout the year to large numbers of people without some form of processing and preservation. For example, peach trees, which yield fruit only once a year, are not normally grown in the same location as cattle. Both fresh peaches and fresh beef are perishable. In warm climates in order to use without spoilage or waste a 1000-pound (450-kilogram) steer in 1 day would require that 1200 people be present to eat the animal. For peaches, 460 people would be required to use the fruit from a single tree. Such numbers of people are found in cities, but not on the land where cattle or peaches are grown. Therefore, what is needed is a system for transporting food from the land to the people.

2. Food from freshly harvested organisms is perishable, and so there must be some system for preserving it while it is being transported to consumers if available supplies are to be used efficiently.

3. Foods differ in the period of time that they may be stored without loss. Seed foods such as rice, wheat, corn, and beans (not listed in Table 9.1), if sufficiently dry and protected from invading organisms such as insects, rodents, and fungi, may be used for long periods of time—months or years—depending on storage conditions (see pp. 217, 218). Consequently, they will make up a greater proportion of the daily diet of persons unable to get more perishable or otherwise preserved foods daily.

From the foregoing considerations, a general concept relating to the need for food preservation follows: Since the organisms man uses for food differ in length of life cycle and in period of usefulness, it would be necessary to have almost daily harvest of diverse types of food organisms if the food were not preserved in some manner. This is an impossible logistics problem for any sizable group of people. Reflect for a moment on the different organisms furnishing your food for a day and you will realize that it would be impossible to have tomatoes, lettuce, chickens, cattle, pepper, peas, corn, and so on growing in your own yard, all ready to eat at the same time. The only alternative then is to find ways and means to process, preserve, and store food as it is produced and then transport and distribute it as needed. Cities—indeed, our whole civilization—simply could not exist without such a pattern of food production and supply. The more compact and sophisticated our urban populations become, the more specialized, sophisticated, and efficient food production, preservation, and distribution must be. So interdependent are the food-producing and food-consuming sectors of our culture that only a few days' interruption in our food supply system would cause havoc in our cities. Even to stop a single phase of our current food supply scheme such as mechanical refrigeration, canning, or delivery by trucks would cause great hardship.

To broaden the base of our understanding of the technical aspects of the fundamental problem of feeding people, we need to begin by focusing our attention on the life cycle of any food (biological) organism. Let the line OHC in Fig. 9.1 represent the time required for a food organism to pass through its life cycle. Point O, *genesis*, represents the beginning of the life cycle. As time goes on, the organism will *grow* and reach *maturation*. The organism will next reach the point of *harvest* and then ultimately that of *consumption*. Of course, it is necessary to allow some organisms to reproduce and start another generation.

For highly perishable foods, the time from harvest to consumption—HC—is very short, and for more stable foods such as rice, wheat, or other dry seeds HC is longer. The longer the time HC, the farther the food can be transported from its site of production to the consumer. This relation, so

GENESIS ⟶ GROWTH ⟶ MATURATION ⟶ HARVEST ⟶ CONSUMPTION

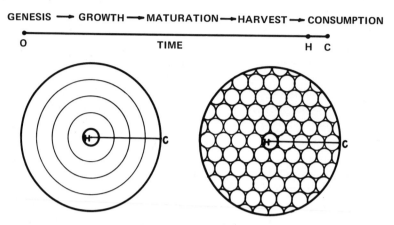

Fig. 9.1. Schematic representation of food production and distribution patterns as affected by the time between harvest and consumption of food organisms.

important in the efficient feeding of people whatever their economic level of life, is also graphically represented in Fig. 9.1. Let the center of each circle represent the location where food is produced and the radius represent the time from harvest to consumption, HC. On the left, all people living within the circle whose radius is HC may be fed from the single center source of supply. Now if by processing, preserving, etc., HC can be increased 3, 5, 7, or 9 times its original value, all the people living within the larger respective circles can be fed from the single point source. If 1000 people live in the smallest circle and if the total area of all circles is evenly populated and all areas have equal transportation and marketing facilities, then as HC increases the single site of food production in the center can feed 9000, 25,000, 49,000, or 81,000 people, respectively. To feed the 81,000 people without food preservation we would need at least 85 production sites without overlap. Still we would find some people not being fed and in order to get food they would have to go to a prescribed area within the small circles. Thus it is quite apparent that extending the usable life of any food will permit its more efficient use.

A corollary follows directly. Given a fixed time between harvest and consumption, more people can be fed from a single food source if the transportation from the source to the consumer becomes more rapid. Figure 9.1 then could be used equally well to represent area or population of consumers in relation to increased distribution velocity. So by increasing time between harvest and consumption as before and doubling the velocity of distribution, the population that could be fed is 26,000, 100,000, 196,000 or 324,000. From such projections, it is easy to understand how, for example, a small production area in California, Arizona, or other location can supply lettuce to the entire United States.

In short, anything that increases the period of usefulness of a food or permits more rapid delivery represents an improvement in efficiency provided, of course, that the process does not destroy or remove the nutritive value of the food. In order to extend *HC*, it is necessary to know something about food deterioration which renders a food unpalatable, unwholesome, or otherwise unfit for human consumption.

A single food as prepared and eaten generally comes from a number of organisms. Consider for a moment the number of different food organisms which are used in the preparation of vegetable soup, fruit cocktail, breaded veal chops, ice cream, pie, or any other food on the dinner table. Bread containing only flour, sugar, yeast, shortening (fat), milk, salt, and water represents at least five food organisms—wheat; sugar cane; yeast; cotton, soybean, or pig depending on the source of fat; and cow. The recipes in a cookbook describing the different ingredients required to prepare common dishes point up the same fact—each ingredient usually represents a different food organism, all of which probably had to be previously preserved and processed into a form suitable for final food preparation. The list of ingredients on the label of a prepared food such as a candy bar, cookies, canned beef stew, or mushroom soup also represents a number of different food organisms. An average nutritious but inexpensive menu for a day for a college student will likely come from over 30 different organisms, each of which required some processing and preservation before being used in final preparation of the food.

There is a detailed and specific story concerning the processing and preservation necessary for each food. Such details are beyond our purpose here, although they may well be of primary practical importance to the farmer, processor, marketer, restaurateur, cook, homemaker, or the person eating the food as well as to the food scientist, technologist, or economist. All such stories have two fundamental similarities. First, foods for human consumption must be protected from attack by many diverse types of organisms—man is always in competition with other animals, plants, and microorganisms. Second, even if protected from biological attack, foods are still subject to chemical deterioration—changes due to oxidation by oxygen of the air and/or reaction of one food constituent with others. Many times these changes are promoted by enzymes, the biological catalysts which make life processes possible and which may remain in the food after harvest. Both these factors can be illustrated with a single common food—bread. Bread, if not eaten, will eventually be consumed by common molds. Although the enzymes of the wheat and yeasts will have been destroyed by baking, even before bread becomes moldy and unfit to eat it will become stale—deteriorate in flavor and become harder even though not allowed to dry out. Stale bread may not have lost its nutritive value, but it is less palatable and may not be eaten, and, of course, to be used food must be eaten. Although both types of deterioration are chemical in the sense that all biological organisms live by chemical processes, it is useful for our purposes to continue the distinction even though at times both types of deterioration may go on simultaneously.

PRESERVATION OF FOOD FROM ATTACK BY BIOLOGICAL ORGANISMS

Some basic characteristics of the biological cosmos of which we are a part are that (1) all organisms are in competition with each other, (2) all organisms begin life, grow, develop, reproduce, die and then are consumed by other organisms, and (3) these processes repeat until the protoplasmic materials made by the autotrophs (green plants or primary producers) from water, carbon dioxide, nitrogen, and minerals using sunlight for energy are eventually converted back to these substances by the heterotrophs (including man), which oxidize consumed protoplasmic substances to supply energy for their vital processes (see Chapter 2). Consequently, in order to feed people it is necessary that the protoplasmic material of food organisms be preserved and channeled to human consumption rather than being used by other organisms competitive with man. We have already discussed the protection of food organisms as they grow and develop (Chapter 8) through the use of fertilizers, insecticides, fungicides, antibiotics, etc. Now let us see how we can protect our food from our ever-present biological competitors.

Protection from Insects and Other Animals

Physical protection of food organisms after they are harvested is much easier if the competitive organisms are relatively large ones—such as rats, mice, monkeys, birds, or insects. It is merely necessary to store the harvested organism in such a way that the attacking animal cannot get to the food. Even so, the loss of food to animals large and small is formidable on a world-wide basis. Losses due to rats and mice particularly may be quite high—10%, 20%, or even much more of all food produced in some parts of the world. Cereal grains and other seeds which are stored for long periods are vulnerable to attack by animals, especially at the local farm level. Not only may grains be consumed by rodents and insects but also the remaining grain will be unfit for use because it is befouled by excrement and other detritis. Grains and seeds may when harvested contain insects or insect eggs which hatch in storage to cause serious damage. To combat this type of deterioration, fumigation can be helpful; i.e., the stored dry grain, seeds, and even spices may be exposed for short periods to gases such as methyl bromide, ethylene oxide, or hydrogen cyanide.

Protection from Microorganisms

The protection of freshly harvested food organisms against attack from microorganisms—which are seemingly everywhere as well as invisible to the naked eye—is much more difficult, particularly for food containing sufficient water at harvest to support their growth, usually more than 10–14%. This is true both for freshly harvested foods and for foods during further processing, distribution, and preparation. So important is this problem that food microbiology is a distinct scientific discipline. Of course, some species of yeasts, molds, and bacteria are useful in food processing and preservation, whereas other species can render food extremely toxic or even lethal (see below). Even though microorganisms present difficulties as far as food preservation and use are concerned, we could not live without them, because they are the ultimate decomposers of sewage, biological residues of all kinds, and even many manmade substances. In this essential way, microorganisms continually restore our biosphere for new generations of all organisms.

Living organisms generally have protective mechanisms against most invading microorganisms. They are both chemical and mechanical in nature. The first line of defense is the skin, which is generally able to prevent microorganisms from entering. Then, if microorganisms do enter, chemical schemes come into play that are often more or less unique to the organism. Man and higher animals have phagocytic cells which digest invading microorganisms or other foreign particles and other cells which produce specific proteins or antibodies that circulate in the blood or other body fluids and can kill invading microorganisms. Plants and microorganisms produce antibiotics or other substances for protection. Also, the hen protects the egg from bacteria in this way (see pp. 32, 65, and 105).

Some plants produce specific chemical compounds to protect themselves not only from microorganisms but also from insects and other plants. However, when a food organism is harvested the biochemical processes of life are altered. When the organism is killed, its source of energy and nutrients is lost and degenerative processes begin. In time, the harvested food becomes more vulnerable to the ubiquitous microorganisms with the result that it becomes unusable and is often described as rotten. Although such food may have a foul odor and be unappetizing or even disgusting in appearance, it may not necessarily be toxic. However, it should not be eaten, because it could be toxic.

The period of time that a freshly harvested food is usable before it becomes unfit or rotten varies with its nature and how it is handled after harvest. The important factors to remember are (1) that upon harvest some dynamic biochemical processes continue, often in a somewhat altered manner, and (2) that most foods are excellent media for the growth of other organisms which can render the foods unfit for human use. Let us illustrate by a few common examples. A ripe tomato, apple, or banana after having been picked may become sweeter and/or softer because of the changing of starch to sugar or of insoluble pectins to smaller water-soluble molecules (see Chapter 6). Eventually the fruit will be almost liquefied by invading bacteria, yeasts, and fungi. A similar course of events occurs with meat, but meat is such a good medium for bacteria that bacterial degradation begins quite early. Other examples could be cited, but it is clear that if a freshly harvested food organism is to be efficiently used both normal biochemical (enzyme-catalyzed) degeneration and microbial degradation must be slowed or stopped. The distinction between these two types of deterioration is made because in processing and preserving food it is possible to arrest one and not the other and *vice versa*. The most familiar method for retarding both types of deterioration is refrigeration. This is merely a reflection of the fact that both are essentially chemical processes and their reaction rates are markedly reduced by lowering of temperature (see pp. 34–40 and 258–261).

Generally speaking, only certain portions of food organisms are consumed, e.g., the pulp of the apple, the muscle of the animal, the endosperm of the wheat, the juice of the orange. So in preparing the edible part for human use, the organism is peeled, skinned and eviscerated, milled, or squeezed, etc. The consequence is that the part of the food organism to be eaten becomes even more accessible to microorganisms. Furthermore, damage to the cells by such manipulations promotes more rapid biochemical deterioration, e.g., the browning of a freshly peeled apple, peach, or potato. The simple procedure of recovering the edible portion of food organisms makes food preservation of some sort even more urgent. This is often the critical period, because if the food is not eaten quickly it is lost. Consequently every practical means is sought to preserve food to this point for commercial marketing. For example, the peas may be canned or frozen, the sugar crystallized, and the meat chilled, frozen, or canned. It is such products that one finds in the supermarket. By such processing and preservation procedures the time from harvest to consumption—HC in Fig. 9.1—can be extended so that food can be more efficiently utilized. Also, food in such forms lends itself to further cooking and preparation for eating with a minimum of waste. Processing and preservation may alter food to some extent, and some foods are more desirable with a minimum of processing—e.g., lettuce and some meats. Cooked and prepared foods as served are also quite vulnerable to attack by ever-present microorganisms, and if not protected in some way they, too, will spoil rapidly—each different food in its more or less characteristic way.

Need for Sanitation

Protection of any food from spoilage due to microorganisms involves two major steps. The first is to prevent insofar as practical contamination of the food with large numbers of organisms. The second step is to prevent or slow down the growth of the microorganisms which get into the food by one means or another. In other words, by having fewer viable microorganisms present in the first place, then whatever is done to control microbial growth is much more effective. Good sanitary practices are essential to the preparation, processing, preservation, and efficient use of wholesome food.

The soil that supports the growth of plants is full of many different kinds of microorganisms. Also, animals walk, defecate, and urinate on the soil as they consume the plants growing on it. They harbor many microorganisms on their hides and in their intestinal tracts. Both plants and animals are usually covered with microorganisms when they are harvested for food. Therefore, washing and cleaning the food organism to remove as many potential spoilage organisms as possible are very important before any additional processing. Also, every step in any subsequent handling of the food should be done so as not to unduly contaminate it. This means that any utensils or equipment which come in contact with the food must also be clean. In industrial operations, this requires frequent or continuous cleaning of equipment so that it does not collect food particles on which microorganisms can multiply and get into the food as it is being processed. Of course, water used for cleaning the food and the equipment or as an ingredient in processed food must be as free of microorganisms as possible. Usually, purified and chlorinated water (20–200 ppm) is used for these purposes.

Milk offers an example of the practical value of cleanliness to avoid contamination. Not too many years ago, most cows were milked by hand—rhythmic manual manipulation of the cow's teats to expel a periodic small stream of milk into an open milk pail—while old Bossie was swishing her tail and stomping her feet to chase flies and thereby stirring up microorganism-laden dust. It was inevitable that many bacteria got into the milk. Milk itself is an unusually good medium for rapid growth of bacteria, and there is no natural built-in antibacterial agent as in the egg. Since normally the calf suckles the teat and takes the milk directly without exposure to many bacteria, there is no natural need for such in milk. Milk obtained by hand is often so badly contaminated that it will remain fresh only for a very short time, and daily home delivery of such milk, even chilled and pasteurized, was necessary.

In many places in the world today this is still so. By mimicking the cow-calf relationship milk will remain fresh in the household refrigerator for two weeks or more. How is this done? The cow's teats are washed clean and sanitized and then fitted with a clean and sanitized milking machine. The milk is collected without exposure to the open air and cooled immediately, after which it is pasteurized and bottled—all in a clean, closed system. The result is that the usefulness of fresh milk is increased almost tenfold and the milk costs less because daily deliveries direct to the home of the consumer are unnecessary. All of this is accomplished largely by preventing incidental bacterial contamination during the milking (harvest).

Food being processed may also be contaminated by handlers. It is important that people who work with food practice good personal hygiene and be free from disease in order not to add to the numbers of bacteria already present.

Foodborne Infections

Food and drinking water can be contaminated in such a manner that they do not appear to be spoiled but nevertheless carry microorganisms which can cause human diseases, including some that cause many deaths throughout the world. One such human disease which is now rare in most developed countries is typhoid fever. Others are tuberculosis, undulant fever (brucellosis), paratyphoid fever (salmonellosis), cholera, dysentery, and tularemia (rabbit fever). In some cases, the microorganism is in the animal which is used for food. Of course, good inspection of the animals can often determine whether an animal is diseased—but not always. In some diseases—particularly cholera, dysentery, paratyphoid, and typhoid—the causative organisms may get into the water or food supply by contamination with excrement from people who have the disease. Rodents as well as people can be the vectors for such contamination. Good sanitation can, of course, greatly diminish the danger from foodborne microorganisms pathogenic to man. Lapses in sanitation and proper preparation of food do occur and cause an occasional outbreak of food- or waterborne disease.

Animals including man are also susceptible to parasitic infections. Some of these parasites, though larger than bacteria, are quite small, such as that causing amoebic dysentery; others such as tapeworms and roundworms are much larger. Even for the larger parasites, their infective stages may be quite small, because many of these organisms have complicated life cycles and may enter the host in more than one way. Fortunately, the eggs or other immature stages of parasitic organisms that are found in food and water are rather easily killed by cooking. The eggs or even adults of parasites are found in the feces of infected animals including man, his pets, and the animals he uses for food. Consequently, when animal and human manures from infected individuals are used as fertilizers in producing food crops, the crops themselves often carry the parasite or its eggs. In order to avoid parasitic infection, the consumer must practice good sanitation and not eat uncooked foods that are likely to contain potential human parasites. A poorly nourished human is far less capable of physiologically protecting himself from parasites than one who is well nourished and healthy. Parasitic infections are much more common in warm climates than in cold climates, particularly where good personal hygiene and sanitation are not practiced.

One parasitic disease that gets a lot of attention in the United States is trichinosis, even though it is of relatively minor importance. It is caused by the roundworm *Trichinella spiralis*. When meat from pigs (or from bears) containing encapsulated larvae of the worm is eaten without proper cooking, it is possible that the consumer may become infected. Few pigs in the United States carry live parasites; they are usually found only in swine fed uncooked garbage. Rigorous inspection of animals at slaughter can detect the parasite. The inspection procedures practiced in some European countries where lightly cooked or even raw pork is customarily eaten assure that infected meat does not reach the market. In the United States, meat is usually not inspected for trichina worms or cysts, and for this reason consumers are urged to cook pork to the well-done stage even though any live larvae in meat are killed at 137°F (58.5°C).

Foodborne Microbial Intoxication

Sometimes foods which give the appearance of being quite safe contain bacteria that, although not the cause of an infectious disease, can produce a toxin to which humans are sensitive. The most important of these is *Clostridium botulinum*, which produces a toxin that is absorbed into the bloodstream. This toxin is among the most poisonous substances known and affects the nervous system often with fatal results. Certain species of *Staphylococcus* produce toxins (enterotoxins) that cause disturbances of the gastrointestinal tract characterized by nausea, vomiting, and diarrhea. Some species of *Salmonella* that may contaminate food can survive in the human gastrointestinal tract. As they complete their life cycle and are digested, toxins are released which can also cause intestinal disturbances somewhat similar to those produced by staphylococci. A victim of food poisoning due to salmonellae or staphylococci may feel that he is going to die, but fortunately recovery almost always occurs.

The conditions that will destroy *Clostridium botulinum* are commonly used in the United States as a guide for determining correct preservation procedures, particularly in the food-canning industry. The idea is that if all *C. botulinum* organisms (spores) are killed, most other dangerous microorganisms cannot survive. *C. botulinum* is a common soil bacterium. It is anaerobic—i.e., it grows where oxygen is absent—and is a spore former. Many microorganisms occur in a vegetative or actively growing stage and then, under certain circumstances, form spores that are somewhat analogous to the seeds of higher plants. The spores may be dormant for long periods of time and may often be carried to other areas by wind. When suitable conditions for growth are available, spores germinate and the microorganisms resume their vegetative form. Often spores are much more resistant to extremes in environment than the organism in its vegetative stage. For instance, the spores of *C. botulinum* are much harder to kill than its actively growing cells. Most of the deaths due to botulism in the United States in this century have resulted from improperly heated home-canned foods. In essence, what happens is that in canning the green beans, peas, or other low-acid foods all vegetative bacteria are killed but a few spores of *C. botulinum* are not. In time, they may start to grow, although the canned food may not appear particularly spoiled, and the food is eaten without reheating. (The toxin of this organism is easily destroyed by cooking whereas the spores are not.) There are well-authenticated cases where a housewife tasted only a little of the contents of a newly opened can or jar of food, decided it was not spoiled, and heated it and served it to her family. She died and the rest of the family, who had not tasted the food before it was reheated, survived. The symptoms of botulism are similar to the horrible symptoms of another generally fatal and much feared disease, tetanus or lockjaw, caused by the toxin produced by a closely related soil organism, *Clostridium tetani*, which can sometimes contaminate a wound.

Many species of *Salmonella* and *Staphylococcus* are common and widespread. The enterotoxins produced by these organisms, unlike the toxins of *C. botulinum* and some other bacteria, are fairly stable to heat. This means that even though mild heating below the boiling point of water will usually kill the living bacteria, the enterotoxins remain. If the organisms grow in a particular food, even though spoilage may not yet be apparent the food, heated or not, may contain sufficient enterotoxin to cause considerable distress.

The most common foods subject to this type of deterioration are cream soups, sauces, custards, filled pies, and pastries, prepared meat salads, and the like. These foods are equally nutritious for humans and most bacteria, and consequently they are excellent media for rapid growth and reproduction of these microorganisms. Such foods may be contaminated with the organisms during preparation, and trouble ensues if the foods are not properly refrigerated. It is this type of food poisoning that is generally the cause of large outbreaks involving the lapse of good sanitation or food handling in hotels, restaurants, other institutions, and catering services that feed large numbers of people. In some areas, an attempt to control this type of food poisoning has been made by the passage of laws against preparation and distribution for retail sale of certain custard and cream-filled foods unless they are kept under continuous refrigeration.

Animals used for food can harbor *Salmonella* organisms without obvious gross infection usually detectable by proper inspection. To combat this problem, animal feeds, particularly those prepared from animal by-products, are tested for excessive contamination by *Salmonella*. Many species of *Staphylococcus* normally reside on the skin and in the oral and nasal passages of man and other animals. Usually there is no evidence of infection, although these organisms are often the cause of pimples and related problems. So it is clear that both man and animal can be the source of microorganisms that cause food poisoning.

In addition to the bacteria just mentioned, a number of other foodborne organisms are of considerable importance. *Clostridium perfringens* (sometimes called *C. welchii*) is a rather common organism which, when it becomes established in wounds, can cause dreaded gas gangrene. This organism also is the cause of a significant number of cases of food poisoning, as are *Bacillus cereus* and some organisms of the genus *Shigella*. Surprisingly, certain strains of *Escherichia coli*, a common intestinal bacterium, are pathogenic and can cause severe or even fatal diarrhea in infants.

Other microorganisms such as molds can be the source of toxicants in foods. Usually excessive growth of such organisms on foods makes them quite unpalatable, although in a few foods such as Roquefort or blue cheese certain molds are desirable and apparently do not produce harmful toxins. However, fungal toxins can cause problems in animal feeds. We have already mentioned that the toxins (aflatoxins) produced by the common mold *Apergillus flavis* can be a problem in peanuts and perhaps in similar foods (see pp. 218 and 227).

PRINCIPLES OF FOOD PRESERVATION AND PROCESSING TO CONTROL MICROORGANISMS

Although harvested food may be protected from competing animals including insects by rather simple mechanical means (see above), foods cannot be so easily protected from the ubiquitous microorganisms—bacteria, molds, and yeasts. To be sure, the numbers of such organisms getting into our foods can be minimized by good hygiene and sanitation but this is not enough for preservation of foods for any appreciable length of time—particularly in the case of foods harvested only once a year. How then can foods be processed and preserved against the growth of microorganisms?

In Chapter 4, it was pointed out that all cells live in water and that the conditions for life are definable in terms of the fundamental parameters of water solutions—temperature, pH or hydrogen ion concentration, water activity or number of dissolved molecules and ions, and kinds of molecules, including nutrients, in the water. These are the factors which can be manipulated to prevent microorganisms from consuming food desired for our own use.

Water Removal

Simply by the removal of water food is protected from microorganisms. In a real sense, food preservation by drying is preservation by lowering water activity (see below), because it is virtually impossible to remove all water from foods. Although this procedure is not useful for a number of foods, it is often not too costly when appropriate. Good examples of foods preserved by natural drying are, of course, the cereal grains, other seeds, and nuts. Many spices are also preserved by natural dehydration. The same principle applies to flours and meals prepared by processing seeds. Sugar refined from processed cane or beet sap and fats and oils refined from plant or animal sources are other examples. Such processed and ready-to-eat foods as crackers, prepared cereals, some cookies, pretzels, potato chips, hard candies, dried milk, chocolate, coffee, and peanut butter are not susceptible to microbial attack because they lack water. More recently, fruits, vegetables, and meats have been freeze-dried; i.e., the food is frozen and then dried by sublimation (distillation) of the water in a high vacuum. This is a rather expensive process. Freezing, wherein all the water is turned into ice, may in a limited sense be considered the preservation of food by the removal of (liquid) water. Dried foods are still subject to chemical deterioration (see below); however, such degradation may be slow or it may be controlled by other processing and packaging procedures.

Temperature

The manipulation of temperature to enhance or inhibit growth or to kill microorganisms is so commonplace that when we practice this food-processing technique we rarely realize what we are doing. Throughout this book, we have repeatedly noted temperature effects in discussing various subjects. For growth, the temperature of the aqueous medium in which the organisms live must be in the biokinetic zone, 0–60°C (32–140°F) (see p. 35).

The use of refrigeration in the home, institution, food-processing plant, transportation facility, or food storage warehouse serves a threefold purpose. Perhaps the most important is to inhibit or decrease the rate of growth of microorganisms present in or on food. Second in importance is to cut down

Table 9.2. Effect of Cooling on
Food Spoilage

Temperature °C (°F)	Days before spoilage	
	$Q_{10} = 2$	$Q_{10} = 3$
40 (104)	1	1
30 (86)	2	3
20 (68)	4	9
10 (50)	8	27
0 (32)	16	81

the rate of metabolism or rate of respiration of food organisms which are still active, including such things as lettuce and apples—fresh vegetables and fruits. Third in importance is to decrease the rate of chemical deterioration (see below). Of course, all three fundamentally are the same—decreasing the rate of chemical reactions (see pp. 34–40). The Q_{10} or temperature coefficient for most chemical reactions and for growth of most biological organisms ranges from 2 to 3 for each 10°C (18°F). Table 9.2 represents the effect of refrigeration on the hypothetical keeping times of foods having Q_{10} values of 2 and 3. Although this table illustrates the value of refrigerated storage of food, each food responds differently. The kinds of spoilage organisms may change with temperature, for some species grow better at lower temperatures than other, competing species.

When the temperature of food is raised above the biokinetic zone, microorganisms are killed. This happens in cooking. This killing of potential spoilage organisms is one of the major effects of cooking. Not all organisms are killed at the same temperature, although when the temperature of the food reaches 100°C (212°F) most microorganisms have been destroyed. However, some viable spores are not destroyed until higher temperatures are reached. Of course, after food is cooled for eating it may be contaminated by other potential spoilage microorganisms. Such cooked and subsequently con-taminated food will spoil microbially after a short time. However, if the cooking is carried out in a sealed container such as a can, bottle, or plastic pouch so that there is no microbial contamination after cooking has destroyed all potential organisms, we have the so-called commercially heat-sterilized or -stabilized food common in the food markets. Many such products will keep for years. This process was discovered by Nicholas Appert, who was awarded a prize offered by Napoleon to whoever developed a process to preserve food for his armies.

A few decades after Appert's process, which was to change the whole food industry, another great Frenchman, Louis Pasteur, discovered that most microorganisms including those pathogenic to man are destroyed at temperatures considerably below the boiling point of water. This led to the process called *pasteurization*, which has found wide usage in the milk, meat, and brewing industries. For example, heating milk at 142°F (61.1°C) for 30 min will kill most bacteria. Even though all bacteria and spores are not killed, heating the milk in this way and immediately cooling it generally eliminates the pathogenic microorganisms and most potential spoilage organisms, so that the milk has an extended keeping time.

Many meat items are processed in a somewhat similar way—cured meats particularly. Neither pasteurization nor the curing salts, sodium chloride, sodium nitrate, and sodium nitrite, produce a sterile product, but both processes can be used together to produce relatively stable meat products, particularly if these products are in sealed containers to prevent contamination and are kept under refrigerated storage. So-called cooked and cured sausages such as wieners, frankfurters, and bologna are common, very popular food products with a long history of usage. Pasteurized and cured canned hams heated to 156–160°F (69–71°C), though not sterile, have long storage life if kept refrigerated below 39°F (4°C). Commercially sterilized canned meat— i.e., meat heated to kill all non-spore-forming bacteria and bacterial spores that would grow during normal nonrefrigerated storage—acquires a texture and flavor markedly different from that of meat heated to the milder normal cooking temperatures of 140–180°F (60–82°C), and many people find this objectionable. These undesirable changes result from excessive thermal degradation of meat proteins.

Commercial heat sterilization is a highly developed technology. The precise conditions of time and temperature of heating depend on the nature and quantity of the specific food, ingredients added, and type of container (can, bottle, pouch, etc.). For example, highly acidic (low pH) foods such as tomatoes and some other fruits require lower temperatures for commercial sterilization than low acid (high pH) foods such as milk, meat, and many vegetables. There are also relationships among temperature and other environmental parameters such as water activity, which may be markedly modified by addition of sugar or salt, and the naturally present nutrients for potential spoilage organisms. Temperatures employed for commercial sterilization range from 212 to 250°F (100–121°C) or even higher for short-time heating processes.

Organisms are killed by heating when the temperature reaches a point that any specific chemical compound necessary for any vital process is destroyed. The most thermally labile compounds normally present in all organisms are the proteins and the nucleic acids. You will recall that enzymes, the catalysts which mediate most of the highly organized chemical processes of living things, are proteins and that the nucleic acids direct all activities of living cells including protein synthesis and transmission of genetic information from one generation to the next. So if any vital protein or nucleic acid molecule of a food spoilage microorganism is destroyed by heating, the organism is killed and thus the food is protected and preserved.

The older, more common methods of using combustion gases, hot air, hot water, boiling water, or steam (cooking) or radiant heat (baking and broiling) to increase the temperature of food have some limitations in our hurry-up world. Many foods to be heated are solid or frozen, and foods in general are very poor conductors of heat. So time becomes a factor when solid foods of any appreciable size are to be heated. If the temperature of the source of heat is too high, e.g., in baking, broiling, or frying, the surface of the food may be burned while the internal portion is still cold and uncooked. This is particularly true for frozen foods. The amount of heat (calories) required just to change ice to water (latent heat of fusion) without raising the temperature of the food a single degree is almost as many calories as needed to raise the temperature of the food from the freezing point to the boiling point of water.[1] In order to overcome this time factor, heating food by means of high-intensity, high-frequency radiowaves, commonly called *microwaves*, has been developed. These electromagnetic radiations can penetrate farther into the food than light or infrared waves, which are electromagnetic radiations of higher relative frequency and shorter wavelength (see Table 9.3). The molecules of the food can absorb these radiations and change the energy of the radio- or microwaves into heat. In other words, by this process heat can be generated within the food rather than being conducted in from the surface of the food. Heating food in this way is rather expensive in terms of equipment and type of energy (electricity) required. Yet in certain special food processing operations and in aircraft, fast-food service restaurants, some institutions, and even homes where time is not available for the customary heating procedures the use of high-frequency (radio- or microwave) heating is becoming more commonplace.

pH, Hydrogen Ion Concentration, Acidity

In Chapter 4 (pp. 40–52), hydrogen ion concentration was shown to be a critical environmental parameter of water solutions controlling life. It is determined by the acidic and basic constituents present in the water in which

---

[1] The heat necessary to change 1 gram of ice (solid water) at 0°C (32°F) to liquid water at 0°C (32°F) is 80 calories. This is known as the *heat of fusion*. To raise the temperature of 1 gram of liquid water from 0°C (32°F) to 100°C (212°F) requires 100 calories. To change 1 gram of liquid water at 100°C (212°F) to steam (gaseous water) at the same temperature, the *boiling point* of water at normal atmospheric pressure, requires 540 calories. This is known as the *heat of vaporization*.

the cells live and is conveniently expressed using the arbitrary pH scale of 0–14 (the lower the pH, the more hydrogen ions; the higher the pH, the fewer). Notwithstanding the fact that quantitation of this environmental factor was not accomplished until this century, the control of pH to promote food preservation is an ancient practice used throughout the world. We need only to mention a number of common foods to illustrate its importance (remember that few organisms can live below pH 4 or above pH 9).

Techniques of microbial fermentation to produce acetic acid and lactic acid from sugars naturally present in fruits, vegetables, and milk have come to us from prehistoric times. Sauerkraut, pickled cucumbers, olives, peppers, etc., and a number of acidified milk and cheese products are examples of foods preserved by lactic acid production. Pickling and preservation by use of vinegar is an old practice. Vinegar is a solution of acetic acid usually produced from fruit juices such as apple or grape. These juices are first fermented by yeasts to alcohol, which then is acted on by acetic acid–forming bacteria. So-called white vinegar can also be produced by similar fermentation of alcohol from other sources. Vinegar is a common acidifying food ingredient and is the preservative in mayonnaise, salad dressings, and condiments of various kinds. The relative stability of citrus fruits—particularly lemons—is due to the low pH resulting from their natural citric acid content. Rhubarb is an additional example of very acidic plant foods that are relatively free of microbiological attack because of their low pH. Other acids naturally present in biological systems are used as acidulants in foods; among them are tartaric acid (grapes) and malic acid (apples). All of these acids are, in the proper amounts, pleasing to the taste even though sour and are used in food for flavoring as well as for their preservative action.

The soft drinks, carbonated and noncarbonated, so common in our culture are preserved by the acids they contain, which may be any of the above or phosphoric acid or carbonic acid (CO_2, carbon dioxide). The pH of many of these products is below that permitting microorganisms to grow and reproduce.

Adjustment of pH to the higher or more alkaline limits as a means of food preservation less common. Foods of high pH are generally not very tasty. However, alkalis such as sodium, potassium, and calcium hydroxides and their carbonates, bicarbonates, and phosphates are used in some processes. Hominy and olives, prior to ultimate pickling, are treated with alkalis and reach relatively high pH values for a time. Soaps and other alkaline salts or alkalis themselves when used in water produce pH values above 9 or 10 and kill most bacteria. When the pH reaches 12, water is almost sterile. In the purification process of a number of municipal water systems, particularly those using the so-called lime-softening process, the pH of the water may go to 11 or slightly above at certain stages. Most of the microorganisms in the water treated in this way are killed. Chlorination of such water to kill bacteria may not be necessary.

Decreasing Water Activity

Decreasing water activity to preserve food is among the oldest of methods used by man. The food organism or part to be eaten is preserved by the addition of salt or sugar or some other water-soluble substance. This has the effect of lowering the activity of the water below that at which spoilage organisms can grow and reproduce (see pp. 52–59). Salt may be added to meat, fish, etc., so that they need not be refrigerated. This ancient technique is still practiced throughout the world. Bacon, ham, and so-called dry sausages such as salami and pepperoni were preserved in this way in early times. They were so salty, 10–15% or more at times, that removal of salt by washing or parboiling was sometimes necessary before eating. Today, because of refrigerated storage and transport such heavy salting is not practiced so widely.

Sugar has been used for food preservation for many centuries. Maple and other types of syrup, sugared fruits, fruit preserves, jams, jellies, cream candies, and some cakes (particularly fruitcake) are rather free from microbial attack because of their sugar content, which depresses the water activity below the level tolerated by most microorganisms. Fruit jellies and jams usually contain 65% sugar and have a water activity of approximately 0.86. Many cakes and confections may have a_w in the range of 0.70–0.88. It is obvious that these products contain water, but their a_w is in the range for properly stored cereals such as rice, wheat, or wheat flour, which have 14% moisture or below and appear quite dry.

Alcohol either produced by fermentation or added to the food can depress water activity to the extent that many potential spoilage organisms may not be able to grow or grow very slowly. Wine containing 15% alcohol, 7.7% simple sugars, and 0.3% tartaric acid has a_w of about 0.92. A lighter wine with 10% alcohol, 5.0% simple sugars, and 0.3 tartaric acid has a_w of 0.95. The decreased a_w together with the lower pH due to the acidic constituents naturally present in the fruit from which the wine is made serves to inhibit or retard the growth of many potential spoilage microorganisms. The liqueurs and distilled alcoholic beverages such as brandy, whiskey, vodka, and rum often typical of particular cultures contain much more alcohol, which acts as a preservative even though such beverages during compounding and aging in wood might pick up enough nitrogen and minerals to support microbial growth. A 100 proof whiskey (50% alcohol by volume, 42.5% by weight) has a_w of 0.77.

Nutrients

Nutrients in the environment make all life possible. Food is the source of all the nutrients for man, with the exception of oxygen, and consequently food can provide the nutrients for the microorganisms which are competitive with man. So while the control of spoilage organisms by changing or altering nutrients in the food may seem easy in theory, in practice it is very difficult. However, the natural avidin in egg white does just that; it eliminates biotin, a necessary nutrient for many competitive microorganisms (see pp. 65 and 105). Other similar examples can be found among the biochemical mechanisms whereby various species of organisms protect themselves from competitive species. However, in food preservation so many organisms are involved that control of spoilage by inducing nutritional deficiencies in only some species of microorganisms is of limited value. Even so, there is one nutrient besides water itself which can be controlled advantageously to prolong the usefulnes of foods. This nutrient is oxygen.

With respect to oxygen requirements, microorganisms fall into four groups—obligatory aerobes, which must have oxygen to live; microaerophils, which must have only limited amounts of oxygen; obligatory anaerobes, which cannot live if oxygen is present; and facultative organisms, which are capable of living in either the presence or the absence of oxygen. Many of the common molds as well as bacteria require oxygen. Processing and packaging in sealed cans, jars, or plastic containers free of oxygen can control these organisms. Some cheeses and fruit products subject to spoilage by mold growth are examples common in the market. Many putrefactive bacteria are anaerobes, and such packaging only promotes their growth. *Clostridium botulinum* is only one of many potentially troublesome anaerobic spoilage organisms. These, of course, can be controlled by packaging in oxygen. Since many foods, e.g., meat, may be in large pieces, oxygen at the surface may not necessarily mean oxygen within the food. By use of a food ingredient which contains a potentially metabolizable source of oxygen, the anaerobes can be controlled. Such ingredients are sodium and potassium nitrates and nitrites—$NaNO_3$ and KNO_3 (saltpeter) and $NaNO_2$ and KNO_2. These substances, which are used by many plants and microorganisms as nitrogen sources for their growth, turn an otherwise anaerobic environment into an aerobic one. They have been used for curing meat—ham, bacon, "corned" beef, fish, etc.—for centuries. Because of their use, botulism poses no problem in cured meats, whereas historically it has been a problem in uncured meats.

You will recall that a fundamental nutrient for all organisms is energy and for the autotrophs this is light and for the heterotrophs energy is obtained from the organic matter consumed. Microbial food spoilage is almost always caused by heterotrophs requiring no light. Of course, food is usually stored in dark places. If any autotrophic organisms were to be found a problem, special precautions could be taken to shield the food from light.

Newer Techniques

The aqueous environment containing the nutrients to support growth, development, and reproduction of spoilage organisms can be altered in ways other than those already discussed. Such changes which retard or inhibit the growth of these organisms will of course retard spoilage and preserve the food. There are two general methods available: one is the addition of substances which will kill the spoilage organisms and the other is the use of energy sources other than heat that are capable of irreversibly changing at least one vital molecule of the spoilage organisms and so killing them. The latter approach, the use of high-energy radiation, is a relatively new one and appears to have considerable potential. For want of better terms, let us refer to these methods as the use of preservatives and the use of radiant energy or irradiation.

Preservatives

As we shall soon see, the term *preservation* must be applied to specific processes of deterioration. There are several types of spoilage. Often a preservative to retard chemical deterioration has no effect in controlling microbial spoilage. Conversely, preservatives to control microbial spoilage are often ineffective in preventing chemical deterioration. Even within these two major types of food spoilage, different preservatives are required for specific purposes. For example, a substance used to retard bread staling is likely to have no effect in retarding rancidity in potato chips, or a substance used to control certain species of microorganisms may have no effect on, or may even promote, the growth of other species. Therefore, detailed discussions of individual preservatives is beyond our purpose here. Yet it is essential to understand that each specific food deteriorates or spoils in its own unique way and that preservation procedures must be designed to fit a particular food.

Any substance introduced into food (or into any ecosystem for that matter) which kills all living things is, of course, of no value, for if it does not kill the consumer, it is likely to ruin the food. Historically there have been attempts along this line. Formaldehyde, the major preservative in embalming fluids, reacts with proteins to such an extent that living things cannot tolerate much of it. Attempts to use formaldehyde in small amounts as a food preservative have been tried in the past and have not been successful. Too much formaldehyde is repulsive to the consumer, and, if enough is used to be a general preservative, the nutritive properties of the food are so impaired that it has little or no value. There are many other substances which fall into the same pattern. Chlorine is a very reactive substance, is an excellent sanitizer, and is extensively used in water purification. However, it is of no value as a food preservative because it would react chemically with the food so that by the time an excess was attained that would have a lasting preservative action the food itself would be useless. Where then do we look for antimicrobial food preservatives? The most promising place is biological organisms themselves. Remember that chemical warfare is a common potent device in the endless dynamic competition among all organisms in the biosphere. As man has studied life processes of individual organisms and their interactions with other organisms in their environment, he has discovered certain practical, more specific preservatives.

A number of foods such as bread, certain varieties of cheeses, and some fruit and vegetable products are subject to attack by molds of various kinds. Many of these molds have difficulty growing in the presence of certain substances produced by other biological systems. Benzoic acid is found naturally in a number of higher plants, as is sorbic acid; and propionic acid is produced in large amounts by many bacteria. All three of these acids when present in small amounts inhibit the growth of many species of common molds and yet are easily metabolized by most higher animals including man. These substances do not interfere with the nutritive value of foods. Consequently, benzoic, sorbic, and propionic acids have been found to be of great value as preservatives in bread and related items, catsup, cheeses, and some other foods.

The antibacterial action of nitrates and nitrites, already mentioned, makes these substances also useful as preservatives. Many microorganisms produce these compounds—which are also produced from nitrogen and oxygen in our atmosphere, particularly by lightning—and other organisms can use nitrates and nitrites as sources of nitrogen for production of protein as well as sources of oxygen (see above). However, for many other bacteria nitrates and particularly nitrites are toxic even in small amounts, and man and most higher animals are apparently not harmed by these compounds at the levels necessary to control the growth of many spoilage bacteria.

Certain cheeses are resistant to specific types of bacterial spoilage because the bacteria used in making the cheese produce antibiotic substances which inhibit the growth of the spoilage organism. It is rather common among microorganisms to produce chemical compounds which control the growth of competitive organisms. Some of these substances, known as antibiotics, have found practical use in medicine and in animal feeding (see Chapter 8). A few of these have been found to have value in delaying bacterial spoilage in certain foods—for example, in meat, poultry, and fish.

Unfortunately, many consumers are wary of the term *preservative* because it is thought to imply inferior quality, when in fact the proper use of preservatives can mean higher food quality. There is nothing inherently wrong with their proper use. However, what may be an effective and legitimate use of a preservative in some cases can be quite improper or even fraudulent in other cases. Sulfur dioxide, or sodium metabisulfite, which yields sulfur dioxide in water solutions, is a valuable preservative of long use in a number of fruit and vegetable products. However, its use in meat is improper because it has the ability to cause meat to appear fresh and free of bacterial spoilage when in reality it is neither. The use of many preservatives does not usually make foods keep for long periods of time under ambient conditions. Other preservation procedures such as drying, freezing, and canning are often superior for such purposes. However, when the detailed nature of the microbial spoilage of a particular food is known and when there is a preservative which is specific in combating the spoilage organisms, it can be quite valuable, provided, of course, that it does not adversely affect the nutritive qualities of the food or produce any type of metabolic disturbance in those eating the food.

Irradiation

The advent of the atomic age brought with it an economic source of high-energy or ionizing radiation. At first, it was thought that the radiation from various radioactive elements could quickly sterilize foods so that they could be kept for very long periods without refrigeration or special handling. This hope has not been realized because sterilization of food by ionizing radiation is not as simple as sterilization by heating. However, it now appears that irradiation with high-energy electromagnetic waves can enhance the keeping time of a number of foods and may become a very important procedure in food processing and preservation.

Earliest man appreciated the life-giving role of the radiant energy (light) from the sun. Sun worship has characterized many cultures. The tremendous energy of the sun is generated by atomic thermonuclear reactions. Hydrogen atoms are converted to helium atoms at temperatures in the millions of degrees with the result that some of the substance (mass) of the sun is converted to radiant energy (according to Einstein's equation for the relation between mass and energy—$E = MC^2$—or energy is equal to the mass which is converted to energy multiplied by the velocity of light squared). Light is only one form of radiant energy or electromagnetic waves, and, in fact, light waves represent only a fraction of the energy from the sun. Although all electromagnetic radiations travel with the speed of light, these radiations occur in many different wavelengths. The shorter the wavelength, the more potential energy there is in the unit of radiation. Fortunately, the atmosphere of the earth acts as a shield and absorbs most of the high-energy radiations of very

Table 9.3. Electromagnetic Spectrum of Radiant Energy

	Wavelength in meters[a]
Ordinary radiowaves	500–100
Shortwave radio, television, radar transmission	100–0.01
Microwave heating	0.3–0.01
Infrared-radiant heating	3×10^{-4}–7×10^{-7}
Visible light—red, yellow, green, blue, violet	7×10^{-7}–4×10^{-7}
Ultraviolet light	4×10^{-7}–1×10^{-8}
Limit of sun's radiation on surface of earth	2.92×10^{-7}
Germicidal range	2.8×10^{-7}–2×10^{-7}
X-rays	1.5×10^{-8}–1×10^{-11}
Gamma rays	1.4×10^{-10}–5×10^{-13}
Cosmic rays	5×10^{-14}

[a] 1 meter = 39.37 inches or 1.094 yards.

short wavelength (Table 9.3), because these are lethal to living things. The radiation that passes through the atmosphere—of wavelength longer than 2.92×10^{-7} meters—is the source of energy for supporting life in the entire biosphere. Electromagnetic radiations are ever present in our environment and most of them are harmless to man. Furthermore, we have learned how to produce and use some of these to our own advantage.

Table 9.3 shows the range or spectrum of electromagnetic radiations in terms of wavelength. You are already quite familiar with many of them. White light is in reality a mixture of light of all colors. When something appears colored such as green grass or a red tomato, it is so because the molecules in the object absorb light of certain wavelengths and reflect or transmit the remaining wavelengths that you see. Each molecule in accordance with the kinds and arrangements of atoms in its structure absorbs certain characteristic wavelengths whereas other wavelengths of radiation are not absorbed and therefore have no effect on the molecule.[2]

The energy absorbed by a molecule may appear as heat, which is simply increased movement of the molecule in its surroundings. If, however, the energy absorbed by the molecule is sufficient, the atoms within it may rearrange to form new molecules or ones with a higher amount of energy. Such

[2] Chemists, physicists, and biologists make much use of this fact as they can determine qualitatively and quantitatively many substances by the wavelengths of electromagnetic radiation they absorb. For this purpose, usually ultraviolet, visible, and infrared light are used.

energy may be transferred to other molecules or cause reactions to occur with other molecules in the immediate surroundings. This is essentially what happens in photosynthesis. If the radiation absorbed by the molecule is of such high energy that atoms or parts of atoms within the molecule are stripped off, the remaining molecule becomes highly ionized or electrically charged and extremely reactive chemically. For this reason, some of the short ultraviolet, X-rays, and gamma-rays from radioactive disintegration or atomic fusion reactions and cosmic rays from outer space are called *ionizing radiations*. Extremely fast-moving particles—electrons, neutrons, protons, atoms—which approach the speed of light have very high kinetic energy and can also produce an ionizing effect. Both types of ionizing radiations are being studied for their application in food processing.

Anyone who has exposed himself to the sun on the beach knows that the ultraviolet rays striking the skin can induce adverse chemical reactions commonly known as sunburn. These ultraviolet radiations also play an important beneficial nutritional role by changing some of the sterols of the skin into vitamin D (see Chapter 5, p. 102). It has been known since the beginning of the science of microbiology that even though sunlight is necessary for green plants many microorganisms are killed by sunlight, especially in the ultraviolet wavelengths. Part but not all of this so-called germicidal effect of ultraviolet light is due to ozone, O_3, produced from O_2 of the air. Ozone has an acrid odor and is toxic in appreciable amounts. Ozone and closely related products are characteristic of smogs. Ultraviolet light artificially produced at wavelengths which produce little or no ozone has been used for almost half a century to retard or completely control growth of spoilage microorganisms on the surface of certain foods.

The basic idea behind sterilization by ionizing radiations is essentially the same as for sterilization by heating—i.e., if any molecule vital to the processes of growth and reproduction can be modified or irreversibly changed (by the radiation or heat), then the organism no longer can live. In applying this principle to food, it is essential that the radiation producing the desired effect be of insufficient energy to induce radioactivity in the food itself. Of all possible sources, the gamma-rays of radioactive cobalt-60 and electron beams of less than 12 million electron volts (meV) have been shown to be the best. Cobalt-60 gamma-rays can effectively penetrate 12 inches (20 cm) into food and a 10 meV electron beam can penetrate 1.25 inches (3.2 cm).

As one might expect, the lethal dose of such irradiation is directly related to the complexity of the organism being irradiated. The measure of radiation dosage is the rad, and Fig. 9.2 is a diagrammatic representation of lethal doses for various organisms. Note that only about 600 rads of whole-body ionizing radiation will kill one-half of a human population. About 10,000 rads is sufficient to inhibit sprouting in tubers such as potatoes. Stopping the

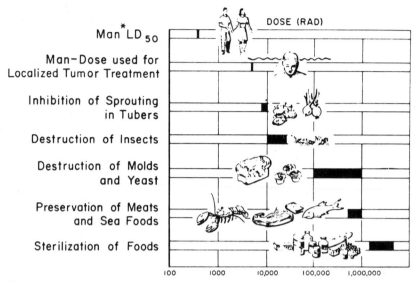

$^{*}LD_{50}$ whole body radiation dose necessary to destroy 50% of people exposed

Fig. 9.2. Approximate dosages of ionizing radiation for specific effects. One rad represents the absorption of 100 ergs per gram of substance. The heat equivalent of 1 rad is 2.39×10^{-6} calorie; 1 million rads or 1 megarad is equivalent to 2.39 calories. (From S. A. Goldblith, *Food Processing Operations*, M. A. Joslyn and J. L. Heid, eds., AVI. Publishing Co., 1963.)

propagation of insects requires only slightly more radiation. These processes have considerable practical value in extending the time from harvest to consumption of certain very important foods and have been approved by government authorities in a number of countries. The killing of yeasts and molds requires up to 1 million rads and destruction of bacteria to produce sterility in foods requires several million. Notwithstanding the relatively high dosage necessary, there appear to be a number of practical possibilities for using ionizing high-energy radiation for commercially sterilizing and preserving foods.

Sterilization of food by heat and sterilization by irradiation differ in the types of chemical reactions producing the effect. Heating, of course, affects the most heat-labile compounds, which are the proteins and nucleic acids (see also Chapters 2, 4, and 6). Furthermore, although such heat treatment (cooking) causes many chemical changes, these are not usually adverse or unacceptable with respect to flavor, texture, and nutritive value. This is not the case for foods sterilized by ionizing radiation at ordinary temperatures above 0°C (32°F), which are particularly likely to acquire undesirable textures and flavors.

When the ionizing radiation is absorbed by molecules of the food, even including the water present, the molecules suddenly have so much energy that they may lose an electron or proton or may even break apart. The resulting highly charged ions or free radicals (molecules not internally balanced electrically, see pp. 44–46) will react with almost any nearby molecule, sometimes even forming new free radicals. Such randomized reactions subside rather quickly when new chemically stable compounds are produced, and most of the energy absorbed is eventually dissipated as heat. Sterilization results when some molecules necessary for life are destroyed either by direct absorption of ionizing radiation or by chemical reactions with the free radicals produced when other molecules are broken apart by the ionizing radiation.

It has been found that the more or less randomized chemical reactions resulting from ion or free radical production can be controlled at low temperatures so that the adverse reactions producing undesirable flavor and texture effects do not occur to any appreciable extent. This is accomplished by irradiating the food at very low temperatures, even down to $-180°C$ ($-292°F$), with 2–4.5 megarads (2–4.5 million rads) using the gamma-rays of cobalt-60 or an electron beam of 10 meV. (The precise conditions vary with the food product.)

Meat and meat products can be quite satisfactorily sterilized in this way. First, the meat or meat product such as ham, frankfurter, boned beef, poultry, or seafood is processed and heated or cooked in the usual manner to about 60–70°C (140–158°F) and then sealed in a package free from air. This heating is necessary to destroy the many enzymes naturally present in uncooked meat because these catalysts are not all inactivated by sterilizing doses of ionizing radiation (see below). The product is then frozen and cooled to about $-40°C$ ($-40°F$) or below and irradiated. The resulting sterile product may be stored at ambient temperatures for extended periods of time. Meat and meat products treated in this way are superior in flavor, color, texture, and nutritive properties to the same foods subjected to high-temperature heat sterilization, according to reports released by the U.S. Army Natick (Massachusetts) Food Laboratories and Medical Research and Development Command. Currently, extensive long-term feeding studies are being conducted to determine the safety of this new food preservation technique.

No fundamentally new food preservation process of the nineteenth and twentieth centuries has been scrutinized more thoroughly before commercialization—not even such commonly accepted processes as heat sterilization and the hydrogenation of fats, both of which entail extensive chemical changes in foods. Certainly high-energy radiation can be used in food processing and preservation wherever a living biological organism must be destroyed, whether it be preventing the sprouting of potatoes so that they can be stored longer, killing insects which might propagate in stored grain, or destroying microorganisms that compete with man for his food.

CHEMICAL DETERIORATION OF FOODS AND ITS CONTROL

To be sure, the biological spoilage of food discussed in the previous section is chemical in nature since all living things are themselves dynamic chemical systems requiring an input of energy. Yet foods will deteriorate chemically in the absence of the living, growing organisms that compete with man for his food. These changes are degradative in that they require no continuing input of energy as do living things. In other words, these chemical changes are due to reactions among the food constituents themselves or with the substances in their surroundings such as the air and containers. Although each particular food will deteriorate in its own characteristic way, there are several well-defined types of chemical deterioration familiar to us all. In many cases, complete inhibition of the degradative process is not feasible. Then, too, chemical and biological deterioration may go on simultaneously. Furthermore, chemical deterioration may or may not alter the nutritive properties of the food. Often because of custom and peculiar standards of quality, foods will become unacceptable solely because of changes in color, flavor, odor, texture, or some other attribute the consumer associates with a particular food.

Enzymatic Degradation

We have already noted many times that almost all the well-ordered chemical reactions that go on in living things are catalyzed by enzymes and that the building processes for tissue growth depend on a continuing input of energy. When a food organism is harvested or killed, its energy supply is taken away and so the direction of the dynamic chemical reactions reverses or otherwise changes. Generally, complex molecules will become simpler ones, for in the presence of residual enzymes and water there is a tendency for them to revert to the smaller molecules from which they were made in a manner similar to the digestive processes discussed at length in Chapter 7. Deterioration of foods resulting from the catalytic action of enzymes is easy to observe. An example is the ripening process of the banana wherein it becomes sweeter, softer, less astringent in taste, and more odorous. One of the reactions in this process is the hydrolysis (degradation) of the starch, essentially tasteless and insoluble in the water of the banana, to simple water-soluble sugars. Another example is the overripe tomato. Here the softening is due largely to the hydrolysis (degradation) of the pectins to their simpler carbohydrate building blocks. Proteins in such foods as cheese, meat, and fish may be hydrolyzed to simpler compounds by the proteases[3] naturally present. Fats in some foods may also be hydrolyzed by lipases and nucleic acids by other enzymes to form the simpler molecules from which they were made. Such chemical changes are often manifested as changes in taste, odor, texture, etc. For example, those who like freshly picked sweet corn are aware that if it is picked too long before cooking and eating it loses its sweetness. This is because the simple sugars disappear and the corn may even acquire a straw-like flavor.

When freshly harvested organisms are processed for eating, the normal cellular organization of the tissues may be disrupted, with the result that residual enzymes may initiate degradative changes at a very rapid rate. One of the most common examples of such changes is the rapid darkening of freshly peeled potatoes, apples, peaches, and pears. Here the oxidative enzymes of the freshly exposed tissues use oxygen from the air to change many naturally occurring colorless compounds (phenols) to colored compounds (quinones). Other oxidizing enzymes induce the common and sometimes intense "hay" flavor of vegetables such as lima beans, corn, and broccoli if they are not cooked soon enough after harvesting. Many of the changes observed when fresh fruits and vegetables are bruised are catalyzed by enzymes, and a great many more examples could be cited.

Chemical reactions catalyzed by enzymes can, of course, be stopped by destroying or removing the enzymes. This is relatively easy to do and was practiced long before modern concepts of biology and chemistry were developed. The method is simply heating or cooking food. All enzymes are proteins, and proteins are easily changed or denatured by heating (see p. 147). The temperatures required to inactivate most enzymes are in the range of 60–80°C (140–176°F), although some enzymes are destroyed below 60°C and some require heating to temperatures above 80°C before they lose their catalytic properties.

Many foods are preserved by freezing. However, freezing does not destroy most enzymes, and many frozen foods can deteriorate enzymatically even though the rates of the reactions may be slowed. Vegetables are the worst offenders. So in order to preserve peas, green beans, lima beans, corn, etc., by freezing, it is first necessary to heat them briefly to almost 100°C (212°F) before they are frozen, to prevent deterioration in frozen storage. Generally fruits do not require such heat treatment, which is fortunate since many of them are adversely altered in flavor when heated. However, enzymatic darkening often occurs in frozen fruits such as sliced peaches, and

---

[3] The suffix -ase denotes an enzyme. Proteases are enzymes which hydrolyze (digest) proteins. Similarly, lipases hydrolyze lipids, sucrase hydrolyzes sucrose, and so on.

to counteract this they are often packed with sugar syrups containing ascorbic acid or similar oxidation inhibitors. Along with proper packaging, this diminishes greatly the amount of atmospheric oxygen reaching the fruit, which is necessary for the darkening reactions.

Although degradation of harvested food organisms by enzymes leads to food spoilage in many cases, in other cases enzymes are used in making and processing foods. Specific enzymes used for such purposes are generally prepared and purified, if necessary, from some biological organism. Since ancient times the enzymes from sprouting barley have been used for brewing and fermentation and rennin from the stomach lining (mucosa) of young lambs and calves has been used to coagulate or clot milk to form a curd which then may be made into cheese. Pectin-hydrolyzing enzymes are sometimes used in winemaking to clarify fruit juice, because the insoluble pectin may cause turbidity. This is the same kind of chemical reaction that produced the overripe tomato mentioned above. Also, the starch-digesting enzymes of sprouting barley act in much the same way as the starch-hydrolyzing enzymes that soften the ripening banana previously discussed.

Nonenzymatic Browning

Nonenzymatic browning is a process whereby foods darken without the catalytic effect of enzymes. It differs markedly from the darkening of fresh fruits and vegetables already described. The food constituents responsible are certain simple sugars which react with proteins, amino acids, or other substances normally present in the foods. Although the darkening in color of the food may be undesirable, the more notable changes are often those of flavor. The rate of the browning reactions increases with temperature, and these chemical changes may contribute to desirable flavor development on cooking. These browning reactions are sometimes called the *Maillard reactions* after the chemist who first characterized them. Maple syrup gets its characteristic flavor from this chemical process as the maple sap is concentrated by boiling. A similar reaction causes the caramelized flavor in canned evaporated milk, which many people find much less acceptable than the taste of fresh milk. But the same flavors of evaporated milk are characteristic of the desirable flavors of caramel candy and certain other foods.

It is this browning reaction in dehydrated foods such as meat, eggs, and some fruits and vegetables that, after storage for a time, often makes them objectionable, so much so that this otherwise promising method of food preservation is extremely limited in its practical uses. These changes are most troublesome in foods, even canned sterilized foods, stored for long periods of time at relatively high temperatures. Consequently, food deterioration due to nonenzymatic browning is a great problem in military and survival rations, which are often stored at relatively high temperatures. Although in a few instances adverse development of the browning reaction can be delayed by certain food ingredients, the best way to diminish nonenzymatic browning is proper food processing and storage conditions.

Oxidative Rancidity

Oxidative rancidity is one of the most important types of food deterioration and has been studied intensively for many years. As its name implies, this type of chemical spoilage consists of the development of undesirable odors and flavors and is due to chemical reaction of oxygen in the atmosphere with fats and foods containing fats. Almost everyone is familiar with the paintlike or acrid odor and flavor of oxidatively rancid fats and oils, nuts, potato chips, frozen meat, salad dressing, cookies, etc., that have been stored in air for a time. Almost everyone has at times also experienced off-flavors from pies, cakes, cookies, fried foods, etc., made with rancid fat or fat that is so near to being rancid that the products made with it quickly acquire undesirable flavors.

Oxidatively rancid fat is nutritionally deficient. The fat-soluble vitamins (A, D, E, K) in the food are destroyed by some of the very reactive substances produced by the oxidation process. Shortly after the discovery of what we now know as the E vitamins (tocopherols) in connection with reproductive performance in animals, it was found that barely rancid fats and fatty foods when fed to pregnant animals prevented them from bearing live offspring. The long and tedious research necessary to isolate and identify the E vitamins was facilitated because the keeping time in air of certain vegetable oils correlated closely with their effect on the success or failure of pregnant experimental animals to bear healthy young. It was found that the fats and oils that supplied the nutrients necessary for a successful outcome of pregnancy also contained natural preservatives that greatly retarded the rancidification process. When the preservatives were isolated, so also was the vitamin. Indeed, the tocopherols, the naturally occurring substances with vitamin E activity, are all *antioxidants*. This is the specific term used for the preservatives that will retard the deterioration of fats and fatty foods by oxygen in the air. (However, all naturally occurring antioxidants do not have vitamin E activity.)

From study of these natural antioxidants, a number of very important substances with antioxidant activity were recognized. Among these, in addition to the tocopherols, are gallic acid and many of its derivatives, other naturally occurring phenolic compounds, citric acid, phosphoric acid, and

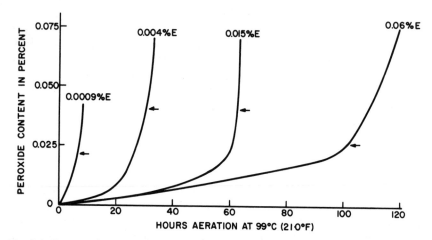

Fig. 9.3. Increase in peroxide content of fat with time. The fat was originally free of peroxide, and the indicated amounts of α-tocopherol, E, were added. The arrows indicate the point of observed rancidity. (Adapted from A. E. Bailey, *Industrial Oil and Fat Products*, 1945; Interscience Publishers, Inc.)

ascorbic acid or vitamin C. Some of these substances have been found to have valuable antioxidant properties when used in foods. Based on extensive research on the oxidation process and how the antioxidants inhibit rancidification, other compounds have been developed such as 2- and 3-*tert*-butyl-4-methoxyphenols, a mixture of which is commonly known as *b*utylated *h*ydroxy*a*nisole or BHA. Sometimes antioxidants are listed on labels as "freshness preservers" or "oxidation inhibitors," or by other descriptive words.

Animal fats generally contain relatively small amounts of natural antioxidants in comparison with vegetable fats. This might be expected. Animals are consumers of oxygen and their active tissues, always using oxygen, have little residual oxygen to react directly with fat *in situ*, whereas plant tissues are producers of oxygen and thus seem to have evolved with built-in systems for preserving their fats and oils.

A unique feature of the oxidation process causing rancidity is that rather than being catalyzed by enzymes the process is a self-catalyzing or an accelerating process once it gets under way. For this reason it is called *autoxidation*. The process is characterized by the early production of peroxides as oxygen gas reacts with fat. Figure 9.3 shows the peroxide contents after aeration at 99°C (210°F) of antioxidant- and peroxide-free oil to which varying amounts of tocopherols have been added. Note the rapid increase in peroxide (consumption of oxygen) once it appears.

Rancidity is evident in most foods when the oxygen reacting amounts only to about 0.05–0.10% of the weight of the fat present in the food. The rate of oxygen consumption is directly affected by temperature. The rancidification process can take place at temperatures below freezing, at room temperatures, and at elevated temperatures. A fat stable for 2 or 3 months in air at 20°C (68°F) may be subject to oxidation by the oxygen of the air after only a few hours or a day at 95°C (203°F). At temperatures of 140–180°C (284–356°F) used for frying various foods, fats begin to oxidize in a very short time, and therefore residual frying fats in foods quickly become rancid unless protected in some manner. The gumming of the fat on the surface of deep friers or on baking pans is evidence of quite advanced oxidation of fats.

The susceptibility of fats to becoming rancid is determined not only by the antioxidants present but also by the degree of unsaturation of the fats, or more particularly of the constitutent fatty acids making up the fats (see pp. 127–131). Whereas the saturated fatty acids are quite stable to oxygen, the unsaturated fatty acids in fats and other lipid molecules readily react with O_2 of the air. The so-called essential fatty acids (Tables 5.1 and 6.4, pp. 75–76, 128), arachidonic and linolenic acids, are most reactive; linoleic is somewhat less so. Oleic acid, the most common unsaturated fatty acid in fats, although less reactive than the others is still quite susceptible to attack by O_2. Thus highly unsaturated fats such as linseed, soybean, corn, and cottonseed oils are much more vulnerable than more saturated fats such as these same oils hydrogenated, beef fat (tallow), pork fat (lard), butter, coconut oil, or cocoa butter (chocolate). The antioxidation process leading to rancid fats is essentially the same chemical process as the drying of linseed oil–based paints when that highly unsaturated oil reacts with O_2 in the air to form a dry, water-resistant resin.

Atmospheric oxygen attacks the fat at or adjacent to the double bond of the fat or fatty acid molecule, $—CH_2CH{=}CHCH_2—$, with the result that peroxides are formed. Once this happens, the molecule becomes very reactive. It may break into other smaller molecules and/or free radicals, which in turn are susceptible to attack by O_2 or may react with almost any other molecule to initiate further oxidation by O_2. Some of the new reactive molecules can react with others to form larger molecules or even resin-like materials as in paint or in the gum formed by frying fats already mentioned. When sufficient small odorous molecules have formed, the fat is rancid; only a very small amount of these odorous molecules is required to ruin a fat for food uses. These small molecules are complex mixtures of acids, alcohols, aldehydes, ketones, and the peroxide derivatives of these substances, and they give the acrid character to rancid fat. A number of antioxidants appear to function by interrupting the chain reaction—i.e., they react with free radicals as they are initially formed and so neutralize these catalysts. When the antioxidant is used up, there is no further preservative action. The fat of cured cooked meat is generally more stable than the fat of uncured cooked meat, probably because of the nitrite which is present in cured meats and acts as a free radical acceptor or antioxidant.

There are also substances which can accelerate rather than retard the autoxidation of fats. In some respects, rancidification of fats is related chemically to smog formation from gasoline, exhaust fumes, etc., polluting the atmosphere of our cities and to the deterioration of rubber. One characteristic common to all these phenomena is the accelerating role of light—particularly ultraviolet light. Light and especially minute amounts of ozone, O_3, produced by ultraviolet light in air will promote rancidification of fats. Other substances such as hemoglobin, chlorophyll, and some of their close derivatives act as accelerators in the rancidification of certain foods. Also, the salts of iron and especially those of copper in minute amounts have pronounced pro-oxidant effects. Such materials can get into foods by contact with equipment made from these metals or from other sources. Many foods naturally contain small amounts of iron and copper, both of which are essential nutrients (see Appendix and Table 5.1 p. 74).

Since oxidative rancidity is a deterioration process involving unsaturated fats and oils and since these are so common as food ingredients, often in small amounts, the autoxidation process is manifested in many foods. The changes in flavor may be described merely as stale or off flavor, and occasionally changes in color may be noted. Frozen or dehydrated foods are quite susceptible to autoxidation, particularly those that have previously been cooked. As with other types of spoilage, any particular item of food must be studied in detail to determine the role that oxidative rancidity plays in its deterioration.

The use of antioxidants and/or the use of hydrogenation to change unsaturated fats to more saturated ones can delay the rancidification of many foods. The general use of completely saturated (hydrogenated) fats, which are very stable, is out of the question for two reasons. First, such fats are too hard and brittle to be palatable; second, and more important, they have only limited value nutritionally—all essential fatty acids are destroyed by the hydrogenation process and such hard fats are poorly digested and absorbed. Perhaps the best way to prevent oxidative deterioration of foods containing fats is to process and store them in the absence of oxygen and away from light, but for the consumer this has practical limitations once the sealed package has been opened.

Staling

Staling is a rather general term applied to all manner of chemical deteriorative changes usually manifested as adverse alterations of taste, odor, and texture in prepared foods which are not promptly eaten. The changes in palatability may lead to rejection of stale food even though it may not be much altered nutritionally. The details of the chemistry of the "steam table" or "warmed-over" flavors encountered in restaurants, dining halls, hospitals, and the home, the changes in flavor and texture of stale bread, and the changes in flavor of cold and even refrigerated cooked meats and other foods can become quite involved. However, all of these changes are related to the fact that foods, being of biological origin, are composed of many substances that are chemically altered by heating. Of course, heating may not impair the nutritive qualities of the food and even may enhance them.

Boiling, baking, frying, etc., produce flavors, textures, and colors that we desire in freshly prepared foods. These effects of heating are the result of the thermal degradation or change of the natural food constituents and are described in some detail in Chapter 6. Many of these new compounds, a few of which are shown in Table 6.6, are themselves unstable and react in time with oxygen from the air or with other chemical compounds present in the food. In other words, by heating many chemically unstable compounds are produced which are characteristic of a particular food. By habit or custom we associate the flavors and textures produced in these unstable chemical systems (freshly prepared food) so strongly with quality that we call stale and reject food in which these desirable compounds no longer exist or are greatly reduced.

Some of the unstable desirable substances are oxidized by oxygen in the air to form more stable undesirable substances. To prevent such changes, many commercially processed foods are heated and stored in the absence of oxygen—for example, canned foods. Another way of excluding most of the oxygen in prepared frozen or refrigerated foods is to have the food covered with sauce or some other liquid and packaged in a container with little residual air. The efficacy of this procedure can be easily observed in the home by simply comparing after a day or two in the refrigerator the flavor of cold sliced roasted uncured meats—beef, pork, poultry—with and without a complete cover of a sauce or gravy.

The desirable flavor intensity of a heat-processed food such as canned soup may decrease during storage even though it is sealed from air. This decrease in flavor is likely due to the reaction of substances produced on the heating of the food with other constituents of the food. Such loss of flavor may be masked by adding stable flavoring substances in the form of spices, essential oils, or pure flavoring compounds. Of course, these materials can also sometimes be used to minimize or mask the undesirable oxidative changes in food stored in air.

Some prepared foods are susceptible to textural changes after preparation. For example, liquid may exude from jellies, jams, custards, sauces, puddings, etc., which have been stored for a time. The initial structure and/or the viscosity of foods freshly prepared results from the fact that some of the carbohydrates, generally starch and sugars, and some of the proteins have the ability to hold large quantities of water within their molecular structures so as to form what the chemist calls *gels*—solid, semisolid, or viscous mixtures containing water. Some of these gels which result when the foods are prepared are not stable for long periods of time—particularly the gels containing the ordinary starches of wheat, corn, rice, etc. To overcome such difficulties, starches that have been slightly modified chemically or other carbohydrates capable of forming more stable gels are used in prepared foods that are to be stored for long periods of time (see Table 6.3, p. 125). The texture of foods whose structure depends primarily on protein—water gels such as coagulated egg, meat, gelatin, or other protein—also may change on heating and subsequent storage. Many times when the deteriorated gels are composed of carbohydrates such as starch, the original texture of the food may be restored by reheating, but this may not be so with protein gels.

The staling of bread and related products such as cakes, rolls, and biscuits is familiar to everyone. Furthermore, no better food can be found to illustrate the changes in both flavor and texture so typical of the staling processes. Desirable flavors in bread and many other foods result from crust formation, where the surface of the food, particularly during baking, frying, and broiling, is subjected to temperatures above the boiling point of water, causing intensified breakdown of food constituents. Upon standing, even in packages which allow no water loss, bread will degenerate in flavor and become harder and crumbly in texture. This latter change is not due to drying out but to rearrangements of the molecules of water and the starches and proteins in the bread itself. This hardening is easily reversed if the bread is reheated to 60°C (140°F) or above. (Of course, if bread is allowed to dry out it will harden simply due to moisture loss.) In other words, bread held at 60°C or above without moisture loss will not harden but it will indeed degenerate quickly in flavor if exposed to oxygen. At temperatures below 60°C the rate of flavor degeneration will decrease but the rate of crumb hardening will increase as temperatures are decreased to 0°C (32°F). Freezing freshly baked bread to temperatures well below the freezing point will retard the hardening rearrangement of water, starch, and protein so that immediately after thawing out the crumb of the bread will be soft. However, it will then become harder, as if it had not been frozen, as the bread sits at temperatures above the freezing point of water. From these considerations, it is easy to understand why bread is often eaten as toast. The toasting process heats the internal part so that it becomes soft and forms a new crust on the surface of the slice so that a new supply of flavoring compounds is made.

A number of years ago it was discovered that certain substances could retard the rate of the hardening which takes place during the staling of bread.

These materials are emulsifying agents such as partially hydrolyzed fats—mono- and diglycerides (see pp. 131–132 and 170)—or close chemical derivatives. Some of these substances are used in bread and related foods as "freshness preservers." Even though these compounds may retard the hardening process, they do not retard the chemical process of flavor degeneration due to oxygen of the air.

CONCLUSION

It is clear that all foods are perishable. Some become unfit for human consumption much more quickly than others. Foods, being of biological origin, are quite unlike hardware such as nuts and bolts or bricks, yet modern food markets with their processed and preserved foods tend to give their customers the impression of similar durability for many food items. Some consumers generally forget that milk comes from a cow which has to be cared for or that the tomatoes or peas have to be grown by very competent farmers knowledgeable in modern, sophisticated agricultural techniques. Purchasers may also be almost completely unaware of what is necessary to make these foods so easily available in a form that can be used when needed. In our highly structured industrial urban society, many people simply have had no contact with agriculture or food production and the processing and preservation that are necessary to overcome the extreme perishability of most foods.

In this chapter, we have seen (1) that foods may be attacked and consumed or made unfit for human use by living organisms which compete with man and (2) that, even if these competing organisms are controlled, foods will deteriorate chemically in a number of well-defined patterns all reflecting the dynamic chemical nature of the organisms from which they come. These discussions are not complete in themselves. If we were to examine the exact nature of the spoilage of any particular food organism or processed food under a specific set of conditions of use, we might find competitive biological organisms and/or chemical spoilage processes other than those covered here. However, it should be clear that food spoilage is not a simple process and furthermore that its control and prevention are complex problems. In our modern society and especially when both husband and wife work outside the home, there is a strong trend toward less food preparation at home. When ready-to-eat foods are prepared in the processing plant prior to marketing, in order to reduce or eliminate food preparation in the home, restaurant, or institution, the problems of processing, preservation, and distribution can become quite difficult. Viewed from this perspective, the commercial use of certain food ingredients, methods of processing and preservation, types of packaging, and conditions of storage and marketing becomes more understandable.

ECONOMIC AND SOCIAL CONDITIONS CONTROLLING THE SUPPLY AND UTILIZATION OF FOOD

Economics, sociological factors, and cultural attitudes of large populations, ethnic or religious groups, and even individuals often greatly affect food supplies and general nutrition and health of people. This is understandable because man is a social animal and lives in a social setting. Societies of men are characterized by division of labor and responsibility, and survival of individuals and of societies as a whole depends on this fact. The more highly developed social and economic systems become, the more interdependent individuals become. The industrial revolution of the past two centuries has led to urbanized societies in which fewer people produce food and more people do other tasks necessary to maintain the population as a whole.

It is easy to perceive that in a small primitive society the provision of food is the source of wealth and that food may even be the medium of exchange among its members. Larger societies soon became structured because the more efficient producers of food had more bargaining power and more influence in the group than the less efficient producers. Food is a form of wealth in any culture and becomes a means of control of some individuals over others. As very complex societies develop, the primacy of food in economic and political affairs is not diminished even though its position in the scheme of things may become somewhat less apparent to most of the population. As primitive societies became more complex, the need soon appeared for some method to facilitate the flow of wealth among individuals and groups as they traded with each other. If some individuals were to serve the society in a way other than the production of food, bartering food for nonfood items or services would soon become too cumbersome. So a medium of exchange—money—was developed and today it is the basis for all types of commerce. In modern highly developed societies, the common notion is that money itself is primary because so much can be facilitated with it. Yet money cannot be eaten nor will it do work or guarantee happiness. Rather, money is a representation of the wealth coming from the production of food or other necessary biological materials needed by the people, from mineral production, from the accumulated labor of the population which may add value to these materials, and from the rental of durable items of wealth.

As any member of a society knows, some individuals and groups possess more wealth than others. There is a spectrum from the very rich to the very poor. Some cultures of the world are characterized by a few very rich people, a few in the middle class, and masses of poor. The countries of the world which are most highly developed have a very high proportion of their populations in the middle class as far as available individual wealth is concerned, and have relatively lower proportions of very rich and very poor. No society seems to be without its proportion of poor and disadvantaged people. Then, too, the middle class of one society may appear rich from the perspective of the middle class or poor of another society. Or even the poor of one society may appear middle class when some absolute measures of wealth are used. Cross-comparisons of societies with respect to the social and economic stratification of people are difficult, and sometimes misconceptions and misunderstandings among groups become quite severe. Yet, in all cultures, all people, at whatever level, must eat to survive. Consequently, knowledge of food and how it is used is fundamental to the understanding of any society.

FACTORS CONTRIBUTING TO THE COST OF FOOD TO THE CONSUMER

The cost of food to a consumer is the sum of the costs of all operations of producing the food, of processing, preserving, packaging, and storing it until needed, of transporting it to the consumer, and of financing all of these operations including carrying the risks involved. The more persons involved in each step of this chain of events and the longer it takes, the more will be the ultimate cost to the consumer. The producer and each subsequent handler must realize a sufficient monetary return (profit) to support themselves and their dependents. Of course, how efficiently each step in the food chain is performed will also be reflected in the cost of food to the consumer. Figure 10.1 should help to visualize some of these intricate interrelationships of costs

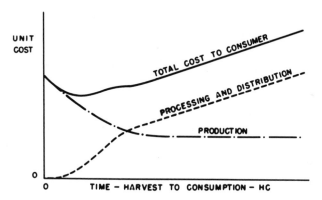

Fig. 10.1. Relationship of period of time between harvest and consumption to food costs.

of production, costs of processing and distribution, and cost of food to the consumer. The curves in this figure are hypothetical and oversimplified but nevertheless are a fair representation that can be applied to any food organism or prepared food. All costs have been related to the time, *HC*, between harvest of the food organism and consumption of the food that is discussed in other contexts throughout this book. In studying these curves, it should be remembered that in various social and economic situations costs take many forms— money directly paid for food and the increased costs resulting from decreased food supplies, poor food quality, deterioration of health because of improper as well as inadequate food, and so on.

Once a food organism has been produced, the cost of production is of course fixed irrespective of how the organism is later processed and marketed. However, if processing and distribution techniques are used to extend the time from harvest to consumption, then it is possible to decrease food production costs to a minimum. This is indicated by the approach of the *production* curve in Fig. 10.1 to a horizontal level and simply means that production efficiency is best when a food can be consumed without wastage for an extended period after harvest. In large measure, this results from the fact that increasing the time between harvest and consumption permits food production to be carried out where the soil, climate, and labor conditions are most favorable for growth of the food organism at the lowest cost.

To illustrate this relationship, consider any series of foods in your own local market and then note where each food organism is grown. Quite likely, many of these—fruits, vegetables, cereals, meat, eggs, spices, etc.—could have been grown locally but were in fact produced elsewhere. Perhaps locally the climate or soil is less favorable than that in more distant regions for growing potatoes (Idaho), wheat (the Dakotas or Alberta), peas (Minnesota), pigs (Iowa), oranges (Florida), bananas (Honduras), and so on. Or even if local soil and climate conditions are very favorable, land costs may be too high because of urbanization and industrialization. Then, too, the land may not be suitable because it is relatively inaccessible to adequate transportation or markets, or even because of an insufficient work force. Costs of land, feed, fertilizers, pesticides, labor, machinery, fuel for farming and transport of supplies, taxes and interest on money (capital) invested in land and equipment, and equipment repair and replacement (amortization) all contribute to the total cost of food production.

The relative importance of each of these individual cost factors varies with the food organism and with the economic and social development of any particular society. For example, in the United States fuel (gasoline, diesel fuel, electricity) and machinery costs are quite high, but in a less developed society such costs may be relatively less because they take a different form. Instead of gasoline for a tractor it may be feed for an ox or food for the farmer himself who may have no tractor but only a hand hoe. A properly equipped sophisticated American farmer can efficiently manage and produce food on a square mile (640 acres) or more of land, whereas in some societies a farmer using only his own body power may toil to operate 1/200 of this amount of land, or even less. In both instances, however, all of the cost factors mentioned above do exist and in each case the farmer must, after paying all costs, realize enough profit to provide for his own welfare so that he can continue to produce food for other members of the society.

Of course, food is produced to supply needed nutrients to the consumer and it must be eaten to be of value. All foods are perishable, but some deteriorate more rapidly than others. We have already seen in Chapter 9 that the time from harvest to consumption may be extended by processing, preservation, packaging, and proper storage techniques and that thereby food supplies can be more cheaply and efficiently distributed to the consumers. Wasted and spoiled food generally represent unnecessary costs to the consumer. But there are necessary costs the consumer must pay for food beyond the price paid to the food producer. These are costs of processing and preservation, which include such procedures as peeling the potatoes, milling the wheat to make flour, squeezing the oranges and freezing the juice, canning the green beans, slaughtering and dressing the chickens, and so on. Then there are packaging costs—the can, the box, the plastic wrapper, the paper sack, the bottles, larger packing cartons for wholesale transport, and so on. Next there are transportation costs—to carry the food organism from the farmer-producer to the processor or food manufacturer and to transport the processed food to a storage warehouse and then to the market and finally to wherever it is consumed. Then there are risk costs—losses because foods are perishable and market prices fluctuate. There are also financing costs for each step in the entire food operation. All of these individual cost factors must be added so that the price that the ultimate consumer pays covers all costs, including, of course, the labor costs of all those who handle the food from the time it leaves the producer to when it is consumed. Such people include truck drivers, railroad trainmen, factory workers, salesmen, clerks, and government food inspectors. We can lump all of these costs together and simply call them *processing and distribution* costs, shown in Fig. 10.1. These costs vary greatly with the type of food and the methods of processing, preservation, storage, and marketing required. For example, most of the cost of a banana to the consumer might well be for transportation and storage, or 60–80% of the price of a meal in a restaurant might represent labor, rent, and other costs of running the restaurant. Each food must be studied individually for specific cost information. Even so, it is possible to see from the curve in Fig. 10.1 how these costs are affected by the time between harvest and consumption.

If someone goes to an orchard or his own garden and picks and eats a peach from a tree, there is essentially zero processing and distribution cost (Fig. 10.1). If the peach is taken to a local market and the consumer buys the peach unprocessed, the distribution cost may be nominal and the processing cost nil. But some peaches may spoil in the local market and this adds to consumer costs. In order to diminish spoilage and allow a longer distribution time for the peaches, refrigerated storage and transport will add more costs. If the peach must be preserved, it must be peeled and canned or frozen. So the costs go up. If the peach is to be kept a very long time, more exacting processing, packing, and storage conditions cost still more. Finally, as long as the peach is stored in a warehouse, storage costs including interest will continue. So it is not difficult to understand that, whereas production costs can reach a minimum, processing and distribution costs always increase with time.

The cost of processing and distribution to extend the time from harvest to consumption may be offset for a while by the decreased cost of production, because the time extension will permit the food to be grown at the most efficient location. This is reflected in the minimum consumer cost at some moderate value of HC in Fig. 10.1, since the curve for *total cost to the consumer* is merely the sum of production costs and processing and distribution costs. In order to satisfy the food needs of a population at the lowest possible cost to consumers, it is clear that the food industry as a whole must strive to operate in the minimum region of the consumer cost curve. This simply means that in a dynamic society there must be sufficient processing and preservation to permit the maximum efficiency of food production and distribution. But for some foods perhaps it is not economically desirable to process, preserve, and package them to keep for very long periods. So in order to avoid spoilage losses everyone working in the chain of supply of food including the consumer must exercise reasonable inventory control to see that foods keep moving and that they are not stored beyond the period of time intended. This is merely a reflection of the fact that even processed and preserved foods are perishable. In special cases of foods required by the military, cached foods needed to sustain populations after natural disasters such as earthquakes, or foods needed in sending men into space, costs become secondary to survival itself and the increased price of long-time storage is unavoidable. However, such special needs are a minor part of the total food requirements. So in any society, region, or country the cost interactions schematically shown in Fig. 10.1 are operative. Consequently, in any particular population, attempts are made to feed people at the lowest possible costs commensurate with the economic and cultural framework of the society.

Anyone who has compared a so-called undeveloped or underdeveloped country with a highly developed country such as we consider the United States immediately realizes that the food cost factors portrayed in Fig. 10.1 operate in different ways. An economically disadvantaged group might be dependent on subsistence agriculture, where perhaps 80% or more of the total population is primarily occupied with food production. In such a situation, people simply do not have money to pay for much in the way of processing and distribution costs and so food production itself may be quite inefficient because the society must depend on locally produced food. Food production cannot be moved to its most efficient location and the cycle of availability of food organisms leads to periods of surplus with waste and spoilage and periods of shortages for more perishable foods. Dependence on seed crops is high in such civilizations, which are more common in warmer climates of the world with reasonable supplies of rain. In contrast, in a highly developed urbanized society food processing and distribution costs exceed production costs. The overall food cost to the consumer would be well beyond the economic resources of the people of the undeveloped countries, but it is well within the resources of the general population of the developed country. However, even in developed countries food costs, particularly for the urban poor, may be relatively so high that there are pockets of underfed people in a general population where many individuals are overfed.

FOOD COSTS, ECONOMIC DEVELOPMENT, AND SOCIAL CHANGE

Figure 10.2 gives a clearer picture of the relation between economic development and food costs in various populations of the world. It shows the average food costs to individual consumers in terms of percent of the total private (personal) consumption expenditures. Such data are very nearly the

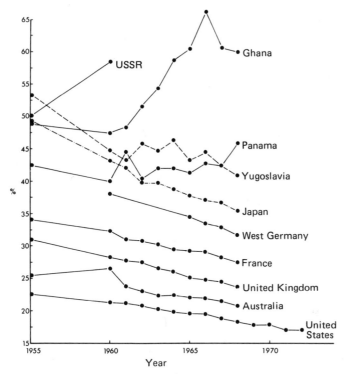

Fig. 10.2. Average food costs to individual consumers of several countries expressed as percent of personal consumption expenditures spent for food. (Based on data in *National Food Situation*, U.S. Department of Agriculture, No. 119, February 1967, and No. 137, August 1971.)

same as the proportion of a consumer's total disposable income (after taxes) which he must spend to feed himself and they reflect the general economic development of a nation. Reliable data from many of the less developed countries are not readily available. Nevertheless, Fig. 10.2 is sufficiently meaningful to show the relative food costs of countries with different degrees of economic development. First note that in the more developed countries the total expenditure for food of the population as a whole are proportionally less than in economically underdeveloped countries. In many countries, people spend more for food than for clothing, housing, medical care, or any other necessities of life. This is not true in the United States, where as much or even more money is spent for personal transportation.

Continued overall economic development in various countries such as Japan, France, Australia, the United Kingdom, West Germany, and the United States is reflected in the decreasing smooth curves of proportional food expenditures. The data begin with 1955, when the major economic, social, and political adjustments after the upheavals of World War II had been made in many countries. Where the curves are relatively horizontal with time, there appears to be no continuing social and economic development. Some of the curves show up and down trends which appear to reflect major political, economic, or agricultural production problems such as adverse weather in a particular country. Major fluctuations of this type are more discernible in underdeveloped countries with large proportionate food expenditures than in highly developed countries where less personal income is spent for food. In considering data of this type showing steadily declining values in developed countries, it is reasonable to ponder what is the lowest possible percentage of income required to be spent for food. Speculation on such matters is difficult. However, if conditions in the United States are any indication, it is worth noting that the minimum percentage of 15.4% was reached in 1971–1972. Data not shown in Fig. 10.2 indicate a slight upturn in 1973 to 15.9 and in 1974 the percentage reached 17.0%.

When the percentage of a population's total expenditure spent for food diminishes, this means that fewer members of the society are required to produce the food for all and that more people can do other things—make automobiles or television sets, build roads, teach school, or perform other services that help society to function. The result is urbanization—migration of people from the rural areas to the cities. The major impetus for such shifts comes from low per capita farm income resulting from too many people trying to make a living by food production and from increasing demand for nonfarm labor. Such population shifts associated with economic development can only come with increased efficiency of food production. Food production becomes more specialized, and the least efficient food producers seek a better livelihood in other pursuits.

A too rapid movement from rural to urban areas tends to contribute to massive concentrations of urban poor because agricultural workers displaced by more efficient food production techniques are generally untrained and uneducated for skilled industrial work. Such migrants often find no employment or only menial jobs in an economy already oversupplied with unskilled labor. But to live these people must eat and to eat they must have work to earn money to buy food and other necessities. When these needs are not met or are met inadequately, grave poverty results with its concomitant social problems. Although magnified in many developing countries, these problems are also common in so-called economically developed countries. When the poor of any society must spend a disproportionate share of their personal income for food compared to the population as a whole, they become severely disadvantaged in their society. The poor spend most of their money for food carrying relatively high processing and distribution costs and have no re-

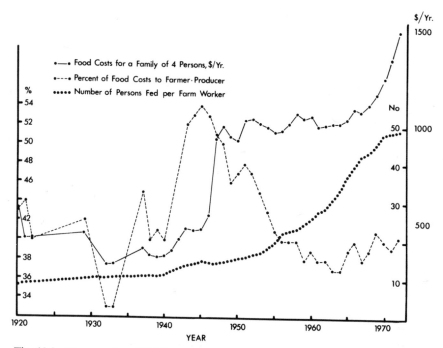

Fig. 10.3. Changes since 1920 in consumer food costs, percent of consumer food costs going to the farmer–food producer, and number of persons fed by each farm worker.

sources left to provide the other things deemed important in the society in which they live—such things as adequate housing, clothing, educational opportunities, transportation, and health care. Such poverty can be self-perpetuating because it prevents the poor from joining the mainstream of society. Rectifying such social and economic handicaps and the hardened attitudes of bitterness and hopelessness of the poor is much more compli-cated than simply improving food production and distribution. Yet these are directly involved, as is apparent from the curves in Fig. 10.2, where personal food costs are clearly shown to be related to economic development.

Figure 10.3 presents a somewhat different perspective on the relationship of economic development to a number of the factors discussed above. Here are shown for the United States the changes since 1920 in the numbers of food producers, the dollar costs to feed a family of four, and the proportion of consumer food costs going to the farmer-producer.

Let us look first at how many food producers have been required to feed our population. In 1920, one farm worker supplied the food needed by 8.3 people. This is only double the number fed by a U.S. farm worker a century earlier in 1820, when one farm worker fed 4.1 people. In 1840, this figure dropped to 3.9, reflecting the beginning of the great migration and settlement west of the Appalachians. This trend was then reversed, and by the outbreak of the Civil War one U.S. farmer supplied food to 4.5 people. By 1900 the figure had risen only to 6.9. As Fig. 10.3 shows, there was a gradual increase until 1940 in the number of people fed by a single food producer to 10.7. The need for increased food production during World War II accelerated its efficiency, as shown in the more rapid rise in the curve. At present, one U.S. farmer feeds 53 people.

What does this curve of numbers of persons fed per food producer mean in terms of population patterns? From 1920 to 1940 when the total U.S. population increased from 106 to 132 million, the numbers of people living on farms remained at about 31 million. But as the U.S. population increased to 210 million by 1972, farm population dropped from 30 million to about 9 million. This represents only about 4.3% of the total U.S. population, in contrast to 29% in 1920. In other words, between 1920 and 1972 rural population dropped from 30 million to 9 million while urban population increased from 76 million to over 200 million. Such rapid shifts in population patterns have caused radical changes in life styles and concomitant social stresses for great numbers of people to a degree never experienced before in human history.

The increased food production efficiency of the last half century is related to another factor relatively new to civilization—the controlled use of stored energy from fossil fuels—particularly petroleum rather directly and coal by way of its conversion to electricity. Until the present era, the work of tilling the soil and husbanding animals was performed by man and beasts of burden such as the horse and ox. They got their energy from the food and feed produced on the farm by photosynthesis of the plants. In fact, agriculture may be viewed as channeling the sun's energy trapped by photosynthesis into the protoplasmic material which man and animals must have for food. Fossil fuels also represent photosynthetic energy, but energy which was accumulated or stored millions of years ago. Whereas the food producer of the past had to concern himself with providing food for people and feed for his beasts of burden, the advent of the internal combustion engine freed the farmer from the responsibility of feeding horses. As agriculture became mechanized, more emphasis could be placed on feeding people.

In 1920 the horse population in the United States was 23 million; now it is only about 3 million and most of these are used for recreation and sport rather than as beasts of burden. In other words, in the last 50 years the land used for the production of feed for 20 million horses to supply energy for work has been diverted to the production of human food. So in a sense we are living on petroleum. Other technological changes have accentuated the role of petroleum, and now for every Calorie of human food produced in the United States approximately 2.2 Cal of energy from petroleum products is consumed directly in its production *and* three times this amount is consumed in the processing, preservation, distribution, and preparation of the food for final consumption. Of course, this situation is good as long as petroleum is available and cheap. The cost of gasoline, diesel fuel, and oil and of the complex machinery to use such energy efficiently in food production is a very significant factor in the costs of food production in mechanized agriculture. Consequently, it is no simple matter in an underdeveloped country to change from the small manually operated or animal-powered farm with limited economic resources to efficient mechanized agriculture. An interesting sidelight here is that whereas a large family is often an economic advantage for the manually operated farm because more children are more hands for work, large families are not such an advantage in mechanized agriculture. Underlying the increased food production efficiency over the past half century that is represented by a simple curve of the numbers of people fed by a single farmer-producer are many profound sociological and economic factors. People often express their concern about these profound effects on human welfare in political actions, as every American knows. Until recent times, the agricultural sector had more political power than city dwellers because of their numbers of voters. This is no longer true. Yet because of the primacy of food to life itself and to the economy as a whole, those who produce our food will likely retain political and economic influence greater than their numbers might indicate.

Processing and distribution costs over the past half century have shown some significant variations, although overall they have remained rather stationary for the last 15 years. This is seen in Fig. 10.3 where the percentage of the consumer's food dollar that the farmer-producer receives is plotted. Over the last half century, the farmer has received an average of approximately 40% of the consumer's food expenditures. Consequently, processing and distribution costs (Fig. 10.1) have represented approximately 60% of what the American consumer spends for food.

The effects of food surpluses and food shortages on the ratio of production costs to processing and distribution costs are plainly discernible in Fig. 10.3. Note that during the great economic depression of the 1930s, which was a time of food surpluses, the cost of production (cash return to the farmer) dropped, to reach a minimum of 33% in the middle 1930s. Then the calamity of World War II created a scarcity of food, and the proportion of the consumer's food dollar going to the farmer reached a high of 54%. It was not until 1955 that the proportion of the consumer's food dollar going to the farmer stabilized again at about 40%. In the depression of the 1930s, many farmer-producers went bankrupt because their costs exceeded their return from the sale of their products. During the 1940s and 1950s, the profit for the farmers was considerably better in spite of some increased cost factors such as labor and machinery. From 1955 to 1972, because of food surpluses, the "cost–price squeeze" experienced in some sectors of food production again created economic problems. However, the ever-increasing need for food is rapidly creating a world market for any U.S. surpluses and this will affect both domestic prices and supplies (see below and Chapter 12).

Besides domestic surpluses of some foods in the 1955–1972 period, there were two other factors contributory to the discernible though slight general decline in the percent of the consumer's food dollar going to the farmer. These were (1) the increase in institutional and restaurant eating, with its higher service costs, and (2) the demand by the consumer for more ready-prepared and highly processed foods, many of which entail increased storage and distribution (refrigeration) costs and increased processing and packaging costs (see Chapter 9). Ovbiously, the grower of the wheat used in making ready-to-bake biscuits requiring an expensive package and refrigerated distribution will receive a much lower proportion of the retail price of such biscuits than of the retail price of the flour which the consumer could use to mix and bake his own biscuits. Thus when more highly developed and sophisticated food processing and distribution patterns develop, farmer-producers tend to get a smaller fraction of the consumer's food expenditures.

Although money is the medium of exchange that facilitates commerce, the real value of money itself is not fixed. Economic and political forces in the rapidly changing world scene have profound effects on the value of money. A study of these interrelationships is beyond our purpose here. Yet these forces do indeed affect food costs and supplies, as is shown in Fig. 10.3. where dollar food costs for a family of four are plotted for the period 1920–1972. In the first 20 years, the real value of the dollar increased and family food costs in dollars decreased, but as the value of the dollar decreased beginning in World War II prices increased. This inflation—a lowering of the real value of money—has continued since then, rather abruptly immediately after World War II but nevertheless continuing even with an increased rate since 1965. Consequently, whereas annual food costs for a family of four in 1935 were only about 125 dollars, in 1975 these costs were about 14 times higher. Since the amount of food needed by a family of four is essentially constant and thus has a fixed real or intrinsic value, and since the amount of money needed to supply this food may indeed vary considerably, we have clear evidence for our earlier statement that money itself is only a representation of wealth and that food or agricultural production is a major source of wealth.

SOME CHARACTERISTICS OF ECONOMICALLY DEVELOPED AND UNDERDEVELOPED COUNTRIES IN RELATION TO FOOD SUPPLIES

Because of the rapidly increasing population in both economically highly developed and underdeveloped countries of the world, it is natural to be concerned about where all the food needed will come from (see Chapter 12). We are aware that currently there are millions of undernourished people in the world—some even starving. So as we look to the filling of future needs for food, we must try to understand why so many people in the world now have insufficient food. Then we can better comprehend the challenge facing man today—a challenge that must be met because undernourished populations of any species of organisms are susceptible to decimation and decline until food supplies become sufficient to permit survival.

Already many pages have been devoted to describing some of the fundamental biological, chemical, technological, and economic aspects of the production, preservation, and utilization of food by man. All pertain to man's ability to adequately feed himself. But because man is a social organism living by his intellect in a milieu of interdependent organisms, sustaining the social system of which he is an integral part is almost as important to his survival as food itself. We have already sketched some economic and social considerations affecting food supplies. But there are still more, and we can only give a glimpse of some of these factors, a number of which may be classified rather loosely as the "attitudes" of individuals and groups in any society. Man has evolved to where he survives by his intellect and by teaching the next generation in the methods of livelihood and the ideas and attitudes which have served the present generation and which will largely govern human activity in the next generation. Sometimes these ideas may not be entirely compatible with new determinants of human survival. An extreme case of this would be a person who would die of starvation rather than eat a food forbidden by his religion. There are many other less simplistic examples of human attitudes which pertain to man's ability to feed himself. Different attitudes are manifest in different cultures and subcultures of the human race. A detailed study of such matters is beyond our scope and is more properly the domain of anthropologists and sociologists. But since our concern here is food for people, perhaps it is appropriate to make some observations relative to economically underdeveloped and developed cultures in the world in order to understand some of the more subtle social attitudes which affect food supplies. In economically developed countries there is sufficient food for all and most of the people are adequately fed. Generally it is the economically underdeveloped or undeveloped areas where there is widespread malnutrition and insufficient food for everyone.

Obviously, a rich man can afford to buy his food and he will pay high prices for scarce food in order to survive. Similarly, a rich economically developed country will buy food for its needs if it does not produce its own food. But where does such wealth originate? This state of affairs cannot continue forever unless there is a real continuing source of wealth—a renewable resource, as the economists say. Food itself is wealth, and food production is a continuing flow of wealth that comes from the sun by photosynthesis. So if we can identify some of those attitudes and characteristics of society which promote or inhibit food production and efficient utilization, we can better appreciate some of the complexities of providing an adequate supply of food for all people.

Besides adequate food and a relatively few citizens working as food producers, developed countries are characterized by excellent systems of rapid transportation for both people and supplies, communications systems, sophisticated manufacturing facilities such as steel mills, oil refineries, textile mills, and food processing plants, marketing facilities, medical and health facilities, and schools and universities. An educated populace is essential, and we have already emphasized the complexities of food production and supply systems. The other systems that operate in developed societies are equally complex. If any one of the complex systems (such as transportation or communications) fails for any reason, the society will collapse because of this high degree of interdependence, just as a man will die who loses a vital organ. All of these systems can be maintained and operated only by trained and educated people. Of course, not all members of the population need the same kind or extent of education. In fact, one aspect of mass production of manufactured goods in developed societies is that the construction of an automobile, computer, television set, or an article of clothing or even the processing of food can be divided into many distinct but highly specialized operations so that the person doing a particular task does not need to perform or understand all aspects of the entire process.

Education and training of any kind cost money—for teachers, buildings, and supplies. Furthermore, a person in school is not earning but has to be supported. All of these educational costs require capital or accumulated wealth or money saved from food production or from services. Governments tax their citizens to provide for necessary services and for capital investments such as roads, buildings, and basic education.

There is, in fact, a counterpart of capital or accumulated wealth in a biological system such as a human being. In Chapter 5 it was noted that food must supply energy for two purposes: (1) basal energy needed just to maintain life (basal metabolism) and (2) additional energy for doing useful work or for storage as fat which can be used later if food is scarce. The basal energy requirement to survive is analogous to a person's earnings necessary to survive, and the additional energy for additional physical activities represents savings which one might invest in education or in other things that might yield additional wealth. Factories, telephone systems, power plants, airlines, entertainment facilities, etc., represent capital or accumulated earnings of people. Large sums of money are required to build a great steel mill, mine, oil refinery, or chemical factory. Products—food or any other items in transit between production and consumption—represent money, so-called working capital. Even art represents accumulated money, although some people would argue that money invested in art is not productive capital investment in the same sense as money invested in education, a factory, or a farm. In general, a major characteristic of developed countries or societies is that large quantities of capital are necessary for them to function properly.

Before going further in our discussion of the role of capital in food production, a corollary characteristic of highly developed countries must be identified in order to better understand the problems of developing countries. This is the use of energy in addition to that coming from the sun by photosynthesis. Some of this energy comes from water power derived directly from the sun but most of it derives from the burning of fossil fuels—petroleum, natural gas, and coal. To a lesser but increasing extent, energy is obtained from nuclear fuels, usually uranium, or in certain volcanic or thermal regions from heat from the earth's interior. However, all such energy sources are relatively small compared to the sun's energy.

Much of the energy is converted to electricity to operate factories of all kinds, to provide light, heat, and air conditioned comfort in homes and commercial establishments, and to power vehicles. Also, some is lost in transmission—in the United States about one-sixth of all electrical energy. The United States is currently using, on a per capita basis, twice as much energy as any other developed country, and this is many times the per capita energy consumption of undeveloped countries where perhaps the burning of animal dung is the only available source of energy for home use. Currently in the United States the annual energy expenditure is equivalent to 24,000 lb (10,900 kg) of coal or 14,600 lb (6700 kg) of petroleum for each person. This is approximately 100 times the food energy needed. Furthermore, at the present rate of increase, per capita energy use will double by the year 2010. Is such an increase possible?

At present, petroleum is the preferred energy source and oil and natural gas account for two-thirds of American energy needs. The petroleum industry is scarcely a century old, and now not only provides fuel but also the raw material for the manufacture of fertilizers, insecticides, plastics, etc. The countries supplying the petroleum on which modern industrial economy depends are in a strong strategic position in world affairs both for the present and in the foreseeable future. Although food itself is primary to human survival under any circumstances, nonphotosynthetic energy appears to be almost as important in supporting the complex structures characteristic of economically developed nations in which many millions of people must live interdependently. When we consider that the United States has only 7% of the world's inhabitants, it is obvious that energy production as well as food production is a great obstacle to economic development for any country or for the world as a whole.

Many economists feel that capital development is so intimately associated with economic and political development that the difficulty of creation of capital is perhaps the greatest deterrent to converting an undeveloped country into one that is developed by twentieth-century standards. Such enormous amounts of capital are needed and so much time is required to generate them that many people take a rather pessimistic view of the possibility for such change. In some ways, capital growth is analogous to the growth of any biological system—the proper environment must be created—and the economic parameters for capital growth are not yet as well defined as the environmental parameters for biological growth. For instance, how did the United States and other developed countries achieve their position? There is no simple answer. But, according to the British economist Barbara Ward, no highly developed society in the history of man has ever been created without the accumulation and judicious use of capital or accumulated savings. She also points out that in order to accumulate the necessary capital or wealth, every highly developed culture in the history of the world seems to have exploited some section of its own population or of other populations in the form of slavery, feudalism, captured wealth, etc. Notwithstanding how capital funds were accumulated, the creation and wise use of that capital appears to be an essential feature of developing countries. Furthermore, in order to remain viable, developed cultures must nurture capital development and learn how to use wealth properly without excessive wastage. This means that societies must establish rational priorities for the use of wealth.

There is normal attrition of wealth or capital as there is normal attrition of a human being with age, so there must be a continuing supply of new capital or wealth. Here again a dynamic economy is analogous to a biological system. One needs a continuing input of wealth and the other needs a continuing input of energy (food). In order to survive, society needs to use wealth wisely and efficiently and the biological organism needs to channel its energy into activities that assure survival and regeneration.

The role of capital in food production for developed and undeveloped countries can be vividly seen by comparing the single U.S. farmer who operates his large farm of 640 acress (259 hectares) and his counterpart who may farm 3.2 acres (1.3 hectares). The land and machinery for the one may represent an investment of $400,000 or more and for the other perhaps $1000 or less.

The one can feed himself and many more people but the other hardly survives and is often undernourished and hungry. One must be well educated—a businessman as well as a farmer, a husbander of both wealth and food organisms —and the other likely is illiterate and lives by tradition. To be sure, these are extremes, but on a worldwide basis there are more of the latter than of the former.

Often the subsistence farmer may not own the land he cultivates—he may be a tenant, sharecropper, or laborer for a large land holder and in such instances lives in what is sometimes called a *patron-oriented* or *feudal* society. In such a structure, the patron, landlord-owner, or even a representative of the government, if it owns the land, makes all decisions and is responsible for the welfare of those who work the land. Utterly dependent on their patrons, workers and their families have no ambition or creativity except for procreating and are locked in, so to speak, with little chance to better themselves.

With considerable oversimplification, developed and underdeveloped countries may be characterized as *enterprise oriented* and *patron oriented*, respectively, with the individuals and groups within the two populations displaying attitudes which tend to perpetuate their way of life. A few comparative general characteristics of the two cultures are as follows:

Enterprise Oriented

1. Any member of society may, by hard work, diligent effort, and/or creativity improve his station in life both economically and socially in accordance with his abilities. This may be accomplished by increased productivity or by investment of part of earnings (savings) in various business enterprises. An individual at least feels there is some opportunity to "get ahead."

2. Social structure is fluid and to some degree is related to income or economic resources of the individual.

3. Education is for utilitarian purposes and schools are available to all. Most people are literate. Higher education is available to those who have the ability. People go to school so that they can earn

Patron Oriented

1. Work assures only a livelihood in accordance with the standards of the patron, who makes all pertinent decisions relative to the economic and sometimes the social life of the individual. There are insignificant rewards which will affect the individual's economic and social status. Personal enterprise is minimal and not encouraged, for these is "no future."

2. Social structure is fixed and there is little opportunity or incentive for individuals to upgrade themselves socially or economically.

3. Education is for status and is limited to the elite. Few people are highly educated and most of these are of the patron class. A large percentage of the population is illiterate and there are few or no

Enterprise Oriented	Patron Oriented
more and qualify for occupations requiring specialized training.	schools available for most people.
4. Human resources are developed more in accordance with the potential of individuals and with diverse needs of society.	4. Human resource development is very limited and so creative and imaginative leadership is scarce.
5. Population is adjustable in accordance with specific needs of the society. A highly skilled and literate populace can accomplish new tasks, meet emergencies, and adapt more readily to economic, social and political change.	5. Population as a whole is rather inflexible and adjusts to new circumstances with great difficulty. Even if the patron or members of the elite see the necessity for change, the inertia of masses of illiterate, poorly educated, untrained people inhibits change.
6. People generally accept change and new ideas rather readily. In part, this is because the people have the resources to cover risks involved with change.	6. People do not adopt new ideas or changes in life style readily as they are unable to understand the need for change. People do not possess the resources to cover risks involved with change.
7. Systems of communication such as newspapers, magazines, telephones, radio, and television, are highly developed. People are informed about economic, political, and social affairs and are more politically active in their own right and display some measure of critical appraisal of their leaders.	7. Communications are poor. Relatively few people are informed about economic, political, and social affairs. Not too many of the population are active politically. Poorly educated and illiterate people tend to be passive followers if politically active. At times, people may be aroused by articulate leaders who may or may not be concerned about the general well-being of the country or people as a whole.
8. Families tend to be less tightly knit and families often move to new communities or may disperse in accordance with economic needs.	8. Family is a very strong social unit, and often many generations live in the same general locale.

Objective evaluation of the role that these attitudes and values of the people in economically developed and underdeveloped countries play in specific situations is difficult. This is so because each individual reflects his own culture in any appraisal. Nevertheless, if we assume that adequate food for the population is the primary requisite of any stable society, some meaningful relationships may be found between societal characteristics and attitudes and the inability of a particular society, country, or region of the world to provide itself with food. And the better these relationships are understood, the more optimistic is the outlook for adequate food supplies.

Since economically developed countries generally have adequate food and may even have surpluses of food, it appears that economic development is a necessary condition for changing a food-deficient country into an adequately fed one. And sometimes the more glamorous aspects of developed countries such as automobiles, expensive public buildings, and air conditioning are emphasized in underdeveloped countries quite out of proportion to their role in building the intricate fundamental supporting economic structure necessary to real development. We have seen that the scientific and technological growth of efficient food production and distribution systems results only from massive inputs of capital for education, research, roads, factories, machinery of all types, and so on, and that capital is accumulated from earnings of food producers and all others who serve the needs of society. It follows, then, that all of these people must have incomes not just for meager subsistence but to support, in some measure, all segments of society. One might say that each member of society should have an income commensurate with his ability to produce what society needs. This means not only material needs and services but also sufficient savings to continually supply the capital[1] to keep the dynamic economy operating smoothly. It does no good to produce food if people cannot afford to buy it or automobiles if there are no customers and no roads to drive on. And so it goes. The problems of economic development may be even more complex than those of keeping an established economic system operating.

As noted above, many underdeveloped countries have a high proportion of the total population engaged in producing food—a primary source of wealth for economic development. However, because many of these people are subsistence farmers barely feeding themselves there is little or no individual surplus for creating capital. If a patron is involved who siphons off whatever wealth accumulates, there will be little general economic development until these accumulated earnings of his workers are wisely used. The personal attitudes of the patron then control the economic development of all. Will the patron educate his workers, or can he even afford the cost of

---

[1] Savings or capital investment may be used for educational expenses, home building, real estate purchases, and insurance premiums, stocks, bonds, taxes, savings accounts, etc. Some of the last types of savings from many individuals may be pooled to provide sufficient capital to build steel mills, factories, etc., which require large sums of money.

education for his workers to better themselves? Are the workers themselves convinced of the value of education for their children? Many subsistence farmers need the labor of their wives and children and consequently see no value in education. Too rigid social class structures may militate against use of education. If education is used only as a mark of social status or if some people are educated for jobs that do not exist, this is waste of badly needed capital. Unfortunately, these things happen rather frequently in under-developed countries. If a large proportion of the population is locked into certain disadvantaged classes with little chance of economic and social improvement for whatever reason, there is automatically a barrier to general human resource development and a limitation on the number of people with qualifications for creative leadership.

The economic and social development necessary for improving the food (and other) resources of food-deficient underdeveloped countries involves fundamental change affecting every citizen. If tradition is strong and if people do not accept change, then development is inhibited. Often uneducated people rely more on tradition than do educated people who understand the funda-mental natural and economic forces affecting their lives. Consequently, among the illiterate any change is suspect if it is of no apparent immediate personal advantage. Resistance to change can take many forms in under-developed countries. A new agricultural technique such as use of fertilizers or of a different plow is not easily accepted by a hungry subsistence farmer who knows that slight natural changes, such as in the amount of rain, over which he has no control can profoundly affect his own already inadequate food supply. Undernourished people often will not accept a new food, even a nutritionally better food, if it is not one eaten by the patron class. Many other examples could be cited of how the population of a patron-oriented under-developed society is far less adaptable to change than that of a highly developed enterprise-oriented society. And it is this lack of adaptability which inhibits the changes that could lead to increased food supplies and con-comitant economic and social improvement.

RELIEF OF HUNGER AND MALNUTRITION

To a well-fed compassionate person, one of the most disturbing sights is that of malnourished and starving people—skin dry and parchment-like drawn over bones to cover emasciated tissues, bloated bellies of young children, and staring hollow eyes of the prematurely aged in pitiful plea for food. It is understandable that many people in societies and nations which are capable of feeding themselves are concerned that ways should be found to provide food for those who have too little to eat. However, observing the problem is far easier than recognizing the causes of food shortages and making the changes that will insure a continuing food supply.

In preceding chapters we have seen that efficient production of food organisms is itself a complex problem. And production is only the first essential step in providing food for people, because there are the problems of processing, preserving, storing, and distributing food to be met. Each step from production to consumption is affected by economic and social conditions, and in most instances of food shortages there are many contributing factors.

In the last two centuries several severe food shortage situations have been rather intensively studied and documented. Although, as we have already noted, dependence on the potato and the invasion of potato blight clearly set off the catastrophic Irish famine in the 1840s, landlord–tenant relationships and a few other social and economic factors greatly accentuated human misery, as the upper and middle classes of Ireland had little difficulty getting food. During World War II, the United Kingdom and several other countries had to contend with a long period of limited food supplies. By rigid rationing of available food by a system which put high priority on conservation of available nutrients to assure a minimum of waste with maximum possible nutrition for all inhabitants and which forced many people to greatly alter their food habits, starvation and malnutrition were avoided. The acute food shortage in the Netherlands in 1944–1945 caused by the German occupation resulted in starvation and widespread malnutrition, for the food available was only about 600 Cal per day per person. As in any situation of this kind, people became listless, without energy to work or even move about, and were always concerned about food that was not there. Vermin and disease increased; the misery was magnified by winter weather and many of the less strong failed to survive. All of this happened in a densely populated country of well-educated, industrious people who had developed an efficient food production and distribution system in an area of limited land resources for growing food. Fortunately, the severe period of imposed shortages lasted only about 9 months. Even so, many died and many who survived continued to show the effects of acute starvation. After liberation these people who had the know-how to feed themselves could quickly recover once sufficient food was made available to restore their health and tide them over until food production and distribution could be renewed. So we see in the recovery from the Dutch famine the importance of a wide base of human resource development and technology. The same types of human resources avoided excessive malnutrition and starvation in Britain and some other combatant nations during this war.

The incidents just described involving acute food shortages in traditionally adequately fed nations present quite a different situation than that found in many regions of the world where food shortages are chronic. Here famines are more frequent and insufficient food and malnutrition are a way of life for large numbers of people. Mortality rates are high for infants and for other age groups as well. Life expectancy is much reduced and the period of individual productivity is short. Usually chronic food shortages have many causes, and some adversity at any stage in the entire food supply system may increase malnutrition and famine. For instance, the soil for growing crops may not be very fertile and consequently not very productive. As food crops are removed, soil productivity drops more and there is no fertilizer. The available land may be limited, and so if for any reason crops are lost in one area the whole region or country suffers. Shifts in climate such as a late rainy season, an early killing frost, or too little or too much rain, and outbreaks of plant diseases or insect infestation can lead to crop failures. Even if crops are adequate, breakdown or inadequacy of processing, storage, and distribution facilities can lead to food wastage and widespread malnutrition, particularly among the urban population.

Of course, many of these same situations can occur in economically developed countries. However, their effects may be minimized by the degree of economic development and the adaptability of the people. For example, with good transportation systems available, foods can be more efficiently distributed or a sudden need in a particular region for an insecticide or other disease-controlling agent can be quickly met. With machinery, a faster-maturing variety of a crop can be planted if late frost kills an earlier planting or a mature crop can be quickly harvested to avoid possible losses from adverse weather. If necessary, wealthier countries can buy from the world market to alleviate local shortages provided that food is available elsewhere.

It appears then that the people of economically developed countries have the ability to identify potential problem areas in their food supply systems and that with their economic and human resources they can act to solve the difficulty before it leads to generalized food shortages. By contrast, in underdeveloped countries with chronic food shortages these capabilities seem to be lacking. Consequently the problem of feeding the hungry in countries with chronic food shortages is directly related to the problem of developing these missing capabilities as well as the more obvious deficiencies directly concerned with food production itself. Simply giving food to the hungry will not meet the long-term needs of any population of any country. Although a country that does not posses the food production resources to feed itself can buy food from food producers with a surplus, it must have the money or the products the food producers want and need—again dependent on economic development. And there is no simple way to accomplish this because of the many interacting factors.

A number of economically developed countries and the United Nations through its agencies the Food and Agricultural Organization (FAO), World Health Organization (WHO), and Children's Emergency Fund (UNICEF), as well as private organizations, are concerned with the problems of feeding the hungry of the underdeveloped countries of the world and have had some success at temporarily alleviating acute famines by gifts of food. But the seemingly intractable problem of developing the capabilities of poor countries to feed themselves is intensified by the increasing population of the world. All of the groups working to improve the food situation are aware that self-sufficiency in food requires general economic development in addition to improved agriculture and that economic development requires capital for education, transportation, industrial development, human resource development, etc. Furthermore, there can be no sustained economic growth without jobs; people must be gainfully employed. Too little is known about how best to create and use capital and to develop human resources. Without such knowledge it is hard to establish priorities for the limited resources available for international development. How much should be invested in roads, in education at various levels, in fertilizer plants, power facilities, in housing? In what order and how rapidly should each of these essentials be developed in order to assure maximum benefits for the population as a whole? Precise answers are not available. The politics and attitudes of both donor nations and receiving nations concerning effective allocation of funds to specific developmental programs often cloud the primary goal itself—general economic development. But this depends on agricultural development to provide the necessary food for all and to initiate the continuing growth of capital for the other needs incident to industrial growth which must provide jobs for the non-food producing urban workers.

The conversion of an underfed, underdeveloped country into a developed one involves profound changes in the population—their attitudes, social structure, and way of life. International developmental programs involve the infusion of one country's resources into another to accelerate change. These resources may take the form of money as gifts or loans, of products such as machinery, or of services of individuals with expertise in areas important to the recipient country. With change there are often bad side-effects—some anticipated and others not. The leaders of the recipient country as well as the leaders of the donor countries must want the changes envisioned. But this alone is not enough. Everyone, local leaders and the general population, must be willing to accept the changes.

Often the larger goal is obscured by personal attitudes. Those responsible for executing a program directed toward alteration of agricultural practices, traditional educational systems, and so on may not have any real comprehension of how these contribute to the developed economy that is the presumed goal. By the same token, the representatives of the donor countries may not appreciate many important aspects of a local situation. So serious communication problems can arise at the administrative level of international development programs between donors and recipients because they, as people, do indeed represent different cultures. At the personal level, if the recipient is not wholly convinced of the overall benefits of a particular change, sometimes an underlying inferiority complex can inhibit progress. Similarly, if the donor representative is not wholly committed to trying to help within the framework of the recipient's culture, then an underlying superiority complex also thwarts progressive change. The old adage that it is more blessed to give than to receive fails to simplify the problems of international development. In fact, it is more difficult to receive and to use effectively than to give. The nature of these problems can be clarified by examples.

For instance, an underdeveloped country may have need for transportation facilities such as trucks. The mammoth vehicles that ply the American freeways may catch the eye of the representatives of the underdeveloped country. Even though there may be funds to buy such trucks and he may want them for reasons of pride, he must be discouraged from using the available funds to buy them, because there are no suitable roads in their country. Or it may be decided that a modest factory must be built and a design furnished by the developed country calls for the latest automated labor saving machinery. However, operation of such a plant would require large amounts of energy in an energy-deficient country and even worse displace people in a country already having a surplus of laborers. Since one of the major problems of developing countries is unemployment, the displacement of labor with machines can cause acute social unrest and disillusionment.

Sometimes a technology useful in one country is not easily transferred to another. This is particularly true of agricultural technology. Many countries chronically deficient in food in are the warm semitropics or tropics. Here clothing and shelter do not pose the problem that they do in colder countries, but wintry weather is not without its benefits as far as food production, preservation, and distribution are concerned (see Chapters 8 and 9). Modern technology of agriculture and food usage common to the developed countries are based on temperate climates of the higher-latitudes. Much of such technological knowledge is not directly transferable to warm countries. Therefore, a great deal of research is needed to develop agricultural and food technologies applicable to the warm climates of the countries with limited economic resources and great numbers of mouths to feed.

Just because the problems of increasing food supplies in underdeveloped countries are complex is no reason for failing to try to alleviate food shortages and malnutrition. It is in the nature of man to continually try to solve these problems just as it is to seek new adventures, to explore the depths of the oceans, to discover new continents on earth or galaxies in the skies, or to seek new knowledge by research. The solution of a complex problem is often achieved through identifying the smaller component problems. When this is done, it is usually found that some of the smaller problems can be solved before others, and as the easier ones are solved others may be approached in turn. Certainly the production of nutritious food is fundamental to providing food to all who need it. Increasing production of a staple highly acceptable food crop per unit of land and per farmer-producer is one logical approach. It involves fewer people and requires less change in traditional attitudes of all people involved. The key person is the farmer himself and the possibility of greater economic return for his labor can be the driving force for change on his part. More and cheaper food is the driving force for the consumer. Within this frame of reference, some progress has been made toward increasing food supplies in some countries.

Since rice, wheat, and to a lesser extent corn are the major foods for much of the world, attention has been centered on these crops. A number of private organizations—particularly the Rockefeller Foundation and the Ford Foundation—and some governmental agencies of a number of countries directed their attention to these crops. Many scientists and other specialists diligently studied the production of these crops to determine the limiting factors in underdeveloped countries. Inadequate soil fertility was found to be one problem. More fertilizer could help if the farmer could be convinced that the additional cost could be overcome with increased yields. But buying and using fertilizer other than manure is a new idea for most subsistence farmers in underdeveloped countries. Furthermore, it requires investment of money many farmers do not possess. So a stumbling block was who would advance financial credit to subsistence farmers. It was also found that the types of seed that for centuries has been saved from year to year for new crops often were quite inferior in terms of yield. It has long been known that different varieties of the same crop produce different yields under the same methods of cultivation and harvest. Therefore, by the study of agronomic conditions new high-yielding wheat and rice varieties were bred that would thrive under growing conditions common in many producing areas in food-deficient countries. Their value was easily appreciated by the farmer. The use of these new varieties of traditional crops together with some fertilizer led to the so-called Green Revolution. For some countries a surplus of food suddenly became available—so much that adequate crop storage was not available in many areas. Varieties of corn having greater nutritional value (containing more lysine; see Chapter 5) as well as greater yield potential have also been developed. All of these accomplishments have had a very significant effect in improving world food supplies. However, as pointed out by Norman Borlaug, who won the Nobel Prize for his key role in bringing about the Green Revolution, these are only temporary solutions in view of increasing populations. Other aspects of the total food supply problem remain to be attacked and solved. But in view of these successes the outlook for still greater improvement in food supplies for all people is more promising.

THE RELATION OF CULTURAL ATTITUDES AND ECONOMIC RESOURCES TO FOOD HABITS

Man has prospered through the ages by the use of intellect, but he apparently has evolved and reached the stage where he has in large measure lost the subtle instincts that assure proper selection of nutritious foods. Man is apparently unique in having this problem. Because people are highly interdependent, cultural forces such as tradition, social status, economic factors, and religion all play a significant role in food selection. Sometimes these controlling factors are not in harmony with fundamental nutritional needs. Also, eating is a very personal biological function, and everyone does not physiologically respond alike to the same food. Then, too, personal preferences control to a great degree which foods are eaten, especially if there is a surplus of many foods from which to choose. There are many other factors which control man's food selection and usage, and we shall discuss a few of them in order to understand the importance of cultural, personal, and economic factors in supplying people with nutritious food and getting them to eat it.

The types of food organisms that different ethnic groups use reflect their habitats. An isolated jungle tribe living in the South Pacific will use different organisms than a tribe living in a warm area of the western hemisphere, even though both groups may live largely on plants. On the other hand, Eskimos, who live where few plants grow, survive by consuming animals such as whales, seals, and walruses. Marine foods make up a large proportion of the food in countries whose seacoast areas are relatively large compared to arable land, but such foods contribute little to the normal U.S. diet, which is less than 1% fish or seafood of any kind. The Laplander depends to a great degree on the reindeer, while the Southeast Asian depends on rice and the Punjabi on wheat. Food habits often become so fixed in a culture that, even though food supplies may be inadequate, the introduction of a new food is extremely difficult and can thwart the best intentions of authorities to improve food supplies. The options for augmenting food supplies in highly populated food-short areas are severely limited when the only solution considered acceptable is increase of traditional foods prepared in traditional ways. Often this course of action may not be possible for reasons of land resources or climate, or some other fundamental scientific, economic, or technological problem.

Food habits based on individual tastes are sometimes easier to change than those based on underlying esthetic, cultural, or religious taboos. This is particularly true in cases of severe food shortages. However, as food becomes increasingly scarce even the strongest taboos may be overcome when the alternative is starvation. In extreme cases of disaster people have been known to consume the flesh of dead comrades. In less extreme cases plants and animals not usually eaten may be consumed. The ancient origins of particular eating habits may have been based on sound reasoning, such as the fear of contaminated meat. When such rules become religious dogma, they may be practiced long after their original basis has lost significance. For instance, the unique and sacred place of cattle in the lives of the Hindus has a profound effect on food supplies in India. Since these cattle are not eaten and are allowed to live and reproduce even though producing no milk and little or nothing as beasts of burden, human food is either lost in order to maintain nonproductive cattle or is not produced at all. The attitude of not killing any animal, or even insects, can seriously and dangerously deplete human food supplies, for in reality such religious tradition puts less value on human life than on the lives of other species. This attitude seems to be counter to a fundamental premise of human existence which antedates recorded history, and that is that man must nurture the organs he needs for food and control or destroy those organisms which would kill him or his food organisms. Of course, changing such religious traditions for masses of uneducated people requires generations of dedicated effort. In the meantime adequate food supplies for increasing millions of people may not be forthcoming because of the secondary position which food holds in the priorities of some stressed human populations.

Introducing new foods or new food organisms to alleviate food shortages in populations may lead to unforeseen difficulties of a nutritional or physiological nature. Certain food taboos which may be a part of social tradition in some societies may well have a physiologic or genetic base. Eggs, milk, meat of certain animals and fishes, fruits, seeds, and leaves of certain plants can as we know, cause today allergic or other toxic reactions in certain individuals. Perhaps because of this, many useful foods in one society are by tradition avoided by another. When a certain food produced an adverse physiological reaction in a leader, tribal chief, or medicine man, the use of that food may have been discouraged or prohibited by custom or law. Perhaps certain food taboos developed in this way. But in the long run it is more likely that the process of evolution has played a part in the establishment of food patterns such that radical changes in foods consumed by a particular population can occasionally result in undesirable side effects from a nutritional point of view.

The Eskimos and non-milk-consuming populations offer two interesting examples.

Eskimos live where there is almost no food of plant origin, and they live almost exclusively by consuming hunted animals. Most of their food is uncooked and they consume almost all parts of the animal, even to the stomach contents in some instances. By these customs they avoid vitamin C deficiency and scurvy, which we usually associate with the lack of fruit and vegetables. When the Eskimos use many foods from other cultures to replace their normal food, cooking their meat and discarding the organ meats as do the people of the more developed countries for so-called "esthetic" reasons, nutritional and health problems may develop because of insufficient intake of some essential nutrients. In turn, people from the warmer climates with their personal tastes and prejudices for food find accommodation to the Eskimo diet very difficult.

In the United States, Canada, and the European countries, milk is often described as the "perfect food." Yet for many people in the world it is not. Where for millennia milk has not been a normal adult food, populations have evolved who are intolerant of milk as adults. Milk, of course, is the natural food of young mammals, including human infants, and the main carbohydrate of milk is lactose. After animals are weaned, they generally do not consume milk, and as they mature they lose the ability to digest lactose because of the disappearance of lactase, the intestinal enzyme to digest lactose to glucose and galactose (see Chapter 7). So in these adult animals intestinal disturbances often with diarrhea may result when milk in significant quantities is consumed. But lactase does not diasppear in all individuals and its persistence in adulthood is genetically controlled. When lactase persists, milk is digested without difficulty. This is the case in man.

In those cultures of the world where milk is an important food for adults, relatively few people are lactose intolerant. However, as Table 10.1 shows, where milk is very uncommon as an adult food, most people cannot digest lactose after early childhood. Among the Thai people and the Ibo and Yoruba of Nigeria, almost all adults are incapable of digesting lactose. Among the Chinese, the Ganda tribe of Uganda, and the Hausa tribe of Nigeria, more than 75% are intolerant of lactose. On the other hand, only a few percent of the Swedes are intolerant and fewer than 20% of the Swiss, Finnish, and U.S. White populations show this trait. The Fulani of Nigeria and the Tussi of Uganda, who, unlike their neighboring tribes, are herders of animals and consume much milk whereas the Ibo, Yoruba, and Ganda do not live by husbanding animals and consume no milk in adult life. In their tolerance for lactose U.S. Blacks closely resemble the Ibo, Yoruba and other West African tribes of similar food habits from which they are derived. Cheese and fermented milk products usually contain little lactose, and so these foods

Table 10.1. Percent of Adult
Populations of Various Cultural
Groups Who Are Intolerant to
Lactose

Group	Percent
Thai	98
Ibo (Nigeria)	98
Yoruba (Nigeria)	98
Chinese	89
Ganda (Uganda)	80
Hausa (Nigeria)	76
U.S. Black	71
Hausa-Fulani (Nigeria)	69
Fulani (Nigeria)	21
Tussi (Uganda)	19
Finnish	16
U.S. White	14
Swiss	12
Swedish	3

Adapted from Norman Kretcher, *Scientific American 227* (No. 4): 70 (1972).

generally are digested well even by lactose-intolerant people. But whole milk, whey, and nonfat milk solids contain lactose and so have limited value in relieving food deficiencies for some populations.

The food habits and attitudes of a population are influenced by their economic resources in the same way that personal wealth influences an individual's selection of food. The poor will not eat caviar and the rich will not eat large quantities of corn or millet. Most people like foods of animal origin—meat, eggs, and dairy products—which are also generally of high nutritive quality. These foods are relatively expensive because they cost somewhat more to produce and are in high consumer demand. Particularly in developed nations, foods of animal origin occupy a primary place in the diet. Figure 10.4 shows that meat consumption in various selected countries is closely related to average personal income. To be sure, it appears that Italian and Finnish eating habits call for less meat and the British for some-what more meat than might be expected from income figures alone, but the relationship of meat consumption to economic resources is unmistakable. Furthermore, as underdeveloped countries improve their economic position the total consumption of meat and other animal products increases and that of cereals decreases in much the same way as when individuals move up the economic scale. This is reflected in Fig. 10.5 and Table 10.2, which show

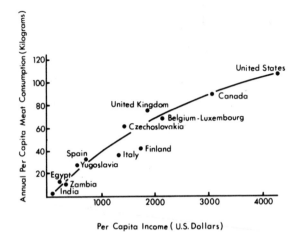

Fig. 10.4. The per capita consumption of meat in relation to the per capita income of the population in selected countries. (Adapted from B. E. Hill, *World Animal Review*, No. 41, 1972, Food and Agricultural Organization of the United Nations.)

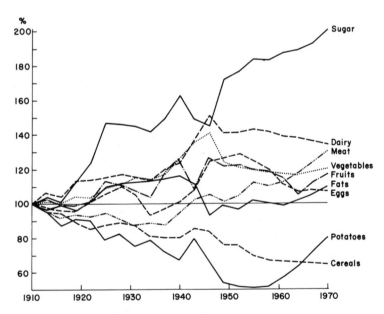

Fig. 10.5. Trends in eating habits. The 3-year average of the 1909–1910–1911 consumption of each category is given a value of 100% at the year 1910. The curves are plots of the 3-year averages from 1909 through 1971. Meat includes fish and poultry. Dairy includes milk and milk products except butter. Fats include butter. Sugar includes glucose, corn syrup, and other sweeteners. Potatoes include sweet potatoes.

Table 10.2. Per Capita Annual Consumption of Major Categories of Food
Shown in Fig. 10.5

Category	1909–1910–1911 Average Pounds	1909–1910–1911 Kilograms	1969–1970–1971 Average Pounds	1969–1970–1971 Kilograms	Change in percent of 1909–1910–1911 average
Meat	175	79.3	229	104	131
Eggs	37.5	17.1	40.3	18.3	107
Dairy products[a]	347	158	467	212	135
Fats	41.5	18.9	45.7	20.8	110
Fruits[b]	136	61.8	183	82.6	135
Vegetables[b]	202	91.7	244	111	121
Potatoes	205	92.6	165	74.5	80
Cereals	295	134	192	87.2	67
Sugar, corn syrup, other sweetners	87.9	40	176	80	201

Prepared from data obtained from *Food—Consumption, Prices, Expenditures*, U.S. Department of Agriculture, Agricultural Economics Report No. 138, July 1968 and Supplement for 1971.
[a] As milk equivalent.
[b] As fresh equivalent.

changes in the consumption of certain types of food in the United States since 1910. Doubtless, other factors such as the advent of the science of nutrition and nutritional education programs have contributed somewhat to the changes in eating habits of Americans. But the changes in economic resources which have led to new technologies seem also to be quite clearly related to changing food habits (compare Fig. 10.5 with Figs. 10.2 and 10.3).

Economic and social development have other effects on eating habits which are not beneficial from the nutritional point of view. Increased consumption of animal protein foods does contribute to improved nutrition, but there are also increases in the consumption of fat (not shown) in addition to that attributable to animal foods—such as fats and oils used to prepare so-called richer foods. This is of concern to some nutritionists and health authorities who believe that fat consumption is related to cardiovascular diseases. Although the precise causes of atherosclerosis and associated problems are still unknown, persons who eat too much and are excessively overweight have a lower life expectancy from a statistical point of view than individuals of normal body weight.

There are other economic factors related to obesity in addition to the fact that people with more money can buy and eat more expensive, rich food than they need. In an economic system where labor-saving devices are emphasized in agriculture, industry, and the home, the need for human manual labor is decreased. Many tasks require only relatively small amounts of energy, and when machines replace manual labor and automobiles and elevators carry people where they need to go, fewer calories of food energy are needed. It is characteristic of highly developed economic systems that many occupations become more sedentary. However, even though people in such a society may need fewer calories, they still require the same amounts of all other nutrients. In terms of eating habits, this means that the consumer must exercise good judgment in selecting foods which supply the necessary proteins, vitamins, and minerals but do not contain more calories than needed. In other words, to be properly nourished a very active person needing many calories has more leeway in food selection than a quiet person working at a desk all day. The same amounts of calorie-loaded foods with little or no other nutrients such as soft drinks, alcoholic beverages, and candy are a greater nutritional obstacle to the sedentary person than to the active one. This is because for the less active person such foods would likely replace other foods that more nearly supply all necessary nutrients.

In more affluent societies there is more to spend on soft drinks, alcoholic beverages, candy, etc. Of all food items, these carry the highest advertising costs. In fact, sales of these are more directly related to advertising than are sales of staple foods. Sometimes these foods even become a sort of status symbol so that poorer people with less to spend on food will often buy them rather than food which would better satisfy their physiological needs. In 1971, 11% of all money used to purchase food in U.S. supermarkets was spent for these foods (see Table 10.3). This figure does not include purchases outside of supermarkets, and large quantities of alcoholic beverages and soft drinks are sold elsewhere.

The family has traditionally been the social unit and the home has been where food is consumed in a social setting. However, twentieth-century socioeconomic development has affected this. We have already noted that developed countries are characterized by mobility of population, and this has changed family life. In a large percentage of families both mother and father work outside the home—many times on different schedules. Children are often at school or in day-care centers and as they grow older are at work themselves. Under these circumstances, family eating patterns become irregular and each member of the family is on his own at times. Little or no breakfasting, eating on the run, and snacking throughout the day and around television in the evening are common alterations of the traditional pattern of set meals wherein the mother was responsible for all meals and for selecting and preparing foods for the family. It was she who influenced eating habits and in so doing was the guardian of her family from the nutritional point of view. Now that more than a third of all food is consumed outside

Table 10.3. Percent of Each Food Dollar Spent for
Major Categories of Foods in U.S. Supermarkets,
1971

	Percent
Bakery foods	8.5
Dairy products	8.3
Frozen foods	6.4
Red meat (beef, pork, lamb)	24.6
Poultry	3.3
Fish and seafoods	0.9
Fresh fruits and vegetables	13.1
Dry groceries (canned goods, flour, cereals, fats and oils, etc.)	23.9
Alcoholic beverages, soft drinks, candy	11.0

Adapted from the 25th Annual Consumer Study of *Super-marketing Magazine.*

the home in the United States and more ready-prepared foods are used in the home in order to accommodate the various schedules of family members, it has been found that these changes can have effects on general nutrition. This is true not only for the poor but also for the middle and upper classes. School lunch programs with emphasis on good nutrition have been started to help counteract this trend. Although these are quite beneficial, they cannot entirely make up for the lack of good nutrition at home. Furthermore, many schools have no programs or inadequate ones. Although economic develop-ment can assure an abundance of nutritious foods, it apparently can diminish the effectiveness with which these food supplies are used. Nutritionists, sociologists, and public health officials are all concerned with this problem, and many technical studies have been done. Also, the popular press, television, and radio often carry information in this area—unfortunately, not always accurate information.

CONCLUSION

In this chapter, some of the economic and social forces affecting the supply and use of food have been described. These factors, distinct from the fundamental chemical and biological principles governing the growth and utilization of food organisms, are so important that they control to a great degree the types of technologies which can be used to feed a particular human population. The way of life of the peoples of primitive, economically under-developed, and developed cultures has a profound effect on how they feed themselves. So if modern science is to improve the quantity and quality of food for any group, the options available for the use of this knowledge are in a large measure determined by economic, social, and even political forces. Consequently, improvement of the food situation in any society or country is a complicated problem to which there is no quick and easy solution. Yet both man and society are by nature dynamic and ever-changing, and so is the quest for food. It is only by trying to understand all aspects of food production and use that viable societies can be maintained.

II

GOVERNMENT REGULATION
OF FOOD

As men established systems of government to serve their needs, food became a governmental concern. Wars have been fought over food supplies and lands for growing food, and control of food has always been used as a military weapon. Armies and countries have been starved into submission. Food has also always been a strong bargaining factor in international trade and diplomacy. The development of modern science and technology, which nurtured the industrial revolution and the economic development of the past two centuries, resulted in more complex societies and governments. The primacy of adequate food supplies to economic growth led to the establishment of ministries or departments of agriculture and institutions for research and education to promote food production.

We have already seen that in complex societies a large proportion of the population is absolutely dependent on other smaller segments of the population for food production, processing, and distribution. In such a situation it is inevitable that the people who have no control over their own food supplies will become concerned about the quality and wholesomeness as well as the quantity of the food available to them. Consequently, citizens are presently demanding that their governments regulate all segments of the food industry from production to consumption to assure not only that there is food in the markets but also that this food is not adulterated, is wholesome and nutritious, and is as advertised and labeled by the seller. Consumers respond more personally to these issues than to the more remote but nonetheless equally important aspects of food supply with which their governments must deal.

Of course, governments are subject to political pressure from diverse segments of their citizenry. Here is the meeting ground of special interests, domestic and foreign political forces, laws, moral and legal philosophies, and the guiding principles of science which determine the survival of man and the organisms he requires for food. In briefly examining some of the scientific aspects of this state of affairs, we will again find that there are no easy answers for many food problems facing us today. So-called experts often disagree even on limited aspects of the total food picture.

CONSUMER PROTECTION

In early civilizations, when the food available for purchase consisted of easily identifiable food organisms—grains, fruits, animals, etc.—there was something of a natural built-in protection for the consumer—the living organism. In other words, if the food organism was alive, then it was safe to use for food. Unsatisfactory food was avoided. A rotten apple was easily identified, as were sand or insects in grain and sick or dead and putrid animals. Milk that had been diluted with water was more difficult to detect. As civilizations became more developed, grains were milled, foods were preserved by fermenting, salting, drying, etc., or otherwise modified, and adulteration became easier for the unscrupulous supplier, processor, or merchant. The consumer's only recourse was bargaining—a weak weapon when foods are scarce. Although without food processing and preservation modern civilizations could not have evolved, these operations make it much more difficult for consumers to protect themselves.

By the late eighteenth century, adulteration and fraud involving food were common, particularly in the larger cities. At this time, the basic foundations of the science of chemistry were being developed. Among the leading scientists of the era was Frederick Accum, a German living in London (see p. 161). He was a perceptive experimentalist and effective teacher who attracted young aspiring scientists, including some Americans. Among other things, he established a firm for making laboratory apparatus and chemical supplies which facilitated research and scientific progress. But more than his contemporaries, he saw the great practical value of the science of chemistry. Through his efforts, gas manufactured from coal was introduced to light the streets and homes of London, a great improvement over candlelight. Other contributions of Accum attracted still more attention. He was often called on to advise Parliament, and he was an articulate writer and lecturer. One of his interests was the applicability of chemistry to food problems. His books on bread, beer, and wine making were so definitive that they were reprinted for many years after their original publication and were translated into other languages. These and his last English book, *Culinary Chemistry*, clearly demonstrated the importance of chemistry in improving food quality.

A TREATISE
ON
ADULTERATIONS OF FOOD,
AND
Culinary Poisons,
EXHIBITING
THE FRAUDULENT SOPHISTICATIONS
OF
BREAD, BEER, WINE, SPIRITUOUS LIQUORS, TEA, COFFEE,

Cream, Confectionery, Vinegar, Mustard, Pepper, Cheese, Olive Oil, Pickles,

AND OTHER ARTICLES EMPLOYED IN DOMESTIC ECONOMY.
AND
Methods of detecting them.

THE SECOND EDITION.

BY FREDRICK ACCUM,

Operative Chemist, Lecturer on Practical Chemistry, Mineralogy, and on Chemistry
applied to the Arts and Manufactures; Member of the Royal Irish Academy;
Fellow of the Linnæan Society; Member of the Royal Academy of
Sciences, and of the Royal Society of Arts of Berlin, &c. &c.

London:
SOLD BY LONGMAN, HURST, REES, ORME, AND BROWN,
PATERNOSTER ROW.
1820.

Fig. 11.1. Title page of Accum's *A Treatise on Adulterations of Food*, Second Edition. (Courtesy of Harvard College Library. Photograph by The Ohio State University Department of Photography.)

Accum was also keenly aware that the principles of chemistry could be used to detect the adulteration and fraud so rampant in the food business of his time. At the pinnacle of his popularity and prestige, climaxing almost 30 years of work in England, he published a most forceful volume, *A Treatise on Adulterations of Food* (see Fig. 11.1). In this book he not only showed how fraud could be detected but also went so far as to identify those involved. The book was a sensation and went through many printings in England and in other countries. Accum's work marked the beginning of the pure food movement—protection of the consumer with respect to the wholesomeness of his food over which he has little control.

Accum was unprepared to meet and fight the wrath and perhaps the cunning of his new adversaries. Criminal charges were brought against him for removing some pages from books in the library of the Royal Institution. (Accum, as did many in his day, removed pages from his own journals and books.) The judge dismissed the case, but he was reindicted. Many of Accum's friends abandoned him. His zeal as a reformer was broken. He forefeited his bond, abandoned his considerable business, and returned to Germany as a professor in Berlin.

The fraud and food adulteration exposed by Accum did not subside, and the first move by the English Parliament for any kind of corrective measures did not come until 40 years later. Some other countries followed with food laws during the period 1860–1875.

Accum's experience was found to be the lot of other pioneers in the pure food movement. In the United States, the career of Dr. Harvey W. Wiley, pioneering chemist of what we now know to be the Food and Drug Administration, was marked by vilification, obstructions, and pressures often arising from Congressional action fostered by special-interest groups of considerable political power. Even today, politics plays a considerable role in matters of consumer protection. It is difficult for most laymen to evaluate the merits of accusations, political pressures, and scientific evidence bearing on consumer food problems. Here we shall touch briefly on the nature of some of these concerns that affect all of us.

Short Weight and Misrepresentation

Common many years ago, short weight and misrepresentation are relatively minor today, although not unheard of. Local and national governments have set up systems for checking the accuracy of scales and other measuring devices. Because so much of our food is packaged and it is required that each package declare, within narrow limits, the weight of its contents (net weight), short weights cannot be easily disguised.

Misrepresentation of foods became more common as processed foods began to appear on the market. In recent years, governmental regulating agencies have made this type of deception more difficult by requiring that ingredients be listed on the label in order of amount present in the particular food.

Another safeguard is the establishment by the government of "standards of identity" for some prepared or processed foods. For example, the basic composition and permitted ingredients for many common foods such as bread, catsup, butter, margarine, mayonnaise, various cheeses, sausages, and some soft drinks have been established through formal procedures stipulated by Congress. In the case of butter and nonfat dry milk, Congress itself established standards of identity. If a food has no standard of identity, its ingredients must be listed on the label. If a standard food has ingredients in addition to those required, only the additional ingredients need be identified on the label.

Standards of identity have contributed significantly to reducing the misrepresentation of foods. But even here there have been abuses resulting from political pressure by special-interest groups. Since it is not required that so-called standard foods be labeled with statements of ingredients, it is possible to hide the presence of certain ingredients from the consumer if the standard of identity can be made to include them. For many decades, artificial colors have been legally added to butter and certain cheeses without declaring such ingredients to the consumer. Until about 25 years ago, similar coloring of margarine was forbidden and even the use of naturally yellow animal and vegetable oils for margarine manufacture was illegal. Certain soft drinks, particularly the cola beverages, contain caffeine as a mild stimulant. Coffee and tea naturally contain caffeine, but caffeine is added to some formulated soft drinks, as are sugar and citric or phosphoric acid. Because standards of identity have been established, manufacturers of certain soft drinks are not required to identify caffeine as an ingredient. There are other examples where standards of identity have perhaps not served the consumer as well as they might. However, in total, such standards have been useful, and, in the view of some people, abuses could be avoided by requiring proper ingredient declarations for *all* processed foods.

Adulteration

As Fig. 11.1 shows, adulteration of food was a matter of great public concern in 1820. This fraudulent practice usually takes three forms: (1) substitution of one ingredient with another which may or may not be harmful, (2) use or disguise of unwholesome, dirty, or filthy ingredients such as those partially spoiled due to chemical deterioration, microbial contamination, disease, insect infestation, or damage by other animal life (rats and mice usually) (see Chapter 9), and (3) inclusion of an illegal substance which may or may not be harmful to the consumer.

Deception by substituting a less expensive ingredient for a more expensive one can be done in many ways. At present, however, by use of highly sophisticated analytical techniques, it is fairly easy not only to detect such adulteration but also to determine how much of the adulterant is present. Since most foods and food ingredients are derived from biological organisms, their correct chemical composition can be defined within fairly close limits, be the organism a cow, orange, onion, or grain of wheat. Thus the adulteration of olive oil or butter with less expensive vegetable oils or even mineral oil can be identified, as can the addition of talc or corn flour to wheat flour, or pebbles to cereals and beans, or cereals or other fillers to ground meat products (sausages and hamburger), or water and sugar to orange juice. Many similar examples could be given.

Water is a common adulterant, although less so now than in the past. Since many foods contain water, the use of it to dilute liquid foods such as milk, beer, and wine or to add weight to solid foods such as sausages and bread is an ancient practice. Now, however, water added beyond proper limits is usually quite easily determined. But since added water is permitted under certain circumstances, governmental regulations are rather complex in this area, as in many others.

Many regulations and their applications involve value judgments and considerations of the problems of feeding people in complex societies. In Chapter 9, the importance of good sanitation in the milking of cows and the processing and distribution of milk was emphasized. To keep milking machines, tanks, pumps, and other items of equipment clean and free from contaminating bacteria, washing with water and cleaning agents and rinsing with clean water is the usual procedure. Inevitably some water remains in this equipment after cleaning. This water, since it dilutes the milk, may be considered an adulterant. Yet without such water milk could not be economically supplied to the consumer. Consequently, 3% added water is permitted in the fluid milk that reaches the consumer. If milk contains more than this amount it is considered adulterated. Added water is permitted for technological reasons in other foods such as fresh poultry, certain meats, and meat products. Of course, water is a normal ingredient in many processed and prepared foods. It is difficult to establish the amount of water necessary or desirable in a particular processed or ready-prepared food and the amount to be considered fraudulent. In the promulgation and enforcement of regulations involving judgments of this kind, regulatory officials find themselves subject to pressure from groups representing conflicting interests.

The inclusion in processed food of parts of food organisms not usually eaten was practised more in the past than at present. The biological origin of foods makes it rather easy to incorporate peels, hulls, stems, pits, hairs, skins, feathers, shells, etc., in foods either intentionally or inadvertently. Furthermore, in food preparation and processing it is easy for a partially peeled fruit, bean stem, or olive or cherry pit (seed) or pit fragment, etc., to get into the canned or frozen food. However, such adulteration is rather easily detected and most such defects can be avoided with proper processing and quality control. Consequently, the standards of identity and food regulations specify the limits of such defects. For example, not more than 3% of pitted prunes may contain a whole pit or pit fragment exceeding 2 millimeters (0.08 inch) in length.

The biological nature of foods quite naturally can lead to contamination of foods by other biological material. We have repeatedly emphasized that a fundamental characteristic of the dynamic biological cosmos is that organisms are in incessant competition with each other. We have noted in Chapters 8 and 9 that a major aspect of food production, processing, and preservation is the prevention of the consumption of our food by competitive organisms whether they be microorganisms, insects, or larger plants and animals. So it is inevitable that in unprocessed and even processed foods there is some degree of biological contamination. Many times this type of contamination is of no consequence and at other times it is a major adulteration. To set and enforce guidelines for protecting the consumer in this difficult area, we look to governmental regulating agencies. Here again value judgments are involved.

There are many kinds of biological contamination or adulteration—for instance, the use of moldy or otherwise rotted fruit; spoiled, tainted, or diseased meat; insect-infested flour, vegetables, or fruits; foods befouled by insect or rodent excrement or other detritus; or meats laden with fecal or intestinal material or soil as a result of unsanitary dressing procedures. Sometimes when contaminated foods are processed or used to make prepared products, their presence can be masked. Indeed, the use of certain spices is thought to have arisen in the ancient past to make more acceptable what we might now consider to be spoiled food. But how can we define unwholesome or spoiled food? Sometimes what appears to be spoiled or unfit food may not be harmful to those who eat it, while at other times food which appears wholesome may in fact be quite harmful. So the establishment of legal definitions of what are and what are not clean, wholesome, safe, and nutritious foods is difficult. Furthermore, cultural, religious, and esthetic considerations play an important role along with scientific and economic factors in the establishment of food standards relating to biological contamination or adulteration. Although the details of the legal guidelines for permitted biological contamination of foods are beyond our scope here, we shall mention some of the allowances which the Food and Drug Administration has established as the highest permitted levels of biological contamination of several foods. These amounts are quite small and often reflect the biological nature of the food itself.

Molds attack many foods in the field and during and after processing and distribution. Colonies of white, green, yellow, blue, and black molds are often seen on fruits, vegetables, bread, cheese, etc. The molds are rather complex multicellular organisms characterized by filaments (mycelia) which are only about 1 or 2 microns in diameter. As the mold grows on the food, the mycelia extending into and above the food form a fibrous mat or bed supporting at their external ends fruiting bodies which are about 20–40 microns in diameter and produce the spores or conidia that are analogous to seeds in higher plants. Individual spores or conidia usually are only a very few microns in diameter. The mycelia tend to entwine to form larger fibers and are easily identified with a microscope even after a food has been cooked, frozen, canned, or otherwise processed. Although many molds are harmless, they often indicate spoilage. So the number of mold fibers or spores in a given amount of food (usually fruits or vegetables) is used as an index of the amount that is spoiled and is called the *mold count*.

The mold count is usually given as a percent and is arrived at in the following manner. A representative sample of the food is spread on a special microscope slide to a thickness of 0.1 mm. One-hundred 0.15-mm^3 portions of the food are examined under the microscope and the portions containing an identifiable mold fiber of specified size or larger are counted. The number represents the percent of 0.15 mm^3 microscopic fields which contain the specified amount of mold tissue. So the mold count given in percent does not mean that that percentage of the food examined consists of mold, but rather it is an empirical index. Actually, the amount of mold in each positive field is likely to be only about 1% or less of the amount of food in the field examined, so the actual amount of mold consumed in the food is more likely to be 1% or less of mold count. For example, canned crushed pineapple with a mold count of 15%, or 15 of 100 microscopic fields of 0.15 mm^3 showing mold, would contain certainly less than 1.5 parts of mold tissue per 10,000 parts of crushed pineapple. However, since the mold count is related to the amount of decomposing or rotting tissue in the food organism, the actual amount of unsound fruit is usually somewhat higher, perhaps about 0.1%. The correlation of mold count with percent rotted tissue is not very close and varies with the observer and with the particular food, yet mold counting is much more accurate than more subjective observation.

The permitted mold count for orange juice is 15% or less, for tomato juice 20%, and for strawberries 55%. Why the difference? Orange juice is prepared from only the internal pulp of the fruit; rinds or peels are discarded.

Tomato juice is also prepared from the internal pulp but ripe tomatoes are more fragile and more subject to attack by molds. By contrast, strawberries are consumed whole. Furthermore, strawberries are of a texture that easily harbors mold spores. Consequently, a mold count of 55% in strawberries does not indicate the same degree of unsound fruit as would 55% in orange or tomato juice, where the peel is removed.

Many foods are spoiled by the invasion of ubiquitous bacteria rather than by molds. Sometimes molds and bacteria are both involved. Bacteria are very small unicellular organisms which are not easily measured and counted except by culturing techniques, which require viable organisms. Molds and mold fragments—even those killed by heat—are easily identifiable microscopically in food. Because similar identification of bacteria in processed food is difficult or impossible, using the bacteria count as an index of unwholesome food or food ingredients is not too satisfactory. Sometimes certain chemical compounds produced by spoilage bacteria can be used as indices of spoilage, but this too is difficult.

For instance, fresh milk and meat are easily spoiled by invading bacteria. These foods may become sour, putrid, gaseous, slimy, or even of different color or texture depending on the species of bacteria. The changes are caused by the products of metabolism of the bacteria such as acetic, lactic, butyric, or other acids, ammonia or other amines, hydrogen sulfide, methane, carbon dioxide, or polysaccharides, most of which may not pose a health hazard. In many cases, such tainted food may contain perhaps 5 or 10 million or more living bacteria per gram of food. Yet on a weight basis the bacterial tissue itself may amount to about only 1 part per million in the food. Counts of viable bacteria in or on food or on equipment may indicate the degree of sanitation or the wholesomeness of a food at a particular time or how long it will take for the food to become spoiled by the multiplying organisms present. However, bacterial counts are poor indicators of whether or not a spoiled or rotted food ingredient has been used in preparing processed food. Many food processing and preservation procedures are based on killing potential spoilage bacteria; therefore, if there are no or few viable bacteria in a particular food, such as a can of soup or a frozen casserole, an ordinary bacteria count will tell little about the bacterial load of the ingredients used in preparing the soup or casserole.

Food organisms during their production, storage, processing, and distribution are subject to attack by insects and higher animals. The mammals are usually rodents such as rats and mice, and the insects may range from very small ones invisible to the unaided eye to much larger ones. Some insects, such as the tiny thrips and aphids, are common competitive organisms during the production stages, whereas weevils and cockroaches are greater hazards during storage and processing. Flies such as fruit files, often known

Table 11.1. Natural or Unavoidable Defects in Selected Foods for Human
Use That Present No Health Hazard

Food	Defect action level
Potato chips	6% of the chips examined show rot.
Chocolate and chocolate liquor	Average of 60 microscopic insect fragments per 100 g when six 100-g subsamples are examined or if any one subsample contains 100 insect fragments. Average of 1.5 rodent hairs per 100 g when six 100-g subsamples are examined or if any one subsample contains 4 rodent hairs.
Eggs, frozen	Shell, 2% by weight. Two cans (of a lot) contain decomposed eggs and subsamples from cans classed as decomposed have counts of 5 million bacteria per gram.
Fish, fresh and frozen	50 cysts per 100 lb of whole fish or fillets provided that 20% of the fish are infested.
Corn meal	Average of 1 whole insect (or equivalent) per 50 g. Average of 25 insect fragments per 25 g. Average of 1 rodent hair per 25 g. Average of 1 rodent excreta fragment per 50 g.
Caneberries (blackberries, raspberries, etc.)	Microscopic mold count average of 60%. Average of 4 larvae per 500 g, or average of 10 insects (larvae or other insects) per 500 g (excluding thrips, aphids, and mites).
Peaches, canned	Average of 5% wormy or moldy fruit by count or 4% if a whole larva or equivalent is found in 20% of the cans.
Raisins	Average of 5% by count of natural raisins showing mold. Average of 40 mg of sand and grit per 100 g. 10 insects or equivalent and 20 *Drosophila* (fruit fly) eggs per 8 oz.
Cherry jam	Microscopic mold count average of 30%.
Black currant jam	Microscopic mold count average of 75%.
Apple butter	Microscopic mold count average of 12%. Average of 4 rodent hairs per 100 g. Average of 5 whole insects or equivalent (not counting mites, thrips, aphids, scales) per 100 g.
Tomatoes, canned	Ten *Drosophila* fly eggs per 500 g, or 5 *Drosophila* fly eggs and 1 larva per 500 g, or 2 larvae per 500 g.
Tomato catsup	Microscopic mold count average of 30%.
Sweet corn, canned	Two larvae, cast skins, larval or cast skin fragments 3 mm or longer of corn ear worm or corn borer, *and* aggregate length of such larvae, cast skins, larval and cast skin fragments exceeding 12 mm in 24 lb (24 No. 303 cans or equivalent).
Broccoli, frozen	Average of 60 aphids, thrips, and/or mites per 100 g.
Peanut butter	Average of 50 insect fragments per 100 g. Average of 2 rodent hairs per 100 g. Gritty, water-insoluble inorganic residue 35 mg/100 g.
Popcorn	One rodent pellet in 1 or more subsamples on examination of ten 225-g subsamples or six 10-oz consumer-size packages and 1 rodent hair in other subsamples, or 2 rodent hairs per pound or rodent hairs in 50% of the subsamples, or 30 gnawed grains per pound and rodent hairs in 50% of the subsamples.

Table 11.1—*Continued*

Food	Defect action level
Cinnamon or cassia, whole	Average of 5% moldy pieces by weight. Average of 5% insect-infected pieces by weight. Average of 1 mg of excreta per pound.
Hops	Average of 2500 aphids per 10 g.
Pepper	Average of 1% insect-infested and/or moldy pieces by weight. Average of 1 mg of excreta per pound (approximately 2 parts per million by weight).

From *Current Levels for Natural or Unavoidable Defects in Food for Human Use That Present No Health Hazard*, second revision, Department of Health, Education, and Welfare, Public Health Service, Food and Drug Administration, Rockville, Md., April 10, 1973.

as gnats, and common house flies can attack food at almost any stage from production to consumption. The same is true of rats and mice. To be sure, good practices at all stages can minimize adulteration by these larger organisms and their detritus, but absolute elimination is almost impossible. Grains, seeds, fruits, and vegetables are cleaned and sorted to remove insects and rodents, their detritus, and food organisms damaged by them, but here again absolute elimination of all such biological contamination is impossible. Allowable tolerances of biological contamination by higher organisms are continuously reviewed by governmental agencies. When biological contamination exceeds acceptable limits, the foods are said to be unsound, filthy, adulterated, or unwholesome and are not to be sold. They may be condemned and destroyed by government action.

Standards of biological contamination become part of the standards of identity and are a matter of public record. Table 11.1 lists some tolerances as published by the Food and Drug Administration. It should be kept in mind that these tolerances represent the maximum contamination permitted and most foods (particularly the better consumer grades) contain much less of the biological adulterants or contaminants listed.

Ingredient Safety

Much public attention is directed at the safety of the many substances which get into our food either as intentional food ingredients or as incidental contaminants such as insecticides, detergents, or other cleaning or sanitizing agents, and packaging materials. One reason for this is that food is of great personal concern, and people are easily alarmed by reports of dangerous ingredients. No one wants to kill or incapacitate himself, slowly and unwittingly poison himself, make himself impotent, or became the parent of a defective child. Public concern about ingredient safety has increased as the industrial and scientific revolutions of the past two centuries have brough rapid social and economic adjustments. Modern sophisticated systems of food production, processing, preservation, and marketing have to a great extent removed food preparation and the culinary arts from the home, and consumers are quite naturally interested in whether these techniques insure the safety and wholesomeness of what they are eating.

As the principles of science and engineering were applied to all manner of food problems, the precise functions of food ingredients, which in times past were largely food organisms or easily recognizable products of food organisms, began to be understood. Eggs, particularly the yolks, are used in many foods for their emulsifying properties, due largely to their lecithin content (see p. 132). So it would occur to a food scientist that perhaps lecithin from soybeans or emulsifying agents of other kinds might serve as well in making cakes, candies, salad dressing, etc. Sour milk is another example of a traditional ingredient used in a number of foods. In a particular food sour milk may be used for the effect of its lactic acid content. Then one might reasonably ask why could not another acidic food be used. Vinegar, lemon juice, or cream of tartar, a product from wine production, might serve as well. And since the acidic principles of these latter substances are acetic acid, citric acid, and tartaric acid, might these not serve as well for a particular food? These are pure chemical compounds which might be more easily obtained from other sources. Pure chemical compounds used in foods are now called *additives*—as are salt, sodium bicarbonate, sodium nitrate, and so on, which have been used since the ancient past. Baking soda (sodium bicarbonate) was discovered to be useful as a leavening agent, particularly in conjunction with an acidic ingredient such as sour milk. Then came baking powder, a mixture of bicarbonate and cream of tartar or other acidic salt. Could not calcium acid phosphate be used instead of potassium acid tartrate (cream of tartar)? After all, calcium and phosphate are essential nutrients. Then what about other compounds such as sodium aluminum phosphate, sodium acid pyrophosphate, and ammonium bicarbonate for leavening of cakes, cookies, etc.? The same type of story can be developed for other foods—notably spices. Is cinnamaldehyde from cinnamon bark any different from cinnamaldehyde from some other source or the pure chemical compound? But how safe is the consumption of these ingredients—biological organisms themselves, ingredients purified from biological organisms, and substances of nonbiological origin produced by chemical manufacture? Unequivocal positive or negative answers to such questions are impossible. Let us see why.

In Chapter 6, the great complexity of the chemical nature of our food was discussed. In Chapter 7, it was shown that the human organism is a dynamic chemical system capable of separating from the chemical milieu called food those compounds it needs and can use and getting rid of those substances which are of no value or even toxic. So if the human organism is designed to handle potentially harmful substances, how can toxicity be defined? To put this question another way, when is a substance harmful or toxic and when is it safe? This is the type of difficult question the consumer looks to the government to answer. Yet there are no easy answers.

All substances, even compounds classed as nutrients, are toxic or harmful under certain circumstances, so the question of ingredient safety relates directly to how the ingredient is to be used. Also, there are different types of toxicity—acute, subacute, and chronic. Acute toxicity refers to the immediate effects of consuming a specific amount of a substance, in contrast to the short-term or subacute effects of ingestion for a period of weeks or the effects of consumption over a long period of time, possibly an entire lifetime or even two generations or more. Acute toxicity is much easier to evaluate in test animals than chronic toxicity, but it is usually the chronic or long-term effects that are of most importance in foods.

Determination of the acute toxicity of a substance often indicates how its subacute and chronic toxicity can be studied. Acute toxicity is usually measured as the amount of the substance given in a single dose that will kill one-half of a population of test organisms. This value is designated LD_{50}— lethal dose for 50% of the population. All individuals in a given population do not respond in the same way to the same amount of material, and so a plot of the percent of the population killed with respect to the dosage gives a sigmoid curve. Figure 11.2 shows the dose–response curves for two substances, A and B. The steeper curve for A indicates that it is more acutely toxic than substance B and that a higher proportion of the population can tolerate moderate dosages of substance B.

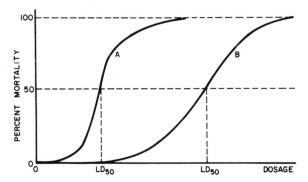

Fig. 11.2. Hypothetical curves relating the percent of animals killed by increasing amounts of two different substances, A and B, given as single doses.

Determination of LD_{50}'s does not give useful information for food ingredients, for food is eaten continuously throughout life. Furthermore, if an ingredient showed acute toxic effects in the amount required for use in a food, we would hardly consider it edible in the first place. Consequently, determining the safety of food ingredients requires long-term studies.

Human beings cannot be used for controlled chronic toxicity studies involving feeding the test ingredient for a whole lifetime. First of all, the human life span is too long. Then in such studies all food consumed and all urine and feces are carefully measured and analyzed, and the test animals are generally killed to permit microscopic study of vital organs or other tests commonly used by toxicologists, pharmacologists, and pathologists. So toxicity studies are conducted using experimental animals which can be grown in confinement under standard reproducible conditions. Commonly used experimental animals for such work are rats, mice, dogs, rabbits, and guinea pigs.

Controlled feeding of the test food ingredient is usually carried out in several different ways using an otherwise adequate test diet. At least two species of animals are used, and only one may be a rodent (rats and mice). The ingredient is fed for extended periods, many times over the complete life cycle or for 2 or 3 years with animals of long life span. Test animals fed the test diet are bred and their offspring are given the test diet, often for several generations. Groups of animals, usually 20 or more per group, are fed a diet without the test ingredient (control) or the test ingredient at the level normally expected in the food, or the ingredient at 10 or 100 times the normal level and sometimes even higher. Such procedures are satisfactory for many ingredients which occur in minute amounts in foods but can present problems for certain other ingredients. For example, if x amount of flavoring compound such as vanillin (from vanilla) or menthol (from peppermint) is used in food and in the test diet, test diets containing $10x$, $100x$, or $1000x$ amounts of vanillin or menthol may be so repulsive in odor and taste that test animals will not eat them. Or if a normal diet contains 99 grams of control food ingredients and 1 gram of salt (1%), animals often will not eat diets composed of 99 g of control food ingredients and 10 g (9%) or 100 g (50%) of salt. Obviously, toxicity testing can become quite involved. It is therefore understandable that the methodology and protocols for evaluating safety of food ingredients are themselves problems of significant importance to the regulating agencies, to those who petition the government for permission to use an ingredient, and to consumers. The last group, particularly, is seldom formally represented in negotiations, but of course consumers are the ultimate test animals. No test on a nonhuman species of animal, no matter how comprehensive, is absolutely predictive of human safety. Also, human value judgments on the meaning of test data and the priorities imposed by society as a whole affect every stage of ingredient testing and toxicity evaluation. Thorough testing can only minimize the risk of hazard to individual humans. This is true for drugs, medicines, and cosmetics as well as for food. When we consider the great number of new medicines and new ingredients for foods and cosmetics that have been tested and put on the market in the past 50 years and the few failures in reducing risks and hazards for the consumer, the efficacy of the testing methodology is evident. But improvements can always be made.

Protocols for appraising potential hazards of food ingredients to consumers have been developed by regulatory agencies. These procedures and methods of testing reflect the application of many scientific disciplines and include many kinds of observations on test and control animals which might indicate physiological impairment of any kind caused by feeding the substance under study. These observations include determinations of food and water consumed, growth rates, general behavior, reproductive performance, and viability of offspring, and periodic detailed examination of blood, urine, and feces throughout the test, along with complete autopsies of the animals which die during the test and similar examination of all animals at the close of the test. Many of the features of the testing protocols as currently followed by the U.S. Food and Drug Administration are briefly outlined in Tables 11.2 and 11.3. These tables indicate that testing a food ingredient to assure its safety is a complicated procedure involving many experimental animals and much time of highly trained scientific personnel. Consequently, the testing cost is quite high, which alone is a deterrent to the introduction of new ingredients having only limited use. If test results indicate that the particular ingredient is safe for specified uses, permission to use the ingredient is granted, usually after public notification and the allowance of time for interested parties to file objections. Commercial use of the new ingredient is monitored, and, based on the findings, the regulating agencies may modify or rescind permission for its use.

All new ingredients are scrutinized in this way, including both intentional ingredients and incidental ingredients which might get into foods during production, processing, and distribution. The latter include pesticide residues, lubricants, detergents, and packaging materials such as paper, films, plastics, inks and chemical used in their manufacture.

It is necessary to understand that all of the thousands of substances which comprise our food have not been tested in this manner. Indeed, our foods likely contain many still unidentified chemical compounds from food organisms. These testing procedures are of relatively recent origin and it is quite impossible to have scrutinized every substance in this manner in a short period of time. However, many of them have passed the rigorous test of time, because most of the food organisms we use have been consumed by the human species generation after generation, often since prehistoric times. Such foods

Table 11.2. General Protocol of Toxicological Tests for Food Ingredients

---

I. Acute toxicity tests—single dose
  A. LD$_{50}$ determination (24-hr test with all survivors followed for 7 days)
    1. Two species of animals, one of which is not a rodent
    2. Two routes of administration—one by feeding and the other by injection, usually into the bloodstream, abdomen, muscle, or under the skin
II. Subacute toxicity tests—daily doses
  A. Duration 3 months
  B. Two species of animals—rats and dogs most commonly used
  C. Three dose levels
  D. Route of administration—feeding
  E. Evaluation of state of health
    1. All animals weighed weekly
    2. Complete weekly physical examination (see Table 11.3)
    3. Determination of sodium, potassium, urea, and glucose in blood; blood cell volume; blood cell counts—red cells, thrombocytes, total and differential white cells—on all animals; analysis of urine for pH, specific gravity, glucose, protein, ketones, crystals, bacteria, and blood
    4. Organ function tests for the liver (bromsulfophthalein retention and serum alkaline phosphatase) and for the heart (serum glutamic-oxalacetic acid transaminase test)
  F. All control animals and those that received the highest doses are sacrificed and autopsied, with histological examination of all organ systems
III. Chronic toxicity tests—daily doses
  A. Duration 1–2 years
  B. At least two species, only one of which is a rodent—selection of species will depend on acute and subacute (I and II) studies and on pharmacodynamic studies on several species, possibly including human trials on single doses
  C. Three dose levels
  D. Route of administration—feeding
  E. Evaluation of health—as in II E 1, 2, and 3 above plus organ function tests at 6-month intervals and on all abnormal or ill animals
  F. At end of test, at least all control animals and all that received the highest doses are sacrificed and complete autopsies are performed, including complete histological examination of all organ systems; similar postmortem studies are made on all animals that died during the course of the test
IV. Special tests
  A. Potentiation or enhanced toxicity with other chemical compounds
  B. Effects on reproduction
  C. Teratogenicity—birth defects or malformed fetuses
  D. Carcinogenicity—induction of cancer

Adapted from *The Chemicals We Eat*, by Melvin A. Benarde, pp. 113–116. Copyright 1971 American Heritage Press. Used with permission of the McGraw-Hill Book Co.

Table 11.3. Signs and Symptoms to Be Observed in Test Animals

Aggression toward experimenter	Inactivity
Stupor or excitement	Spontaneous convulsions
Convulsions to touch	Exploratory behavior
Coma	Abnormal excreta
Paralysis	Shortness of breath
Changes in pupil of eye	Involuntary eye movement
Cataracts or opacities of cornea of eye	Sedation
Sensitivity to pain	Cyanosis—oxygen deficit
Altered muscle tone	Saliva—too little or excessive
Changes in heart rhythm and rate	Discharge from nose
Reflexes—grasping, righting, placing	Erection of hair
Death	Vocal sounds
	Unusual positions—body or tail

Adapted from *The Chemicals We Eat*, by Melvin A. Benarde, pp. 113–116. Copyright 1971 American Heritage Press. Used with permission of the McGraw-Hill Book Co.

and their natural ingredients are generally considered safe in the amounts customarily consumed. When the U.S. Congress made safety of food additives the responsibility of the Food and Drug Administration in 1958, the so-called GRAS (*generally recognized as safe*) list of additives came into being. These substances had not been tested using current methods; however, the safety of some had been studied by concerned groups such as manufacturers, trade associations, food processors, and merchandisers. Although the hundreds of substances on the GRAS list are considered safe by knowledgable specialists, the list is under constant review and is continually revised by the Food and Drug Administration in collaboration with groups such as those mentioned above and particularly the Committee on Food Protection of the National Academy of Sciences–National Research Council, an independent body of scientists who often advise governmental agencies on matters of national importance. This group publishes a quasi-official guide for food ingredients. *The Food Chemicals Codex*, second edition, 1972, NAS-NRC, Washington, D.C., gives detailed specifications for 639 food ingredients. The *Codex* is a direct outgrowth of a Food and Drug Administration request occasioned by the 1958 action of Congress.[1]

One of the most useful guidelines for evaluating ingredient safety over the years has been what some have conveniently called the "1 to 100 rule." The idea is that any nonnutrient substance should not be added to food in any

[1] For additional information, see *Handbook of Food Additives* (Thomas F. Furia, ed.), Chemical Rubber Co., Cleveland, 1968.

amount to exceed 0.01 part or 1% of the amount which produces any adverse effect in test animals in chronic or subacute feeding trials. This premise is related somewhat to another old pharmacological rule of thumb that if a substance is required in x amount to produce a physiological response, either beneficial or deleterious, then the substance is likely to be toxic or even lethal in the amount of $10x$ and of no effect or consequence in the amount of $0.1x$. For example, let us say a human being requires 2800 grams (3 quarts) of water per day. Then 280 g (0.30 qt) will not suffice and 28,000 g (30 qt or 7.5 gal) of water drunk daily is likely to be an almost impossible physiological burden. The same relation holds for salt—3 g per day may be needed, 0.3 g is not sufficient, and 30 g is much too much. This idea also seems to apply roughly to other physiologically active substances such as hormones. To a degree, the accuracy of this guideline of 0.01 or less of the minimum physiologically active amount for establishing safety tolerances of added food ingredients is a reflection of the fact that man (and other organisms as well) is a dynamic chemical machine capable of consuming a complex mixture of chemical compounds, sorting out and using those substances he needs, and getting rid of those for which he has no use.

Carcinogens in Food

One of the most dreaded human diseases is cancer, and it is quite natural to wonder if there is some relationship between what foods are eaten and cancer. Certainly if there is a cause-and-effect relationship, the consumer must be protected. There are many types of cancer affecting different tissues of the body. The common characteristic of all forms of cancer is that somehow the normal processes of tissue growth and repair go awry. Normal cells change to cells which proliferate uncontrollably, often forming tumors; these abnormal cells eventually interfere with some vital function or organ activity and result in death. Cancer appears to be due to some alteration of the metabolic control mechanism residing in the cell nucleus or perhaps some change in the complex nucleic acid molecules of the cell nucleus which control all cellular activities (Chapter 3). Scientists of all persuasions have for many decades sought to find what causes cancer and what will cure it; research has been greatly intensified in recent years as the more common infectious and nutritional diseases of the past have yielded to research. The actual cause or causes of cancer are still unknown. However, it has been shown that the tendency to develop cancers of many types is inherited and that, under certain circumstances, cancers can be induced by topical application, injection, or feeding of a great many naturally occurring substances, some of which are normally present in food. Also, many synthetic chemical compounds not known to exist in nature are capable of inducing cancer. Whether natural or synthetic, cancer-inducing substances are called *carcinogens*.

Coincident increasing public interest in cancer and food additives resulted in the incorporation of a clause in the 1958 revision of the food law prohibiting the addition to food of carcinogenic substances. This stipulation is commonly called the Delaney Amendment after its author, Congressman James J. Delaney. It says that "no additive shall be deemed to be safe if it is found to induce cancer when ingested by man or animals" and "that no such additive may be used in animal feeds unless no residue of it can be found in food products obtained from the animal." The legal effect is that *no* carcinogen in any amount is permitted legally in food. Here the law deals in absolutes—*no* amount. Certainly no one argues for the intentional addition of cancer-causing substances in food, but the legal philosophy of absolute *no* and the scientific realities of living things come into conflict on two counts. First, a great many substances normally present in food can, under certain circumstances, induce tumors, particularly in experimental animals specifically bred to be cancer prone. Second, biological organisms and the food made from them contain minute amounts of many substances coming from their environment which are of no physiological importance. In the past such minute amounts could not be detected. Since 1958, however, the accuracy of analytical techniques has increased a thousandfold or more. Whereas analyses to 1 part per million were achieved 25 years ago, now procedures are available to measure 1 part per billion or even 1 part per trillion for some substances. With such analytical sophistication, zero levels from the legal standpoint are unrealistic, and many scientists and legal professionals feel that tolerances for carcinogens should be established in the same manner as tolerances for other ingredients.

A current dilemma arising from the Delaney Amendment illustrates the kind of incongruities that can develop between well-intentioned manmade laws and effective use of scientific knowledge for more effcient production of nutritious food. Selenium is a widely distributed element even though in soil it is usually present in small amounts. It has long been known that animals foraging in certain areas of the world do not thrive and in some cases are poisoned and die from eating vegetation known not to be toxic in other areas. In the 1930s, it was found that a cause for this situation was the presence of selenium in certain plants growing in high-selenium soils, including a large area in the United States area with the so-called badlands of western South Dakota as the focal point. Many investigations were made on selenium metabolism and toxicity in plants and animals. Some studies seemed to show an increase in the number of liver tumors in animals fed toxic levels of selenium, 5–10 parts per million of the diet, after liver damage (cirrhosis) had been induced by protein-deficient rations.

In the 1950s, research on the nutritive requirement for and physiological functions of vitamin E indicated that minute amounts of selenium were related to the use of vitamin E in the body. Subsequent work established unequivocally that selenium is a required nutrient for higher animals in the range of 0.1–0.25 parts per million of diet. Much of the soil in the eastern half of the United States has so little selenium that plants do not contain that much. Animal nutritionists observed that animal and poultry rations composed almost entirely of feed produced in the eastern United States often were inadequate and that the problem occasionally seemed to respond to high vitamin E supplementation. Further work clearly showed that the cause was selenium deficiency rather than vitamin E deficiency and that the problem could be corrected by increasing the selenium content of otherwise adequate rations to about 0.25 part per million. But when the FDA was asked in early 1970 to permit supplementation with this concentration of selenium, a proved nutrient, under the Delaney Amendment the FDA could not do so because of the presumed carcinogenicity of selenium, notwithstanding the fact that proper dietary levels of selenium have also been found in some studies to decrease the number of cancers. Thus dietary use of a required nutrient was prevented by the Delaney Amendment with the result that the efficiency of animal and poultry production was impaired when their rations were made up of plants grown in low-selenium soils.

To try to reconcile the selenium dilemma, the Food and Drug Administration in 1974 agreed to permit supplementation of animal rations with selenium while the possible role of selenium as a carcinogen was being reexamined in the light of newer knowledge about the causes of cancer. Also, some aspects of the Delaney Amendment itself are under scrutiny. In time, the law may be modified by Congress so that scientific and legal philosophies are more compatible.

NATIONAL REGULATORY AGENCIES

Every human being is concerned that he have an adequate supply of food, that food be available at a price he can afford, and that it be clean, nutritious, and free of any ingredients which might impair his health and well being. In modern complex societies the role of governments with respect to these fundamental concerns is clear, although as we have seen consumer protection is no simple task. It is a focal point of concern for individual citizens, for economic interests representing discrete sectors of the agricultural and food industry, for national interests, and for politicians representing every segment of society. Rational protection of the consumer by governmental agencies must rest on the fair application of regulations, standards, and procedures based on sound scientific principles. These agencies are politically created to perform legal functions and so are subject to the forces of scientific, political, and legal philosophies which are not always complementary. In spite of the difficulty of their mission, governmental agencies perform an essential service to all. However, like any segment of a dynamic, ever-changing society, these regulatory agencies must evolve along with society as a whole. They must protect the consumers not only by preventing harm and minimizing risks in foods but also by constructive encouragement of newer developments for increasing supplies of wholesome nutritious food for the increasing population of the nation and of the world.

On June 30, 1906, Congress established by two separate acts the two federal agencies having primary responsibility for regulating the food industries. Both were originally part of the Department of Agriculture, but they developed in quite different fashions. Today the Food and Drug Administration is in the Department of Health, Education, and Welfare, whereas the Meat Inspection Service has remained in the Department of Agriculture. In later years other agencies have been established: regulation of fish and other seafood was made the responsibility of the Department of the Interior, and the Federal Trade Commission was set up in the Department of Commerce to handle so-called fair trade practices such as advertising and price fixing in all commerce, not just food. The Constitution requires that these agencies operate only with respect to interstate and international commerce. Consequently, there have been established state and local regulatory agencies to handle intrastate and intracity food problems. The local and state agencies often look to federal agencies for guidelines, as do many smaller nations. Consequently, we shall limit our discussion to the federal agencies and to only two of these—the Food and Drug Administration (FDA) and the Meat and Poultry Inspection Program (MPIP) of the Animal and Plant Health Inspection Service of the U.S. Department of Agriculture. FDA and MPIP operate in quite different ways—sometimes with conflicting responsibilities.

Since the inception of FDA and MPIP there has been a remarkable growth of knowledge in the fundamental sciences and in the technologies that result from such advances. Our society today rests on these developments. Without agricultural technology to produce our food and the technologies of food processing, preservation, and distribution to insure efficient utilization of foods, our culture would collapse. Yet it is fair to say that this prodigious scientific development took place essentially outside the regulating agencies.

The excitement of scientific research and development of this century tended to minimize and even downgrade the regulatory functions of science. This plus lack of government funding and public support did not attract many scientists to the agencies, where legal and political philosophies tended to thwart the natural curiosity of scientists. The result was that the agencies were often called on to regulate industries which were advancing technically more rapidly than the scientific manpower of the agencies could handle. At times, the agency's lack of scientific sophistication led to dependence on the competence of petitioners for new products or processes or for quality control to meet promulgated standards. At other times, this same deficiency led to disallowance of new processes or new products because a negative action resulting from technical insecurity was easier than positive action from a position of scientific strength. Though the agricultural and food industries have often produced developments beyond the technical expertise of governmental regulating agencies, these agencies have control over food for the consumer. This is as it should be for protection of consumers, yet the result of this situation has been that research and scientific progress in the food industry have not been stimulated but rather inhibited by many FDA and MPIP actions. Consequently, progress has lagged in the technical solution to many food problems (see Chapter 12). This is particularly true when the food industry is compared to other segments of the economy which are less rigidly controlled such as the electronics industry.

There is little doubt that the scientific deficiencies and slow technical growth of the regulating agencies have made them vulnerable to public criticism, which in recent years has often become emotional and even sensational. This criticism clearly reflects some of the political difficulties and scientific dilemmas faced by FDA. Notwithstanding their many deficiencies, certainly stemming in some measure from insufficient funding and lack of public support for many decades, it is fair to say that FDA and MPIP have achieved a considerable degree of success in achieving their mission of assuring wholesome and safe food to consumers. In the past 10 or 15 years, FDA particularly has received greater public support and is becoming better able to cope with its responsibilities. FDA and MPIP have somewhat different responsibilities. They have different operating philosophies and so their methods are not similar, as the following brief descriptions will indicate.

Food and Drug Administration

The responsibilities of FDA are quite broad in that it is responsible for the safety of drugs, medicines, and cosmetics as well as food. However, although its jurisdiction over drugs, medicines, and cosmetics is complete, it has only control over some foods. For instance, meat and poultry are the responsibility of MPIP. FDA is responsible for the safety of all food ingredients—even those in meat and poultry products—and for standards of identity, wholesomeness, labeling, sanitation, etc., for all other food products.

In order to enforce its regulations, FDA has inspection and compliance divisions to continually check on products, processing, procedures, etc. FDA inspectors have legal rights to inspect food plants, including inventories of ingredients, supplies, food processing records, and products sold, and to sample ingredients and products to determine whether all legal standards are met. Inspectors are supported by technical laboratory personnel of many diverse specialties. If findings warrant them, FDA can take actions of several kinds ranging from warnings and negotiated correction of the problem and recall from the market of suspect food to recommendations that legal action be taken by the Department of Justice to effect condemnation and destruction of filthy, unwholesome, adulterated, or substandard food; cease and desist orders and close the plants of offending food handlers, processors, and marketers. The guilty parties are subject to fines and imprisonment.

When formal legal or criminal proceedings are instituted, the matter becomes the subject of public record. Yet it is fair to say that only a few of these FDA actions receive public attention. Emergency seizures of suspect or illegal food that are not challenged do not become part of the public record. Critics have pointed out that FDA can apprehend only a few of many potential offenders and that fines and condemnations are not commensurate with profits made on cheap, substandard products. There are no doubt cases of this kind of abuse of food laws where court-assessed fines for noncompliance are simply a cost of doing business for the repeating offender. In such a frame of reference, the punishment appears not to fit the crime. In other words, there is some question of whether the penalties for violators are punitive or a deterrent to repeated food law violations. Light penalties for criminal violations of food laws are more understandable when it is noted that the American legal system derives philosophically from the old Anglo-Saxon law, wherein it is expected that the citizenry will obey the law and only a few will find it advantageous to disobey and to risk conviction. To some people such concepts may be a weak base upon which food laws are enforced. Some legal scholars will also point out that swiftness and certainty of imposition of sanctions of whatever form upon the offender is a greater deterrent than the severity of the sanction which tends to placate the victims. The small, unscrupulous obscure food operator may for a time be able to function outside the law. However, for the big food grower, processor or merchandiser, the fine or other direct penalty is far less a deterrent than is the loss of a large volume of sales due to possible adverse publicity or to large numbers of dissatisfied consumers. Successful food companies depend on repeat sales for a profitable and relatively stable business and they strive for uniform quality to build customer confidence in the products that carry their brand names, company names, or trademarks. If customers lose confidence because of faulty products or bad publicity, their business suffers. So for large food operators the potential penalty for flagrant disobedience of FDA regulations is far more the loss of business than any court-assessed punishment, as long as the consuming public has confidence in the credibility of the FDA and the courts.

Quality control and compliance with FDA standards are the responsibility of the supplier, processor, and marketer. Food manufacturers now have sophisticated techniques available for maintaining strict quality control over all aspects of their operations. Food technologists, chemists, microbiologists, and engineers are often involved in building compliance with FDA regulations into the manufacturing operation. Although some food processors can request continuous FDA inspection, most do not. In policing the food industry FDA relies for the most part on periodic unscheduled detailed plant inspections and on systematic, sound statistical sampling and examination of food products. Some critics believe this spot-scheck technique for policing the food industry to be inadequate. But to put FDA personnel into every food establishment on a continuing basis, to police every facet of food manufacture and marketing, would cost enormous sums and tend to relieve industry of its own quality control responsibilities. The spot-check technique of enforcement puts food law compliance and quality control responsibility squarely on the food operator. He is responsible for the safety, soundness, quality, and legality of the food he sells to the public. If the food is faulty in any way, he is subject not only to prosecution by government authorities for food law violations but also to civil action by those who buy and use the food.

FDA has a formidable and complex task in fulfilling its mission to society. It is expected to promulgate standards of all kinds, to determine the safety of all ingredients whether intentional or incidental, to evaluate petitions for new food ingredients and processes of all kinds, and to enforce its regulations to serve the best interest of all segments of society. In addition, this same agency has similar functions with respect to medicine and all manner of health services and the cosmetics industry. Criticism from some groups is inevitable. It is a surprise to many people that for most of its history FDA has struggled to perform its broad mission with a budget of only one-tenth or less of that appropriated for meat inspection. That FDA has done so well with limited public support is due in large measure to the dedication to public service of its founder, Dr. Harvey W. Wiley, those who worked with him, and their successors.

Meat and Poultry Inspection Program

MPIP, which regulates the meat industry, has operated under several names since the original Meat Inspection Act of 1906. Only recently has poultry been included in the program. Since originally it dealt only with meat going into interstate commerce, many states set up meat regulatory programs of their own. Some of these state programs operated with inferior standards, some were equal to the federal programs and some were definitely superior. This multiplicity of federal, state, and sometimes municipal regulations in the same industry presented jurisdictional problems. Since the 1967 "Wholesome Meat Act" there has been a very strong trend toward putting all meat and poultry inspection under the jurisdiction of the U.S. Department of Agriculture's Animal and Plant Health Inspection Service. All jurisdictional problems and agency rivalries have not been solved, and there are still litigations in the courts. Recently courts have held that state regulations and sanitary and product standards, even those superior or more stringent, shall be superseded by federal regulations.

The operation of MPIP is quite different from that of FDA. Meat and poultry under the regulatory jurisdiction of MPIP represent about 28% of direct food costs, whereas all other foods under FDA represent 72% of food costs (Table 10.3). For many decades the cost of operation of MPIP has been more than 10 times that of FDA, and it still is many times more even with recent increases in FDA appropriations. Since much of FDA's function is not related to food, meat inspection costs are 20 times the costs for federal regulation of all other foods. On the basis of amount of money spent for meat and for non-meat foods, costs for inspection of meat are almost 50 times higher than those for actual inspection of other foods. In view of this and the central role of meat in the American diet, it is rather surprising that the public is far more aware of the regulatory function of FDA than that of MPIP. Some of this disparity in regulatory costs has to do with the biological nature and perishability of meat as well as its prominent role in the U.S. diet. However, it likely also reflects the history of the meat industry itself, and the philosophy and mode of operation of the meat inspection agency through the years.

In many cultures, slaughter of animals and sale of meat are local operations, as they were in earlier times in this country. Long ago local butchers or farmers devised ways of smoking, salting, and curing meat and making sausages of various kinds, all with the purpose of prolonging the period of usefulness of this perishable and highly nutritious food. Years ago in northern European countries slaughter of animals was seasonal—following the completion of the summer growing season. The colder weather of autumn and winter coupled with preservation by salt and saltpeter (sodium chloride and sodium nitrate) permitted such meat to be available throughout the year. With the beginning of the industrial revolution and urbanization, the source of meat supply remained local but fraud and sale of unwholesome meat in certain meat products increased.

The advent of mechanical refrigeration about a century ago brought a profound change in the meat industry, more so in the United States with its warmer climates than in the northern European countries. Carcasses cooled immediately after slaughter would keep longer and the meat could not only be sold fresh over a longer period of time but also could more easily be processed and preserved by traditional methods or the newer canning techniques. All of these developments took place almost before the sciences on which current meat technology is largely based became established. But to take advantage of mechanical refrigeration required larger operations than just local butcher shops, and the necessary equipment cost money. The result was the rapid concentration of the industry into the hands of large operators. The hub of the meat industry was Chicago—a place where live animals could be transported by rail from all the major animal-producing areas of the country. Meat suddenly became big business for both domestic and foreign markets. The industry pioneered in the mass-production and assembly-line techniques later adopted by Henry Ford, the automobile pioneer, and manufacturers of other types. This revolution in the meat industry took place in the latter part of the nineteenth century and resulted in the formation of the first major corporate organizations in the food industry.

Such rapid change and concentration of industrial power led to many abuses—exploitation of immigrant labor including lack of safety devices for those who worked with the new machines, use of unwholesome and sometimes spoiled meat or meat from diseased animals in manufactured items such as sausages and canned meat products, fraudulent use of nonmeat ingredients in processed meat products, and use of what we consider today to be highly unsanitary practices. Attendant on these changes in the meat industry were abuses of political power such as coercion, bribery, and collusion between politicians and meat industry personnel. These problems of the burgeoning meat industry affected the growth and character of Chicago itself. There was increasing public concern not only about these abuses mentioned but also about the powerful economic control exerted over the farmer-producers who had to sell their animals to the large meat packers. At the turn of the century, this concern was increasingly manifest in many quarters including Congress, which was also being made aware of food adulteration of other kinds.

In 1905 a young novelist, Upton Sinclair, wrote and published serially a story entitled *The Jungle*, which was about exploited immigrant packinghouse workers and the sordid abuses of the newly dominant meat industry. This vivid portrayal with its easily identifiable characters was an immediate sensation not only in the United States but also in many other countries which were grappling with problems of industrial, social, and economic change and which were importing U.S. meat. The call for reform was clear. Congress passed the Meat Inspection Act of 1906 in the same year that *The Jungle* appeared in book form. This novel has been reprinted many times and

Fig. 11.3. MPIP stamp identifying meat as coming from Establishment No. 38.

is considered a classic in literary work directed toward social and economic reform.

The goal of the Meat Inspection Act of 1906 was to see that meat offered to the consumers through interstate commercial channels came from healthy animals and that animals were slaughtered and meat was prepared and processed in a sanitary manner. Having these purposes, the law made meat inspection the province of the veterinary profession. Systems have been instituted whereby every animal slaughtered is inspected for disease as it is dressed. Any diseased, suspect, or damaged tissue is removed and meat considered by the inspector to be unwholesome is condemned and destroyed or diverted to nonfood uses. The carcasses and parts of carcasses deemed suitable for food are identified by stamps affixed by the inspectors (Fig. 11.3). Inspectors also make sure that condemned animals and tissues are turned into tankage and fertilizer.

To fulfill their responsibility with respect to sanitary handling of meat, resident government inspectors in slaughtering, packing, and processing plants follow meat through various steps whether it leaves the plant fresh or processed into sausage or other cured or canned products. If the inspected meat or meat products go to another food plant for incorporation into processed foods in which meat is used as an ingredient such as soups, stews, chili, and pizza, government inspectors reinspect the meat, follow these operations, and affix their inspection stamp on the ultimate products. Government meat inspectors assume regulatory control over every stage of processing, which has led to MPIP approval for equipment design and plant layouts, which are essentially engineering problems. The MPIP, although leaving ingredient safety to FDA, also has division of standards, labeling, etc., for meat and meat products as the FDA does for nonmeat foods.

From this brief résumé, it is clear that MPIP, through its thousands of resident inspectors, supervises as well as regulates almost every operation of the meat processing industry. It is somewhat like having traffic officers on continuous duty as every intersection and periodically stationed along highways to continuously inspect each car or truck while at the same time other officers oversee and supervise all aspects of road construction. From this perspective, it is easy to understand why MPIP costs so much more than FDA. In some meat operations MPIP costs alone may exceed 1 cent per pound of fresh meat inspected, and with new consumer protection programs being added there is increasing concern on the part of both industry and government officials over compounding regulatory costs. In its method of operation, MPIP assumes a large measure of responsibility for quality control in the meat industry, whereas the FDA spot-check operation places major responsibility for compliance on the food industry itself. To illustrate the difference in philosophy of operation between the two systems, MPIP inspectors read thermometers on meat in smokehouses to tell meat processors when the meat is pasteurized or cooked, whereas in a milk processing plant, not under MPIP jurisdiction, this chore is done by the milk processor, who assumes complete responsibility (with attendant liability) that the milk is properly pasteurized.

Meat is slaughtered, processed, and marketed today in much the same way as it was in 1906. To be sure, the supermarket may be brighter and more attractive in appearance than the old butcher shop, but the less attractive aspects of meat handling are merely moved out of the sight of the consumer, who only sees the meat already cut, packaged, weighed, and priced. Compared to the progress made in the dairy industry, which is based on an animal protein food as perishable as meat, one naturally wonders whether progress in the meat industry is fostered or inhibited by meat inspection practices. It is fair to say that even now consumers shopping for meat display much less confidence in their purchases than in other food items such as milk, canned or frozen fruits and vegetables, and bakery products. Shoppers want to see their meat, although they accept most other foods in the supermarkets sold in blind packages. Technically some aspects of the meat industry are more advanced in other parts of the world than in the United States. Through the years, the giants of the industry have spent far more for research and development on by-products such as soap, fertilizers, and chemicals outside of MPIP control than they have on their primary product, meat. There has been much less basic and applied research concerned with meat done by industrial, government, and university scientists than research on other foods. (Foods in general have received less scientific attention than other sectors of the total U.S. economy.) This certainly appears to be related to the policies of MPIP, which, although they effectively eliminated the gross abuses of yesteryears, have failed to stimulate progress in the industry. There are a number of factors which may have contributed to this situation.

MPIP and its predecessors essentially assumed quality control over the meat industry. Since the entire meat industry operated on the same standards, methods of operation tended to settle to the least common denominator of compliance. There was little or no incentive for improvement by industry because there was little competitive advantage to be gained. Furthermore, research and development leading to new processes or equipment were hampered by the high cost of litigation for change of MPIP standards. Since MPIP approval was by no means assured, research and litigation costs of proposed improvements might go for naught. An example of this is seen in animal fats.

For centuries, animal fats were the most desired cooking fats for frying, baking, and so on. In fact, animals were often bred and grown for fat rather than for lean meat. However, animal fats have the disadvantages of a flavor not always compatible with certain foods and more particularly of the tendency to become oxidatively rancid (see Chapter 9) in a relatively short time. As the chemistry of fats and oils began to be understood, it became feasible by relatively simple processing techniques to correct these defects, because animal fats such as lard and beef fat can be rendered into more than 99% pure triglycerides (Chapter 6). MPIP refused to permit the use of such processes to improve these products and retain their identity as far as the consumer was concerned. In three decades, these fats fell into disrepute and had such low economic value as to be used for soap or discarded. To replace them, the vegetable oil industry arose, based largely on the use of cottonseed, palm, soy, and corn oils. These oils in crude rendered form, unlike butter, lard, and beef fat, are generally unacceptable for food. However, by such techniques as refining, hydrogenation, and bleaching these fats were given many of the desired consumer characteristics of animal fats without their deficiencies. The vegetable oil industry was under the jurisdiction of FDA— not MPIP, which would not permit similar or even simpler processing of animal fats. Since World War II, however, MPIP has permitted more refining of animal fats, and these have come back into food uses largely in cake mixes and similar products and as commercial shorteners rather than as home shorteners and frying fats.

Notwithstanding the difficulties just mentioned between MPIP and the meat processing industry, some critics feel that the agency is sometimes in league with the industry in its position as arbiter of quality control. Critics have cited changes in product standards, including permitted increases of fat in sausages such as wieners and addition of water to cured meat products such as ham and corned beef, as evidence of collusion. Many consumers are also rightly concerned that even though MPIP is an expensive regulatory agency it does not go far enough. Its control of fresh meats essentially ceases as carcasses and wholesale cuts leave the packing plant. This perishable food can still be easily abused and adulterated in the retail and institutional markets.

Nationwide, MPIP requires a tremendous number of inspectors. There are simply not nearly enough veterinarians to perform the agency's work as now practiced. The result is that a majority of meat inspectors are lay inspectors who function at the discretion of the veterinary administrators and supervisors. This in itself is not necessarily poor practice; however, many of the lay inspectors are inadequately trained technically and do not understand the principles of the modern technological processes for which they are responsible. In large measure the sciences on which food processing and preservation are based developed outside the veterinary profession. The emphasis of veterinary school curricula is on animal disease, and the food sciences and technology receive little or no attention. Consequently, the veterinary–lay dichotomy within MPIP has in the past made this agency even less attractive to the scientific professions than FDA because chances of advancement in MPIP are limited for nonveterinarians. MPIP veterinarians are quite competent at inspecting animals and carcasses for diseases. But MPIP personnel are largely at a technical disadvantage in many other aspects of the meat industry as it serves the public. The personnel responsible for the actual operation of the meat industry are often much better trained to do their jobs than are the MPIP inspectors.

MPIP also differs from FDA in a significant way as regards assuring compliance with its regulations. Both agencies, when the court so decrees, may condemn, seize, and destroy illegal products. However, because MPIP carries out continuous in-plant inspection and, as animals are slaughtered, passes healthy ones and condemns and destroys unfit ones, as well as following the carcasses through each step of processing, it has a very effective enforcement tool. Resident MPIP inspectors have the authority to stop or delay processing on the spot to enforce compliance with their decisions. Delaying or stopping production in a complex meat processing plant is very costly, not only because of the perishability of the product but also because idle workers must be paid. The fear of immediate financial loss if inspectors are not satisfied is perhaps the most effective deterrent to noncompliance. Inspector–operator interactions are personal and sometimes strained because of differences of opinion on the meaning of regulations.

Conclusions

It is clear that FDA and MPIP have developed quite different inspection technologies by which they regulate the safety of the food reaching consumers. Although these agencies carry the major burden for consumer protection on a national level, enforcement of food safety standards, particularly at the retail market and in the institution or restaurant—the ultimate consumer level—is often a responsibility of state and local agencies with sometimes limited capabilities. In spite of the problems with which regulating agencies have to contend, they have achieved a considerable degree of success in assuring wholesome and safe food for the consumer. Certainly, compared to the not too distant past, Americans are fortunate in the quality, quantity, and variety of foods available at the lowest cost relative to personal income of any nation in history. But in view of increasing population and demand for food, we must continually evaluate our present situation and seek ways of improving our food and protecting the consumer that do not lessen the supply or unwittingly make it too costly.

ROLE OF GOVERNMENTS IN THE CONTROL OF INTERNATIONAL COMMERCE

Some scholars have stated that if the history of man had been recorded and studied from the point of view of food rather than wars and battles our present understanding of past cultures might be more complete. Certainly food supply has been and will continue to be fundamental to international diplomacy and commerce as it is a basic concern of national policy itself. As world population grows without commensurate increase in food supplies (Chapter 12), food will play an even more critical role in international relations.

The United Nations, which is often thought of simply as a supranational organization for debate and settlement of disputes between nations, established shortly after its inception a division concerned with increased food production in underdeveloped food-deficient countries and with international commerce in food. This agency is known as the Food and Agricultural Organization (FAO) and has its headquarters in Rome. Another branch of the United Nations, concerned with international health problems, is the World Health Organization (WHO), operating out of Geneva. Somewhat more visible to many Americans through its popular appeals for funds to aid the underprivileged and underfed children of the world is the United Nations International Children's Emergency Fund (UNICEF), headquartered in New York. All of these divisions are concerned with various aspects of world food supplies. The main thrust of FAO is, as its name implies, improvement in production, processing, preservation, and distribution of food. WHO's interests in foods are related to the maintenance of nutrition, the wholesomeness of foods, and the health of the consumer. UNICEF activities are directed toward child welfare, and, of course, food is primary to its mission.

Consumer protection is quite properly a concern of all countries, although many do not have organizations of their own comparable to those in the more developed countries. Small or less developed countries often look to the regulatory agencies of other countries, particularly the United States and West Germany. As individual countries set up their guidelines for sanitation, adulteration, ingredient safety, and so on, it is inevitable that differences in standards develop. National standards often reflect internal political and national interests. The result has been some confusion of standards which at times has worked against increasing world commerce in foods. Accordingly, FAO and WHO have cooperated toward the development of international standards for the food industry in establishing the joint FAO/WHO Food Standards–Codex Alimentarius Commission. This group has committees dealing with such problems as meat hygiene and ingredient safety which are attempting on an international level what MPIP and FDA do in the United States with regard to food standards. The FAO/WHO commission does not have enforcement and compliance divisions as do national regulating agencies. Consequently, the Codex Alimentarius Commission's main impact is the establishment of standards acceptable to member nations for international trade in food. Development of this type of program is tedious and slow, and the full fruits of these efforts have not yet been realized.

We have seen in Chapter 10 that the social and economic development resulting from the industrial revolution and the advent of modern science led to great shifts of population from rural food production pursuits to urban employment in other sectors of the economy. As these changes occurred, governments found it politically expedient to subsidize food production in some manner, for two reasons. First, urban dwellers had to have food at prices they could afford and, second, farmer-producers had to receive enough for their efforts to stimulate food production. Farmers must be able to produce enough and sell it at a price that assures reasonable economic stability for the farm unit—generally a family-operated unit. This stability is also required if government agricultural policies result in surpluses of food, so that farmers still receive sufficient return for their labors to prevent bankruptcy, which might result if they had to sell food in a glutted market where prices fell below production costs. Government subsidies can serve to protect the less efficient farmers or other sectors of the food industry, but many political leaders feel that this effect of subsidization of food production is much more desirable than inadequate food supply with skyrocketing prices. Farm prices and hence consumer food costs have traditionally fluctuated greatly. This is because of the biological nature of food itself. Crop yields vary with weather, disease and insect pests, and so on (Chapter 8), and consequently the amount of food entering the market is variable. On the other hand, demand for food is less variable since is is directly related to the numbers of people to be fed. Populations of people, having long life cycles, do not fluctuate like populations of their short-lived food organisms. So government subsidization has the effect of stabilizing markets. From these brief remarks, it is not difficult to understand (1) that government involvement in agricultural and food affairs is very complex and (2) that each nation develops its own agricultural programs including subsidies in diverse forms to serve its own national goals. Food exporting, importing, and self-sufficient nations quite naturally differ with respect to national food policies.

Because of their natural resources and perhaps because of subsidization programs, some countries have surpluses of certain food types—grain, butter, meat, sugar, etc. The outlet for surpluses is the world marketplace, provided that buyers can be found with money to pay. International trade is of course controlled by the trading nations, and their dealings are made from the standpoint of their own national interests. To control international trade, countries establish export and import duties, embargoes, and such. Some nations profess a policy of free trade without tariff restrictions. Of course, national manipulations of the types just mentioned apply to all items of trade —minerals, manufactured goods, etc.—not simply to food. For example, nonindustrialized countries which export food for most of their foreign trade often have very high duties on manufactured goods such as automobiles. Conversely, highly industrialized food-deficient countries have low or no duties on food.

In addition to the more direct and obvious controls of duties, tariffs, and outright embargoes, there are more subtle means whereby nations control international commerce. One means particularly applicable to foods is the use of standards: standards of identity; health, sanitation, and hygienic standards; standards for acceptable levels of biological and other contamination; and ingredient safety standards. Sometimes special interest groups within a country put pressure on their national food regulating agencies to set standards for imported food which can diminish or remove foreign competition. Or they may try to exclude another country's food by pressing for an embargo on the basis that it carries some disease or pest. A particularly sensitive situation occurs when an importing country insists that it have its own resident inspectors to control the food operations of the exporting country. This happens most often in the international markets dealing in meat. It is hoped that internationally acceptable standards can minimize this use of national food standards as trade barriers.

One of the major problems in the development of the European Common Market was food. Although since World War II close economic cooperation of the Western European countries has seemed desirable in order to minimize many problems among them and enhance their political, economic, and even military strength vis-à-vis Eastern Europe, food problems arising from such a union even now persist to a degree. Each nation in the market had its own food subsidy programs and special food and agricultural interest groups.

Some highly industrial nations depended on food imports; other nations, although less efficient agriculturally, were almost self-sufficient through subsidy programs; still other nations depended to a very great extent on food exports for economic vitality. As a result of coalescence into an economic community of nations free of trade barriers, food prices went up in some countries while in others farm prices dropped markedly. In all countries, large population groups of diverse economic interests had to make major adjustments. It is clear, therefore, that a workable common market requires a great amount of time for gradual adjustment and a great deal of perseverance by the governments and citizens of the member countries. Nevertheless, in today's highly industrialized and scientifically oriented world with its increasing population and economic versatility, the European Common Market epitomizes recognition of the fact that international cooperation is essential for continued growth, for political stability, and perhaps even for survival.

All people need wholesome and nutritious food throughout their lives. But the fulfillment of this fundamental biological requirement is a formidable task for any group of people—the more highly developed the society, the more difficult the task. In any complex society, it is beyond the capability of the individual to meet his own personal needs for food. So individuals must look to their government to develop programs to assure an adequate supply of food and at the same time to assure its wholesomeness and safety. To effectively meet these responsibilities, the diverse agencies of government—dealing with different aspects of the overall food needs of their people—must be responsive to an ever-changing world.

12

WORLD POPULATION GROWTH AND FUTURE FOOD SUPPLIES

Food enough—there is no greater problem facing the world today. Food requirements are absolutely related to the number of people living on the earth. And this number is increasing at a rate never before experienced in human history. If the world's population growth is maintained at its present rate, the number of people who must be fed will quadruple in the average lifetime of the readers of these words. This means that in 70 years four times as much food as is available now will be required to feed the hungry. The enormity of the problem is hard to grasp for most Americans, who, as a nation, have never experienced hunger. Such has been the bounty of U.S. agriculture and the efficiency of the food industry—both fruits of the advent of modern science coupled with good fortune in natural resources. But even in the United States food problems of diverse kinds will continue to get more and more public attention—rising prices; crop failures due to drought, floods, and freezing weather; food shortages, famine, and starvation in various parts of the world; international trading of food on a scale never before seen; food trade and the balance of payments. Such attention is evidence of increasing public awareness of the need for more food—the challenge facing all of us.

POPULATION

Already many references have been made to the rapidly increasing number of people living on the earth. We discussed briefly some ecological aspects of this situation in Chapter 8. The number of people living in the United States plus Canada is approximately equal to the population of the

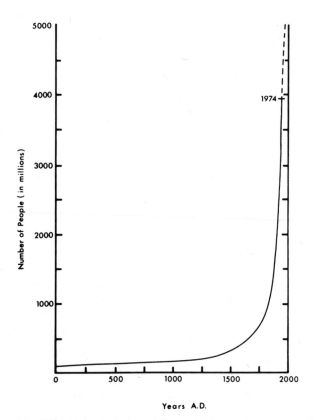

Fig. 12.1. Estimated population of the world for the years A.D. 1–2000.

Table 12.1. Number of Years Required
to Double the World's Human Population

Year	Population (billions)	Number of years to double
1	0.25	1650
1650	0.50	195
1845	1.0	85
1930	2.0	45
1975	4.0	35
2010	8.0[a]	35[a]

Adapted from Harold F. Dorn, *Science* **135**:283
(1962).
[a] Projection.

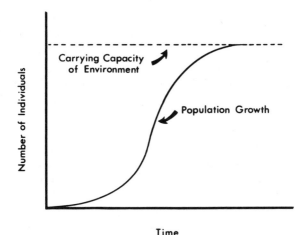

Time

Fig. 12.2. Initially the population of a species increases rapidly in a favorable environment, but the rate of increase declines and the population stabilizes at or below the carrying capacity of the environment.

entire world in the year A.D. 1. The number of people living at any particular time since man appeared on the earth is not precisely known, and even today many countries do not have very accurate census data. However, demographers (specialists in population studies) have been able to make quite meaningful approximations of the number of people living on the earth in the past. It is much more difficult to accurately predict the size of future human populations, and many past predictions of world population have proved to be underestimates.

Figure 12.1 shows the estimated population of the world for the years A.D. 1 to 2000. Comparison of this curve with Fig. 12.2, the general population curve for biological systems (see also pp. 233–234), shows that the human population is in the very rapid (logarithmic) phase of growth. A similarly striking but different perspective on population growth is seen in Table 12.1, where the years necessary to double the population of the world are shown.

For centuries the number of births was almost in balance with the number of deaths. War, pestilence, and limited food supplies were the controlling factors. Beginning in the 1500s the death rates became somewhat less than birth rates and this disparity has gradually increased so that today on a global basis there are each year about 17 more births than deaths for each 1000 population. This great change is due to increased food supplies and effective control or elimination of many diseases which in the past were common causes of death. The result has been a longer life span, which, of course, leads to more births. In the 1500s in Europe only about 30% of all female babies lived to age 20 and fewer than 10% lived to age 50. Today about 97% of all U.S. white females born are alive at 20 years of age and almost 90% are still living at 50. Even in underdeveloped countries female survival rates far exceed those of Europe 400 years ago. Thus the number of women of childbearing age has vastly increased in the past two centuries and so has the number of births. More births of people of longer life span leads to greater and greater population.

Cultural attitudes and social and economic considerations can have an important effect on the number of children a family will have. Agrarian cultures have traditionally placed a high emphasis on large families. Basically this is because large families are an economic asset—more children are available for work. And more working family members can better support the aged. In our own history this was true and is still true for many underdeveloped countries. And yet these are the countries where food shortages to support the increasing numbers of people are most severe. By contrast, in the highly developed countries children are not generally considered an economic asset. The long years of education necessary to prepare a child for gainful employment together with the other costs of living in an industrialized society represent a sizable economic burden on the family. Increasingly, people look to their government rather than to their family for support of the aged through social security and pension programs. Present birth rates in Western Europe, Japan, Australia, New Zealand, Canada, and the United States are generally less than half those in many countries in Asia, Africa, and Central and South America. Even though economic and social development depresses birth rates, people in developed countries live longer, so their populations increase, although not as fast as in many underdeveloped and developing countries.

The growth of the population in the United States since 1750 is shown in Fig. 12.3. The projections beyond 1970 represent the high and low estimates to the year 2010. This country now has the lowest birth rate in its history, less than half of what it was 75 years ago. If the present birth rate in the United States is maintained, by 2040 the much-heralded level of no increase in population will have been attained. In addition to socioeconomic factors, the development of reliable, relatively cheap and easy methods of birth control has contributed greatly to this situation. Now people have available effective and cheap means of controlling their own reproduction without sexual abstinence. It appears that in the more developed countries people are becoming concerned both at the personal level and on the global level about the so-called population explosion and the limits of the earth to support all of the people. In these countries birth rates have dropped. In contrast, similar concerns are not so apparent in many developing countries where there are masses of people who are unaware of means of birth control and who do not comprehend the impact of unchecked increases in population. Although at

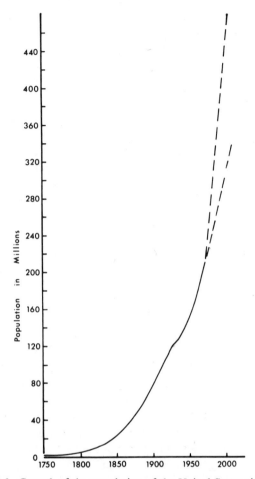

Fig. 12.3. Growth of the population of the United States since 1750.

present it is the people of developed and underdeveloped countries who display these contrasting attitudes about family size, population control will in time probably be a major governmental concern.

Population control can only come through high morality rates or low birth rates. Although it appears that for the present the people of highly developed countries have opted for fewer births and hence some measure of population control, these people represent a minority of the world's population. As Fig. 12.1 shows, the fantastic growth of the world's population is a fact; it cannot be abruptly arrested except by mass extinction. So the big question is whether enough food can be produced to feed the masses or whether the rapid population growth will be controlled by high mortality—starvation. Already many people in the world are hungry—they are not getting enough food energy (calories). Furthermore, many more people in the world are malnourished—they are not getting enough protein—which means inability to work, sickness, and early death for many. Where is the necessary food for all these increasing numbers of people to be found?

INCREASED PRODUCTION OF TRADITIONAL FOOD ORGANISMS

The production of more wheat, more rice, more corn, more beans, more milk, more meat, etc., implies (1) that more food can be produced per unit of land available and (2) that land not now used for food production can be put into production. There is no doubt that both avenues might yield more food, but, recalling the cost factors discussed in Chapter 10, it is also clear that economics will control how large will be the increase in the amounts of traditional foods produced.

The predictions of Malthus in 1798 concerning the dire consequences of expanding population in the face of the limited food supplies of that day have not materialized, although hunger and famine have afflicted peoples of the world in the past two centuries. The real crisis predicted by Malthus has not yet materialized largely because new lands have been brought into food production in the last 200 years in the Americas, Australia, and New Zealand and because the advent of modern agricultural technology has increased food production efficiency. But now there are essentially no great new land masses available for conventional agriculture. So our hope is in better use of land and water. The Food and Agricultural Organization estimates that on a worldwide basis land used for food production is producing less than 30% of what is currently possible.

As we take a brief look at the possibilities for increasing the production of the more conventional foods, it is well to keep in mind a few fundamental considerations developed in earlier pages. All energy for biological systems comes from the sun, and the primary producing organisms are the autotrophs—the plants. Animals including man are consumers, and so production of human food rests first on plant production. The environmental factors which promote optimum growth of human food or animal feed are water, temperature, pH, water activity, and nutrients (Chapter 4). For plants, these translate into rainfall or irrigation water, ambient temperature, the pH and activity of the available water, and mineral nutrients. Although plants can be grown in light without soil by using artificial supports and bathing the roots in nutrient solution by immersion or spray,[1] the natural and cheapest support for plant growth is soil. Soil determines the pH of the water available and serves as a reservoir for water and for the mineral nutrients. In addition to furnishing the energy for growth, sunlight as well as geographical latitude and elevation above sea level determines ambient temperature. In a practical sense, then, sunlight, temperature, rainfall or other source of water, and soil conditions determine how efficiently food plants can be grown.

In the world there are large areas where temperature, sunlight, and soil are sufficient but there is insufficient rain or other source of fresh water. Many deserts will bloom when there is adequate water, but often the cost of providing water would result in prohibitive food costs. And desert soils may be too high or too low in pH and contain too much salt or other water-soluble materials. Also, the available irrigation water may have too much mineral matter, including salt, resulting in too low water activity.

There are large areas in the world, particularly in the warmer parts, where all environmental conditions are met except mineral nutrients. Plants are scrubby and stunted in these areas, called *savannas* or *campos cerrados*, because of lack of mineral nutrients for plants or low soil pH. These areas have rainy and dry seasons. In the tropics, large rainforest areas have rain, sun, and warm temperatures throughout the year. Although lush in appearance, these rainforests are maintained as a stable ecosystem on soil poor in plant nutrients compared to many soils in temperate climates. We have not yet learned how to use these soils of the tropics effectively. In both rainforests and savannas, it seems that most mineral plant nutrients reside in living plants and the soils themselves have little nutrients in reserve; as one plant dies and decomposes, a living plant quickly assimilates the mineral nutrients released. If these vast areas are to be devoted to intensive food production, much research is needed to learn how to use them. Successful use will likely depend on heavy fertilization—supplying the necessary phosphate, potassium, nitrogen, calcium, and trace elements required by the food plants. And such fertilization will add greatly to food costs.

In the temperate regions of the world, a casual traveler along the highways has little difficulty observing that soils appear to differ in productivity even in areas with the same temperature and rainfall. This can be due to differences in mineral nutrients, differences in soil pH, too rapid drainage of rainwater or too slow drainage which inhibits plant root development, or some other factor producing an unfavorable environment for plant growth. The productivity of some of these soils can be improved by fertilization or by drainage or irrigation systems. Such improvements cost money and hence increase production costs. So it is obvious that which lands are used for food

---

[1] Plant culture of this type is sometimes called *hydroponics*.

production and the kinds of crops grown are directly related to economic factors. As the need for and price of food go up, it may become profitable to put so-called marginal land into production. However, the amount of marginal land which can be used with only moderate increases in costs is not great.

In terms of the limited land resources and restrictive climatic conditions in some populated areas of the world, it is truly remarkable what some societies have accomplished to enhance food production. The terracing of hills and mountainsides done by the ancient Inca and Maya cultures of South and Central America and similar terracing that has been done in various parts of Southeast Asia and Indonesia and even in Tibet and Nepal have provided land to support food production for these peoples. Such earthworks along with water management provide the proper environment for growth of food plants. From ancient times, irrigation systems have been developed in water-short areas and drainage systems in swampy areas to provide land for crops. The Dutch for centuries have worked to reclaim farmland from the sea and are still at it on a grand scale. All of these are examples of making otherwise useless land productive. These works required large inputs of manpower and money but were necessary for the survival of the people who built them. But in the rapidly changing world with international trading in food the traditional agricultural techniques of some of these societies are slowly yielding to economic pressures for more efficient agriculture.

The small fields characteristic of terraced mountains and hills are generally unsuitable for mechanized agriculture and therefore such plots must for the most part be cultivated manually, supplemented with animal power in some instances. Irrigation and drainage systems can be adapted more easily to modern agricultural techniques.

A great deal of research has been devoted to the production of fresh water from seawater; however, the cost is prohibitive at this time for food production on a large scale. The reason is that most food plants require in excess of 300 units of water to produce 1 unit of organic matter, of which only a part is useful as food. Nevertheless, food production based on irrigation water rather than unpredictable rain can be very efficient because the total environment of the plants can be managed more effectively. In such instances, some of the costs of water and fertilizer can be offset by greater productivity per unit of land, water, energy, labor, or other economic input.

With respect to land productivity in different locations and the agricultural techniques used in crop production, it appears that the proper use of fertilizers and improved farming techniques offer the best hope for increased food production of traditional foods in the immediate future. Much of the land now farmed in the world could yield more by increased fertilization and the use of newer techniques for weed and pest control. This requires education and money or credit to buy fertilizer, if it is available. Many underdeveloped countries with rapidly increasing populations do not have fertilizer production facilities and supplies of the necessary ingredients. In the more developed countries, the productivity of some lands also can be improved by fertilization. Of course, not all soils or all plants need the same kinds and amounts of fertilizer, and improper fertilizer usage represents money wasted. With adequate soil testing to guide application of fertilizer at the correct time in relation to planting and cultivation practices for a particular crop, the productivity of poor and marginal land can be improved if temperature and rainfall are sufficient. In large measure this is the essence of modern scientific agriculture. Some agricultural scientists believe that the limit of productivity of good supporting soil with adequate water resources, fertilization, and ambient temperatures has not been reached. Others are less optimistic. In any event, cost factors in relation to market demand will control fertilizer usage to increase food production whether in an underdeveloped or a developed region of the world.

The crops which contribute the greater part of the food consumed by the population of the world as a whole are seed crops—wheat, rice, soy, corn, beans, etc. Besides calories these crops supply a large part of the protein for a majority of the people in the world. Seed crops also make up a large part of the rations fed to animals to produce meat, milk, and eggs. Each crop has its own life cycle and environmental requirements for growth, and even within crop species different varieties have somewhat different growth characteristics. Consequently, some crops grow well in some locales and poorly or not at all in others. But a common characteristic of all the seed crops is that, for best yields, they all must reach maturity as rains diminish and the climate becomes relatively dry. Seed crops are generally difficult to harvest in the rainy season, and much of the crop is lost if harvest is attempted when plants and ground are wet. Furthermore, wet seeds spoil quickly and cannot be stored well. So the major seed crops are grown in the temperate zone or in the semitropics and tropics with distinct rainy and dry seasons. Rainforest areas are generally unsuited for production of seed crops. The most familiar commercial crops of warm rainy areas which supply a significant proportion of the food of some populations are bananas, cassava, and sugar cane—all essentially carbohydrate, low-protein foods. Because of this, many people in tropical areas have difficulties in getting an adequate diet unless there are some supplies of seed foods or meat, fish, milk, eggs, etc.

The production of animals for meat, milk, and eggs can of course be increased if sufficient feed is available. As pointed out in Chapter 8, animals can live in a wider range of climatic conditions than plants and furthermore can adapt to using feedstuffs from different plants. Yet many aspects of feed production for animals are the same as food production for humans. This is even true for the ruminants, which can thrive on forages or pasture rather than seed crops because they are physiologically adapted to consuming and utilizing whole plants—leaves and stems as well as flowers and seeds. To be sure, cattle have traditionally grazed drier and less productive lands than those used for cultivation of seed crops. Nevertheless, improved yields of forage for grazing require improved pasture management, fertilization, water, etc.—the same types of inputs necessary to improve the production of traditional crops whether they be seeds, fruits, or vegetables.

Increased production of traditional food organisms may be the best hope for increasing world food supplies in the near future for two very good reasons. First, traditional foods are more acceptable to people, and there is no problem of learning how to use new foods. Second and perhaps more important, the conditions required for production of traditional food organisms are known better. The essence of scientific agriculture which has typified developed countries with plentiful food supplies is that it is soundly based on detailed knowledge of the physiology, chemistry, nutrition, and genetics of food organisms and competitive organisms, including an understanding of the fundamental environmental requirements for growth, development, and reproduction of these organisms and the chemical, physical, and biological composition of fertile soil. A great deal of research has been and is continuing to be conducted to improve productivity even at the local or regional level, because, for example, corn production in Iowa may require somewhat different techniques and varieties than in Ohio, Missouri, or Central America. Such technical refinements of food production have come from university, government, and industry research efforts and educational programs aimed at getting this useful information to the farmer producing the food crop.

Most of the scientific effort that has produced the sophisticated agricultural industry characteristic of the economically developed countries has been directed toward agriculture in the temperate zones of the world. Relatively little scientific effort has been put into the agriculture of the semi-tropics and tropics where the greatest population growth is taking place today and where the demands for more food are most urgent. Even such an important crop as rice has received relatively little scientific attention until recent years. The results of even a few years of research on rice have shown promise in increasing the supply of this major food crop. However, the question remains—can other traditional crops be adapted to economic production in the warmer lower latitudes of the earth? Many pasture crops used for animal production in temperate climates cannot be produced in the warmer climates, but agricultural scientists are having some success in finding and developing other forages more suited to the semi-tropics and tropics. For the more traditional seed crops, much work must yet be done in order to learn how to produce them more effectively in those warmer areas where increase in population is leading to worse food shortages.

POTENTIAL BUT UNUSED FOOD ORGANISMS

The quest for new food sources is as old as man himself and the search will continue with even greater intensity because the demand for food is increasing at a rate never before seen. Although the exploits of the great navigators and explorers in their search for gold and new empires have been emphasized and romanticized in our history, a major concern of expeditions from those of Marco Polo to the more recent ones of Columbus, Magellan, Cook, and others was the search for new food organisms. To understand that these were among the riches brought back to their homelands we need only to observe what we eat. Potatoes, tomatoes, corn, beans, cassava, and turkeys come from the New World and sugar cane, rice, wheat, peas, and pigs came from the Old World (see Chapter 8), and now all of these foods are relatively common throughout the world. The idea of processing otherwise unusable organisms to provide food is also old, although we are doing more of it today. The production of sugar cane, the milling of seeds, the processing of toxic cassava to produce less toxic stable foods, and even cooking itself which have come down to us from centuries past are in a real sense the same approach that has brought us in recent years such important foods as cottonseed oil from toxic cotton seeds, corn and soybean oils, corn syrup and glucose from corn, and purified proteins from soy beans. So the search for newfood organisms is governed by two factors: (1) whether the new organism is edible and nutritious and (2) if it is not, whether it be made so by processing and preservation.

Of the many hundreds of thousands of species of organisms on the earth, only a relatively few contribute significantly to the world's food supply. It is only natural in the present era of scientific, social, and economic inquiry to wonder if there are organisms which can serve as sources of human food other than those that man has already selected through the millennia of his evolution.

In reflecting on the biology of food organisms (Chapter 8), the necessity and requirements for food processing, preservation, and distribution (Chapter 9), and the economic, social, and cultural factors affecting food supplies (Chapter 10), it is clear that there are many conditions which must be met for a hitherto unused organism to become a significant source of urgently needed food. Although these complex considerations may engender some pessimism, the scientific era has given us new tools—information—to guide this quest for more food. From our knowledge of the chemical constitution of biological systems (Chapter 6), the nutrient requirements for ourselves and the organisms which sustain us (Chapter 5), and how we utilize our food (Chapter 7), it is possible to develop and apply new technologies to produce new sources of nutritious human food. Furthermore, it is no longer necessary that a food organism be edible without processing and with little or no preparation. By refining and processing, inedible or even toxic biological material may yield useful food. So from this point of view the number of potential food organisms is vastly increased. Whether or not new organisms will be useful as food sources is dependent on the same criteria that apply to any traditional food organism: Can the new food be economically produced, processed, and distributed? Will it be accepted? Will it be safe and nutritious?

We shall take a brief excursion into this fascinating and very broad subject of potential new food organisms,[2] emphasizing foods rich in protein. As foods become scarce on a global basis, protein deficiencies may arise even before deficiencies of calories, which are largely supplied by carbohydrates and fats. Consequently, at present most attention is being given to increasing the availability of protein foods, because it is a more urgent problem.

The mention of protein foods brings to mind meat and milk—both from animals. Of all the species of animals which inhabit the earth only a few have been domesticated, and yet many species are hunted and eaten. In certain parts of the world wild animal populations are fairly large. In our own country millions of buffalo roamed the vast central plains until the land was settled and diverted to agriculture. Now this once majestic species is limited to several thousand, mostly in captivity. Some of these herds are being studied to determine whether the buffalo can once again be used as a food source. Wild animal production and harvest in certain African areas has shown promise in a limited way. The numbers of animals in an ecosystem multiply in accordance with their food supplies, and if animals could be husbanded within the limits of their natural food supply a certain proportion would be available for harvest. Perhaps this approach, which is still the basis for the fishing industry, holds less promise than determining whether certain selected species can be produced economically in traditional ways. As discussed in Chapter 8, the animals selected would have to be herbivorous (plant-eating or first-order consumers ecologically), fast growing, manageable, and have good reproductive performance. If they were to be raised in captivity, the plants on which they depend or to which they can adapt would have to be produced in plentiful supply.

The eland or African elk has received some attention as a possible new source of meat. It grows to over 1000 pounds (450 kilograms) as do cattle and it is farily adaptable as to climate. Other species of antelopes and deer have also shown promise. Where these are protected and fed, as in national parks, their numbers occasionally increase so much that some must be slaughtered— a source of useful meat. Since antelopes and deer are ruminants as are cattle

---

[2] N. W. Pirie has made numerous poignant observations in his excellent book *Food Resources, Conventional and Novel* (Penguin Books, 1969). See also *Scientific American* *216*, No. 2, p. 27 (1967).

and sheep, they can thrive on cellulosic plant material and nitrogen compounds unusable by man. The capibara, native to warm climates of South America, has been suggested as another likely species for development. It grows to 44 lb (20 kg) or more and is something like a large guinea pig. Even the manatee—a large freshwater mammal which lives on aquatic plants—has attracted attention, not only as a source of meat but also as a consumer of profusely growing aquatic plants which often clog up irrigation canals.

Although there are many species of aquatic plants which grow in fresh and salt water, only a few are harvested for human food. Some of the seaweeds are harvested and eaten directly or processed for use as ingredients in prepared foods. There are relatively few fully submerged flowering and seed-producing plants and only one is known to be harvested for food. This is eelgrass (*Zostera marina* L.), which is eaten by the Seri Indians in Mexico, who harvest the seed and grind it into flour. It has been suggested that eelgrass be studied more completely as a possible new food organism because it thrives in ocean water and so is one of the few plants now cultivated for food that does not require fresh water. Perhaps in the marine swamps, salt marshes, and high-saline soils there are other salt-tolerant plants which might be cultivated. Determining whether the culture in quantity of eelgrass or other plants is economic using seawater in the warm arid areas near the oceans will require considerable research effort. However, the rewards of success would be great, for there are unlimited quantities of seawater and vast areas of desert land near the oceans.

Fishes of many species have been used as food since prehistoric times. Some species are preferred over others, and some not generally eaten are edible. There are even a few species of fish eaten which are poisonous if improperly prepared. In Japan, where seafoods comprise a major part of the diet, certain fish can be sold only by licenced vendors, chefs, or restaurants. Generally both fresh- and saltwater fishes and other marine foods are simply hunted. The danger is that overfishing or overhunting can deplete the desired fish or animal to the point of extinction. There is much concern about overfishing, but regulation is difficult since most fishing is done in international waters of the open sea. In recent decades there has been an increase in the use of some species of fish not previously used. There are even examples of renaming fish or processing them in a new way in order to make them marketable as human food. The seas will not yield an unlimited supply of foods, as the commercial fishing industry is finding out. Reserves are being depleted. We know almost nothing about culture of fish in the open sea, and there are many obstacles to such a technology.

The Chinese for centuries have grown freshwater fishes (usually of the carp family) on so-called fish farms. Such cultivation of catfish and trout has been increasing in other countries, including the southern United States.

There is no similar cultivation of marine fishes. A considerable amount of attention is being given to cultivation of oysters and shrimps, but these are expensive foods which supply an insignificant percentage of the total food supply. As with cultivating an organism of any kind, it is necessary to create the environment which can promote growth and reproduction. This usually means adding nutrients for sustained cultivation. So in a sense fish culture is not unlike growing chickens and pigs, where feed must be produced by the cultivation of other crops.

In recent years, fish "flour" or bland fish protein concentrate (sometimes called FPC) has been developed for use in foods. This product has a high nutritive value, as have most animal proteins. The idea behind it is to convert hitherto unused fish into human food. This concept could also be usefully extended to other animals and parts of animals not now consumed as food. Such animals and many fishes of the sea have been traditionally converted into fertilizer and to some extent into animal feed—poor uses of protein of high nutritive quality.

The fish protein concentrate is prepared from whole fresh fish by removing both fat and water by suitable solvent processes, and the residue is converted into a flourlike material to be incorporated into prepared foods such as cereal products. FPC improves their nutritive properties because it contains many of the minerals of fish. Some health authorities have objected to the fluoride content, but proper use offers no problem because fluorine itself is essential to human nutrition. More particularly, government authori-ties have objected for so-called esthetic reasons—doubts about human consumption of products made from whole fish (including eyes, bones, organs, etc.) and from fish usually considered inedible. This reasoning seems a bit shaky since oysters, clams, some sardines, and certain other animals are eaten whole, and in many cultures consumption of almost entire animals is the usual practice. Recently, the U.S. Food and Drug Administration has, after more than two decades, approved limited use of FPC in human food, although such action is not as significant in the United States as it would be in less developed countries with protein deficiencies.

In some respects, the use of FPC in human food is similar to the use of whey—the fluid remaining when milk is coagulated—to produce a curd which is then made into cheese. Formerly, the whey was usually thrown away; however, in the last few decades the dairy industry has salvaged whey for processing and drying to be used in foods. Dried whey contains about 13% high-quality protein, about 73% milk sugar, and the minerals and vitamins common in milk.

As already mentioned, the process used to make fish protein concentrate is adaptable for preparation of other high nutritive quality animal proteins now diverted to fertilizer or low-grade animal feed. Presently the amount of such high nutritive quality proteins lost to the human food supply in North America is of the same order of magnitude as the protein produced by all the wheat grown in the United States and Canada. Because of the peculiarities of our own culture, a large proportion of the tissue of domestic animals slaughtered for meat is discarded—internal organs, bones, adipose and connective tissue, blood—all rich sources of high-quality protein, all edible, and all presently eaten by people in other societies of the world. The technology is available for processing these materials discarded by our packing plants into useful, safe, highly nutritious human foods. These materials composed of the high-quality protein, vitamins, and minerals needed by man in this era of increasing need for food and can be converted into bland products similar to FPC for incorporation into processed foods of many kinds. We have seen the technological conversion of inedible soybeans into edible oil, lecithin, and more recently edible protein concentrates for use in many foods. We have now in the market purified soy proteins spun into fibers to simulate meat. So it seems entirely reasonable to salvage animal proteins now thrown away and process them as we do other protein sources to increase our own food supply. We cannot long afford the luxury of ignoring such a large source of badly needed protein food also naturally rich in minerals and vitamins.

Techniques similar to those used in the preparation of soy proteins can be applied to other seeds or seed meals which are residues from vegetable oil production. Considerable study is being made of the possibility of using cottonseed protein for human food. Sometimes current processes for recovery of oil from oil-bearing seeds result in much damage to the quality of the protein in the residue. However, oil extraction procedures can be modified to preserve the nutritive quality of the seed proteins if economic conditions warrant it.

Grasses and leaves of plants are the natural feeds of many animals and some of these are excellent sources of proteins. During World War II, some of the grasses were studied quite extensively with respect to their potential of supplying protein for human consumption in food-short nations. Although we do eat some leafy vegetables such as spinach, lettuce, and cabbage, these are not particularly good sources of protein. Human beings are not adapted to eating grasses and leaves, even those having a relatively high content of good-quality protein; these materials contain so much fiber (cellulosic material) that they are not tolerated in any great amount in the human gastrointestinal tract. However, if these materials could be grown, harvested, and properly processed, it is possible that suitable proteins for human consumption could be produced.

The possibility of using leaf protein in tropical, less developed countries where protein deficiencies are chronic has been investigated by N. W. Pirie.

In the process, the leaves are mashed and the cell sap of the leaves is pressed out. The strained green sap is heated to coagulate the protein, which is then collected by filtering. Although this simple process gives a greenish paste with some flavor, Pirie reports that the final product is quite edible and nutritious. Furthermore, the process is so cheap as to be within the economic capabilities of most underdeveloped countries. By suitable refinements, leaf protein might be used in a wide variety of prepared foods. Of course, not all plants have leaves which are sufficiently rich in harvestable protein and much work remains to determine which plant species are most promising. The leaves of some legumes are relatively rich in protein, because these plants, through a symbiotic relationship with nitrogen-fixing bacteria living on their roots, are capable of converting nitrogen of the air into protein. Alfalfa has long been cultivated as a pasture crop or a harvested crop for animal feed because of the high protein content of its leaves.

Many of the unicellular organisms—algae, protozoa, bacteria, and yeasts—have characteristics which might be exploited for food production. Some species are fast growing in the proper environment—a generation time often in terms of minutes. Many species are capable of living on substances not usable by man as food—cellulosic materials, effluents from paper mills and breweries or other industrial plants using biological materials, cheap carbohydrates, carbon dioxide, inorganic or simple forms of nitrogen, methane (natural gas), petroleum, and so on—and converting them to the carbohydrates, proteins, and lipids of their own protoplasm. Could these microorganisms be used as food? The answer is that some are used already in certain ways. Many herbivorous animals and especially the ruminants are highly dependent on the microorganisms produced in their rumens as a source of protein; in fact, cattle can thrive without external sources of protein and many of the vitamins provided that their rations contain sufficient carbo-hydrate, nitrogen in simple forms, and minerals to promote the growth of the unicellular organisms living in their rumens. Yeasts and bacteria have been used since prehistoric times in processing and preserving food. In view of these considerations, it is reasonable to find out whether unicellular orga-nisms per se can be used as human food rather than just some of the molecules they produce such as alcohol, acetic, lactic, citric, and other acids, vitamins, antibiotics, and so on. Much work is now being done to determine if its possible to use these organisms for the production of protein for human food. SCP, for *single cell proteins*, is the term commonly used to denote such potential protein foods from microorganisms.

Human beings have difficulty tolerating enough algae, yeasts, or bacteria as whole organisms to supply their daily protein requirements, notwith-standing the facts that SCP can supply the necessary amino acids and the cytoplasm of these organisms contains essential vitamins and minerals. It is the other constituents of these microorganisms that cause difficulty, so the problem is not only to produce the organisms in quantity cheaply but also to process the organisms to yield a usable food free of the objectionable constituents. The technology needed to produce the single cell organisms in quantity is well understood—the fermentation industry as we know it today evolved from the ancient technologies of wine and pickle making and later use of yeasts for beer and bread making. The remaining problem in using these organisms as a major source of food is to develop processing and refining methods.

The efficient cultivation of microorganisms differs quite markedly from agriculture as we know it. It is more like a factory (brewery) operation than a farm; the microorganisms must be produced in a controlled environment not subject to climatic variations. They are grown in aqueous media, at the correct temperature, pH, and water activity, to which have been added required nutrients. All nutrients must be transported to the plant where the microorganisms are to be grown. As organisms grow, they are harvested by filtration and the growth medium is replenished with nutrients, which as noted above are simple compared to those needed by man. Efficient production of microorganisms on a large scale requires a considerable amount of engineering skill and technical control and large capital investments for equipment, which tend to counterbalance to a degree the low cost of nutrients for the organisms. Since freshly harvested organisms are perishable, as are most other freshly harvested food organisms, facilities for further processing must be located where the organisms are grown.

Some species of algae have been harvested, dried, and eaten in a small way by groups in Africa living near Lake Chad, and ancient Aztec tribes in Mexico followed a similar practice. Whole algae are not the tastiest dish and have some of the characteristics of fresh leaves. Algae are photosynthetic and get their energy from the sun; besides water, they require carbon dioxide and a supply of essential minerals. Although these nutrients are cheap in one sense, the need for sunlight as the energy source and carbon dioxide as the carbon source presents some limitations on algal culture as far as engineering is concerned.

Yeasts can be produced easily from carbohydrate sources of many kinds —starch, sugar, wastes from their production, paper mill wastes, and wood and paper itself after digestion to simpler carbohydrates. Certain species of yeasts do better on some carbohydrates than others. Yeasts require no protein—only nitrogen as ammonium salts or some other simple form which can be manufactured easily and at a low price compared to protein. Yeasts cannot use nitrogen directly from the air. Although yeasts and processed yeasts are used in many foods in a small way and can improve the nutritive properties of many foods (usually cereal-based foods), yeasts or yeast proteins have not been developed which are capable of supplying a reasonably large proportion of the human protein requirement. However, progress is being made and yeasts are being produced commercially for use as protein sources for animals which can tolerate reasonable amounts.

Among the bacteria are species that can live on almost any kind of organic or carbon-containing compound as an energy source and use simple forms of nitrogen to produce their own protein. Various segments of the petroleum industry are putting considerable effort into the possibility of using bacteria to produce human food from petroleum. Some species of bacteria can thrive on natural gas (methane) and various fractions of partially refined petroleum—gasoline, kerosene, and fuel oil or even heavier oils. Part of the difficulty in culturing such bacteria is to get the petroleum to disperse in the aqueous phase containing other bacterial nutrients. However, such problems have been solved and production of bacteria from petroleum can be quite efficient. Production of protein for human food from bacteria growing largely on petroleum has not yet been achieved, but use of such protein for feeding chickens and swine is reported to be practical in the present state of world markets.

In spite of extensive favorable results from animal testing of these new protein feedstuffs produced by bacteria growing on petroleum, commercial development of such radically new ventures has been impeded by governmental regulations or lack of them. One problem that has arisen illustrates once again that regulatory agencies are being asked to deal positively with unfamiliar concepts (Chapter 11). Most petroleum products are mixtures of hydrocarbons. It seems that the bacteria which grow on petroleum prefer what the chemist calls normal paraffins, with branched-chain paraffins being less rapidly consumed and some aromatic hydrocarbons poorly consumed. In the last class of compounds making up many petroleum products, including those fed to the bacteria, there may be minute amounts of carcinogenic hydrocarbons such as 3,4-benzopyrene. The petroleum products used in the culture of the bacteria contain more than 98% paraffin hydrocarbons—mostly of the normal variety—but when the bacteria are harvested and processed into feed, minute traces, fractions of a part per billion, of the offensive hydrocarbon can occasionally be detected in the finished product. The unresolved question is whether the eggs, meat, and perhaps milk produced for human consumption by feeding animals such rations will contain the supposed carcinogen in amounts, however small, having adverse effects on the consumer.

At this time, single cell proteins (SCP) whether from yeasts or bacteria have been found in certain countries to be a potentially practical source of proteins for animal feeding. As more soy and other oilseed proteins are refined for human use, SCP can be used to sustain production of the traditional high-quality protein foods coming from animals. If the new technology now being developed makes SCP useful as a human food, it seems most likely in view of present eating habits that such products will be used as ingredients in processed and prepared foods rather than as separate foods. In this way the nutritive properties of high-carbohydrate foods with limited protein value could be considerably improved.

Although increased production of traditional food organisms even at higher cost may be the most likely source of increased food supplies in the immediate future, throughout his history man has found and cultivated new organisms to augment his food supply. So as the demand for food doubles and perhaps quadruples in the coming century, the outlook is rather good that new organisms can contribute significantly to our food supply. These new foods may require some readjustment of traditional ideas and customs concerning food, but if the alternative is to eat less or even starve people will surely opt for survival.

POPULATION GROWTH AND CHANGING
NATIONAL AND REGIONAL FOOD SUPPLIES

In the preceding chapters, various fundamental aspects of the complicated problems of feeding people have been described. The more complex a society or nation becomes, the more complex will be the feeding problem. The ever-increasing numbers of people to be fed increase the urgency for each nation to solve its own food problems. The production and use of traditional foods must be increased and made more efficient. Hitherto unused organisms can add to the total supply, but such new ventures will take much research and time to develop. And all foods must be processed, preserved, and distributed so as to minimize waste and maximize nutrition. As we analyze each of these phases of the human food supply from the scientific point of view, it seems that we may have the capability of feeding more people—but not an unlimited number. Generally there is a considerable gap between what is scientifically and economically possible and what actually is accomplished by social and political action. How well nations meet their own food obligations will be determined by how well they understand their food supply problems and how effectively they can marshall their human, natural, and economic resources to the task. Although it is impossible to analyze the food problems of individual nations here, we shall discuss some comparative factors in order to gain perspective on the nature of the challenge facing the United States and other countries.

The most populous nations of the world have temperate or warm climates with sufficient fresh water and soil resources to produce food for their people.

Yet there are some contrasts of importance. It is probably true that civilized man first appeared in the warmer parts of the world, and certainly early civilizations of recorded history developed in warmer climates. Early man gathered his food—the first requirement for survival. Success at gathering or hunting daily food, not preservation and storage, determined the food supply. As man learned to cultivate his crops, he found that certain lands and climates were superior for growing foods. In the warmer climates these locations were those with rainy and dry seasons, because crops could be grown in the rainy season and dried, stored, and used during the dry season until the next crop year. Even so, many perishable foods could not be used all year. Food spoilage is always a big problem, especially in warm climates, and man early recognized the need for some kind of food preservation. When he learned how to provide better clothing and shelter for himself, man found that in the temperate climates the warm summer months could be used for growing food and that the cool or cold winters greatly facilitated food storage, preservation, and distribution. So in later history more complex civilizations developed in the cooler or more temperate climates because food could not only be produced abundantly but also could be efficiently utilized over a longer period of time. Methods were developed in later years for preserving fruits and vegetables—largely seasonal foods—to the extent that today people in the United States and other temperate-climate developed countries often eat much more fruit and vegetables than people in the warmer climates that have longer growing seasons. Basically this is because the foods that are more easily grown in warm climates spoil more easily there, too. In terms of modern food technology, the costs of refrigeration and other types of food preservation and storage are higher in warmer countries than in cooler countries.

Although many people do not enjoy cold winters, such weather besides making food storage and preservation easier has desirable effects for growing crops. For instance, many potential insect pests cannot survive winter weather and those that are adapted to temperate climates generally survive only in the form of eggs or larvae which require a period of growth in the spring before they become fully destructive. When crops mature and are harvested, crop residues—roots, leaves, stems, etc.—are left. The cold winters retard the decomposition of these biological materials until the next growing season, when decomposition resumes, releasing the residual plant nutrients to the new crop. The residual organic plant material also improves the physical characteristics of the soil for growing the new crop. In warm climates, these agronomic advantages of residual plant material are often lost because the decomposing organisms in the soil continue to live, converting most of the residual plant material to carbon dioxide and water, and then the rains come and leach out residual mineral plant nutrients.

Populations in the warm countries are increasing even more rapidly than those in the temperate countries. Yet, as already discussed, what we have come to call efficient scientific agriculture has been directed toward food production in temperate rather than semitropical and tropical countries. Some research and development effort has been directed toward improved production of semitropical and tropical crops, but usually only those crops desired by the economically developed countries such as sugar, bananas, coffee, and cacao.

The vast rainforest lands of the warm countries are often thought of as simply unused but easily used lands for food production. But they are not. The traditional method of the indigenous populations (tribes) of slash, burn, clear-off natural vegetation and cultivate for a few years until the land will not produce sufficient food and then abandon for 50 years or more and repeat the sequence in a new locale is not the answer to intensive food production. We have yet to learn how to use rainforests and other lands in warm countries without depleting or even ruining it for future generations. And these are the kinds of lands that exist in economically underdeveloped and developing countries with rapidly growing populations and very limited resources for discovering how best to grow food (see Chapter 10).

Many of these developing countries are already not producing sufficient food and have not kept pace with their population increases of the last quarter century. How can the future food demands for these already densely populated countries be met? This is perhaps the most disturbing food situation in the entire world, for these countries have few resources and little to sell on the world market to gain needed capital for development.

There are a number of underdeveloped and developing countries in the warm climates which have become food suppliers for the more developed countries. In fact, export of sugar, coffee, etc., has been the major source of foreign exchange for the purchase of goods from industrial nations, for buying petroleum or other fuel, and for financing economic development. These food exports are in a large part from high-labor crops which are marketable because of local low or even subsistence labor costs. But with increasing populations can the export of food be maintained by these countries without diminishing their already limited local supplies of nutritious food? Will it become necessary to divert lands from export crops to more nutritious foods for local people? If so, how will industrial imports be financed? Just the maintenance of the status quo in these countries means increased food production for both export and local use. We can only speculate about the outcome of diminished food for export. And even if the food-exporting developing country becomes successful in its economic development, this means industrialization and likely higher agricultural labor costs. If the costs of of producing export foods go up, food economies of developed buying countries which have come to depend on less well-developed countries for significant amounts of food will be affected. So in a real way the population explosion is beginning to rapidly alter world food markets and the economies of all nations.

Even though population growth in the economically developed nations is significantly less than that of developing nations, the populations of many economically developed countries are still rapidly increasing (Fig. 12.3), and consequently their food requirements are increasing. However, some countries such as Sweden and Japan have shown considerable progress in stabilizing their populations. A characteristic of the economically developed countries is that they have been able to adequately feed themselves through the years by either producing their food or importing it. Today two highly industrialized and economically developed countries have reached the point where they are dependent on other nations for a large part of their food—Japan and the United Kingdom. They must manufacture industrial products and sell them to other countries in order to be able to buy food for themselves.

As a group, the developed nations have in the past been able to feed themselves through modern scientific agriculture, although the efficiency of food production varies greatly among these countries. The demand for food resulting from increased populations in their own countries as well as worldwide will affect the developed countries in different ways. This is because coupled with food requirements are the increasing energy (largely petroleum and electricity) needs also characteristic of economically developed industrial nations.

Rapid readjustments with respect to food supplies caused by increasing populations and industrial development on a global scale are observable in the United States. Traditionally we have been a nation of plentiful food and have been in a position to contribute our food surpluses to alleviate acute food shortages and famine in other countries. Today the efficiency of the U.S. agricultural and food industries rests on abundant supplies of petroleum. In the United States as well as other developed countries and even some developing countries, petroleum demands are increasing even more rapidly than food requirements. The result of these changes is that this country, which once had sufficient petroleum for its needs, must import a significant part of this vital energy source. Furthermore, there is now more worldwide competition in the sale of manufactured goods, and profits from the sale of industrial products to other countries are less able to supply the exchange to buy the necessary oil. However, now that the world food demand is greater than ever before the United States is becoming a much more important supplier of food to other countries on a continuing commercial basis than ever before in its history. More and more the United States will probably be selling food to buy oil—a situation formerly more common in underdeveloped nations. The result is likely to be that, even though the traditionally efficient

U.S. agricultural and food industries have the capacity to more than adequately feed our increasing population for some time, population pressures here and abroad will place American consumers more directly into the world food market. Our position is becoming more directly competitive with that of the people of other countries who have the resources to bid for food produced, processed, preserved, and packaged in the United States.

The effect of this new dynamic dimension to our own food supplies is seen in higher food prices in relation to expendable income, fuel shortages, changing patterns of food supplies in the supermarkets, and increased use of foods made from soy and other nontraditional food ingredients. There will certainly be other changes affecting the food supplies available to American consumers. What they will be is at this point speculative. But whatever they turn out to be, it is likely that by the year 2000 the food supplies and food habits of the average American will change more than they have since the beginning of this century—and the period from 1900 to the present has itself been a period of great change. It is reasonable that the food situation will also change as much or more in other developed and developing nations.

If we learn from history, it is clear to us now as it has been to many philosophers, scientists, and political and military leaders of almost all ages that the destinies of nations are determined by how well they are able to feed themselves. With the awesome increase in population raising the demand for food to an unprecedented level, the magnitude of the food problem facing the nations of the world is enormous. But this does not relieve the nations of the world of their obligations to their people—find a way to feed themselves or die. Although each nation's food problem is great and its solution may be complex and different from that of other nations, the fundamental scientific and economic principles on which these solutions will rest are the same for all.

In the long run, people will survive through the application of collective ingenuity. Our unique human biological attribute is intellect—an inquiring nature and a problem-solving ability. And here there is hope. We can find a way to feed ourselves in the future if we direct our creative abilities not toward our destruction but toward our survival. But sometimes recognizing the difference is difficult. Let us hope that the expansion of knowledge concerning the nature of our human species and the world we live in will help us find the way to provide sufficient nutritious food for everyone.

APPENDIX

COMPOSITION OF FOODS

The food composition tables on the following pages are reproduced from *Documenta Geigy Scientific Tables*, 7th edition, by permission of the publishers, CIBA-GEIGY Limited, Basel, Switzerland.

For the recommended daily allowances to meet human nutritional requirements, see Tables 5.1, 5.3, and 5.10 (pp. 74–75, 79, and 93).

The values given in the tables are the contents in 100 grams of edible portion, uncooked unless otherwise noted. Zero values are shown by 0, and where information is unknown — is shown.

The data are representative values selected from a large number of sources. As do all biological organisms, foods show within certain limits some variation in composition. Prepared foods also show variations.

Some nutrients may be altered during food preparation such as cooking. The most common losses are in water-soluble vitamins and minerals, some of which may be thrown away in cooking water or drippings. The table below indicates percentage losses of some vitamins which may occur during cooking.

Vitamin A (including β-carotene) and vitamin D amounts are given in International Units (IU). All other vitamins are in milligrams (mg). 1 IU vit. A = 0.0003 (or 0.0006 mg β-carotene); 1 IU vit. D = 0.000025 mg vit. D_3, activated 7-dehydrocholesterol. In IU the recommended daily allowance for the adult for vitamin A is 5000 IU and for vitamin D 400 IU. The tables indicate whether the foods give an acidic (A) or a basic (B) reaction after digestion and utilization by the body (see pp. 183–185).

To determine from these tables the nutrients consumed in a particular situation, it is necessary to know the weight of the food consumed. Meaningful estimates of the weights of portions eaten can be made in various ways. Information on food packages can be helpful. For example, if a loaf of bread

	Thiamine (B$_1$)	Riboflavin (B$_2$)	Nicotinic acid (Niacin)	Ascorbic acid (C)
Meats	35	20	25	—
Meats plus drippings	25	5	10	—
Eggs	25	10	0	—
Legumes	20	0	0	—
Vegetables, leafy	40	25	25	60
Vegetables, other	25	15	25	60
Tomatoes	5	5	5	15
Potatoes	40	25	25	60

weighs 1 pound or 454 grams and is cut into 18 slices, then each slice weighs approximately 25 g. Also, the following list of approximate equivalents by weight and volume will help in making estimates. With some care in estimating foods consumed and a little practice, it is not difficult for anyone to make useful evaluations of the nutritional adequacy of his diet by using these tables.

American–Metric System Equivalents

Weight: 1 kilogram = 1000   grams = 2.20   pounds = 35.3   ounces
                      100   grams = 0.22   pounds = 3.53 ounces

                     28.5 grams = 0.0625 pounds = 1.00 ounce
                     454   grams = 1.00   pound = 16.0   ounces
Volume: 1 liter = 1000 milliliters = 1.05 quarts = 4.2 cups

Volume	Approximate volume and weight of water in milliliters and grams[a]
1 teaspoon	5.0
1 tablespoon = 3 teaspoons	15.0
1 fluid ounce = 2 tablespoons	29.6
1 cup = 8 fluid ounces	236.5
1 pint = 2 cups	473.0
1 quart = 2 pints	946.0

[a] One liter of water weighs 1.000 kilogram or 1000 grams.

Content per 100 grammes edible portion (unless otherwise stated)	Water g	Proteins g	Fats Total g	Fats Poly-unsaturated g	Carbohydrates Total g	Carbohydrates Fibre g	Calories* kcal	A** IU	B₁ mg	B₂ mg	B₆ mg	Nicotinic acid mg	Pantothenic acid mg	C mg	Other vitamins*** mg	Malic acid mg	Citric acid mg	Oxalic acid mg	Excess acid A / Excess base B	Sodium Na mg	Potassium K mg	Calcium Ca mg	Magnesium Mg mg	Manganese Mn mg	Iron Fe mg	Copper Cu mg	Phosphorus P mg	Sulphur S mg	Chlorine Cl mg
Fruits, Fruit juices																													
Apples (sweet) (*Pirus malus*)	84.0	0.3	0.6	–	15.0	0.9	58	90	0.04	0.02	0.03	0.1	0.1	5†	E 0.3; biotin 0.001; FA 0.002	270–1020	0–30	1.5	B	1	116	7	5	0.07	0.3	0.08	10	5	4
per lb as purchased (refuse 18%)	*312*	*1.1*	*2.2*	–	*56*	*3.3*	*216*	*335*	*0.15*	*0.07*	*0.11*	*0.4*	*0.4*	*19*	*E 1.1; biotin 0.004; FA 0.007*	*1000–3790*	*0–112*	*5.6*	*B*	*4*	*431*	*26*	*19*	*0.26*	*1.1*	*0.29*	*37*	*19*	*15*
dried	20.4	3	0.7	–	73.6	4.0	281	–	0.05	0.08	0.16	0.5	–	10†	–	–	–	–	B	5	557	31	29	–	1.6	–	52	19	19
Apple juice, fresh	86.9	0.1	trace	–	13	–	47	–	0.01	0.02	0.03	0.5	0.02	1	FA 0.001; biotin 0.0005	700	230	0	B	2	100	6	–	–	0.6	0.35	9	–	–
Apple sauce, sweetened	75.7	0.2	0.1	–	23.8	0.5	91	60	0.01	0.01	–	trace	–	1	–	–	–	–	B	0.3	55	4	5	–	0.5	–	5	–	–
Apricots (*Prunus armeniaca*)	85.3	0.9	0.2	–	12.8	0.6	51	2700	0.03	0.05	0.07	0.7	0.3	7	FA 0.003	–	–	–	B	0.6	440	17	9	0.2	0.5	0.12	23	6	2
per lb as purchased (refuse 6%)	*364*	*3.8*	*0.9*	–	*55*	*2.6*	*217*	*11500*	*0.13*	*0.21*	*0.29*	*3.0*	*1.3*	*30*	*FA 0.013*	–	–	–	*B*	*2*	*1880*	*72*	*38*	*0.9*	*2.1*	*0.51*	*98*	*25*	*9*
canned, sweetened	76.9	0.6	0.1	–	22.0	0.4	86	1740	0.02	0.02	0.05	3.0	0.1	4	–	330	1060	–	B	2	256	11	7	0.08	0.3	0.05	15	1.0	2
dried	25.0	5.0	0.5	–	66.5	3.0	260	10900	0.01	0.16	0.25	3.3	0.7	12	FA 0.005	810	350	–	B	26	1700	67	65	0.28	5.5	0.4	119	164	35
Avocados (*Persea gratissima*)	73.6	2.2	17.0	2	6.0	1.5	171	290	0.11	0.20	0.61	1.6	0.9	14	FA 0.03	–	–	–	B	3	340	10	30	0.3–4.2	0.6	0.4	42	25	10
Bananas (*Musa sapientum*)	75.7	1.1	0.2	–	22.2	0.6	85	190	0.05	0.06	0.32	0.6	0.2	10	E 0.2; biotin 0.004; FA 0.01	500	150	6.4	B	1	420	8	31	0.64	0.7	0.2	28	12	125
per lb as purchased (refuse 32%)	*233*	*3.4*	*0.6*	–	*68*	*1.9*	*262*	*586*	*0.15*	*0.19*	*1.0*	*1.9*	*6.1*	*31*	*E 0.6; biotin 0.012; FA 0.03*	*1540*	*463*	*19.7*	*B*	*3*	*1300*	*25*	*96*	*1.97*	*2.2*	*0.6*	*86*	*37*	*386*
Blackberries (*Rubus fruticosus*)	84.5	1.2	0.9	–	12.9	4.1	58	200	0.03	0.04	0.05	0.4	0.25	21	biotin 0.0004; FA 0.012	160	trace	18	B	4	181	32	24	0.59	1.0	0.12	19	17	15
frozen, sweetened	74.3	0.8	0.3	–	24.4	1.8	96	140	0.02	0.10	–	0.6	–	8	–	–	–	–	B	1	105	17	12	–	0.6	–	17	–	–
Cantaloups (*Cucumis melo*)	91.2	0.7	0.1	–	7.5	0.3	30	3400²	0.04	0.03	0.036	0.6	0.26	33	biotin 0.003; FA 0.007	0	–	–	B	12	230	14	17	0.04	0.4	0.04	16	11.7	41
per lb as purchased (refuse 50%)	*207*	*1.6*	*0.2*	–	*1.7*	*0.7*	*68*	*7710*	*0.09*	*0.07*	*0.08*	*1.4*	*0.6*	*75*	*biotin 0.007; FA 0.016*	*0*	–	–	*B*	*27*	*522*	*32*	*39*	*0.09*	*0.9*	*0.09*	*36*	*27*	*93*
Cherries (*Prunus avium*)	83.4	1.2	0.4	–	14.6	0.5	60	1000	0.05	0.06	0.05	0.3	0.08	10	biotin 0.0004; FA 0.006	1250	10	trace	B	2	260	19	14	0.03	0.5	0.07	19	8	3
per lb as purchased (refuse 10%)	*340*	*4.9*	*1.6*	–	*59.6*	*2.0*	*245*	*4080*	*0.20*	*0.24*	*0.20*	*1.2*	*0.32*	*41*	*biotin 0.0016; FA 0.024*	*5100*	*41*	*trace*	*B*	*8*	*1060*	*78*	*57*	*0.12*	*2.0*	*0.29*	*78*	*33*	*12*
Cranberries (*Vaccinium macrocarpon*)	87.9	0.4	0.7	–	10.8	1.4	46	40	0.03	0.02	0.06	0.1	–	12	–	260	1120	–	B	1	65	14	8	0.3	0.5	0.09	11	7	5
Cranberry sauce	62.1	0.1	0.2	–	37.5	0.2	146	20	0.01	0.01	–	trace	–	2	–	–	–	–	B	1	17	6	2	–	0.2	–	4	–	–

* To convert to kJ (kilojoule) multiply the values given by 4.1855.
** Vitamin A activity due to vitamin A + carotenes; 1 IU vitamin A = 0.0006 mg β-carotene.
*** FA = folic acid; E = α-tocopherol unless otherwise stated.
† Wide variations between different varieties.
² Highly coloured variety.

Content per 100 grammes edible portion (unless otherwise stated)	Water (g)	Proteins (g)	Fats Total (g)	Fats Poly-unsaturated (g)	Carbohydrates Total (g)	Carbohydrates Fibre (g)	Calories* (kcal)	A** (IU)	B1 (mg)	B2 (mg)	B6 (mg)	Nicotinic acid (mg)	Pantothenic acid (mg)	C (mg)	Other vitamins*** (mg)	Malic acid (mg)	Citric acid (mg)	Oxalic acid (mg)	Excess acid A / Excess base B	Sodium Na (mg)	Potassium K (mg)	Calcium Ca (mg)	Magnesium Mg (mg)	Manganese Mn (mg)	Iron Fe (mg)	Copper Cu (mg)	Phosphorus P (mg)	Sulphur S (mg)	Chlorine Cl (mg)
Currants																													
red and white (*Ribes rubrum*)	85.7	1.4	0.2	–	12.1	3.4	50	120	0.04	0.02	0.05	0.3	0.06	41	biotin 0.0026	50	2300	19	B	2	275	36	15	0.6	1.0	0.12	23	29	13
black (*Ribes nigrum*)	82	1.0	0.1	–	16.1	5.7	62	220	0.05	0.03	0.08	0.3	–	136	–	400	3030	4	B	3	336	17	10	–	0.9	0.12	28	–	–
Dates (*Phoenix dactylifera*) dried	22.5	2.2	0.5	–	72.9	2.3	274	50	0.09	0.10	0.1	2.2	0.8	0	FA 0.025	–	–	–	B	1	790	59	65	0.15	3.0	0.21	63	65	290
Elderberries, black (*Sambucus nigra*)	80.9	2.5	0.5	–	15.9	6.8	42	600	0.07	0.08	0.25	1.5	0.18	18	biotin 0.002; FA 0.017	–	–	–	B	0.5	305	35	–	–	1.6	–	57	–	–
Figs (*Ficus carica*)	81.7	1.2	0.4	–	16.1	1.4	65	75	0.09	0.08	0.13	0.63	0.4	2	FA 0.01	trace	340	–	B	2	190	35	21	–	0.8	0.06	22	12	14
dried	23.0	4.3	1.3	–	69.1	5.6	274	80	0.10	0.10	0.32	1.7	0.5	0	FA 0.03	–	–	–	B	34	780	126	82	0.35	4.0	0.35	116	69	105
Fruit cocktail canned	79.6	0.4	0.1	–	19.7	0.4	76	140	0.02	0.01	–	0.4	–	2	–	–	–	–	B	5	160	9	8	–	0.4	0.03	12	2	3
Gooseberries (*Ribes grossularia*)	88.9	0.8	0.2	–	9.7	1.9	39	290	0.15	0.03	0.02	0.3	0.15	25	biotin 0.0005	500–2080	–	–	B	1	210	35	9	0.04	0.5	0.08	31	15	9
Grapes (*Vitis vinifera*)	81.4	0.6	0.3	–	17.3	0.5	67	100	0.05	0.02	0.1	0.3	0.08	4	biotin 0.002; FA 0.006	650	–	–	B	2	250	12	7	0.083	0.4	0.1	20	9	2
Grape juice	82.9	0.2	trace	–	16.6	trace	66	–	0.04	0.02	0.021	0.2	0.04	2	biotin 0.0003; FA 0.003	310	20	–	B	1	120	11	4	–	0.3	0.02	12	–	–
Grapefruit (*Citrus decumana*)	88.4	0.6	0.1	–	9.8	0.5	39	80	0.04	0.02	0.02	0.2	0.25	40	E 0.25; FA 0.003; biotin 0.003	80	1460	0	B	2	198	17	10	0.01	0.3	0.02	16	5	3
per lb as purchased (refuse 51%)	*197*	*1.3*	*0.2*	–	*22*	*1.1*	*87*	*178*	*0.09*	*0.04*	*0.04*	*0.4*	*0.56*	*89*	*E 0.56; FA 0.007; biotin 0.007*	*178*	*3250*	*0*	*B*	*5*	*441*	*38*	*22*	*0.02*	*0.7*	*0.04*	*36*	*11*	*7*
canned, sweetened	81.1	0.6	0.1	–	17.8	0.2	70	10	0.03	0.02	–	0.2	–	30	–	–	–	–	B	2	135	13	11	–	0.3	–	14	–	–
Grapefruit juice, fresh	89.2	0.4	0.1	–	9.8	0.1	41	10	0.03	0.02	0.014	0.2	0.16	45	biotin 0.0007; FA 0.001	–	–	–	B	2	150	8	12	–	0.4	–	14	5	2
Lemons (*Citrus medica*)	90.1	1.1	0.3	–	8.2	0.4	27	20	0.04	0.02	0.06	0.1	0.2	45	FA 0.007	trace	3840	–	B	6	148	26	9	0.04	0.6	0.26	16	8	4
Lemon juice	91.0	0.5	0.2	–	8.0	–	25	20	0.03	0.01	0.039	0.1	0.1	50	FA 0.001	290	6080	–	B	1	130	14	7	–	0.2	0.13	11	2	4
Lime juice (*Citrus aurantifolia*)	90.3	0.3	–	–	9.0	–	26	10	0.02	0.01	0.05	0.1	–	32	–	–	–	–	B	1	100	9	–	–	0.2	–	11	–	39
Loganberries (*Rubus urinus* var. *loganobaccus*)	83	1.0	0.6	–	14.9	3.0	62	200	0.03	0.04	–	0.4	0.4	24	–	200	–	0	B	1	170	35	25	–	1.2	0.14	17	18	16
Melons, water (*Citrullus vulgaris* var. *colocynthoides*)	92.6	0.5	0.2	–	6.4	0.3	26	590	0.03	0.03	0.033	0.2	0.2	7	biotin 0.004; FA 0.0006	–	–	–	B	0.3	100	7	8	0.02	0.5	0.07	10	9	8
Nectarines (*Prunus persica* var. *nectarina*)	81.8	0.6	trace	–	17.1	0.4	64	1650	–	–	–	–	–	13	–	–	–	–	B	6	294	4	13	–	0.5	0.06	24	10	5

* To convert to kJ (kilojoule) multiply the values given by 4.1855.   ** Vitamin A activity due to vitamin A + carotenes; 1 IU vitamin A = 0.0006 mg β-carotene.   *** FA = folic acid; E = α-tocopherol unless otherwise stated.

Content per 100 grammes edible portion (unless otherwise stated)	Water g	Proteins g	Fats Total g	Fats Poly-unsaturated g	Carbohydrates Total g	Carbohydrates Fibre g	Calories* kcal	Vitamins A** IU	B₁ mg	B₂ mg	B₆ mg	Nicotinic acid mg	Pantothenic acid mg	C mg	Other vitamins*** mg	Malic acid mg	Citric acid mg	Oxalic acid mg	Excess acid A / Excess base B	Sodium Na mg	Potassium K mg	Calcium Ca mg	Magnesium Mg mg	Manganese Mn mg	Iron Fe mg	Copper Cu mg	Phosphorus P mg	Sulphur S mg	Chlorine Cl mg
Olives (Olea europaea) green	78.2	1.4	12.7	1.0	1.3	1.3	116	300	0.03	0.08	0.02	0.5	0.02	0	FA 0.001	–	–	–	B	2400	55	61	22	0.05–1.0	1.6	0.46	17	32	3750
Oranges (Citrus sinensis)	87.1	1.0	0.2	–	12.2	0.5	49	200	0.10	0.03	0.03	0.2	0.2	50	E 0.23; biotin 0.001; FA 0.005	trace	980	24	B	0.3	170	41	10	0.025	0.4	0.07	23	8	4
per lb as purchased (refuse 27%)	288	3.3	0.7	–	40	1.7	162	662	0.33	0.10	0.10	0.7	0.7	166	E 0.76; biotin 0.003; FA 0.017	trace	3250	79	B	1	563	136	33	0.083	1.3	0.23	76	26	13
Orange juice, fresh	86	0.6	0.1	–	12.9	0.1	49	100	0.07	0.02	0.026	0.2	0.14	50†	biotin 0.0003; FA 0.002	370	–	–	B	0.5	190	11	11	–	0.3	0.08	17	8	4
Peaches (Prunus persica)	86.6	0.6	0.1	–	11.8	0.6	46	880	0.02	0.05	0.02	1.0	0.12	7	biotin 0.002; FA 0.004	370	370	trace	B	0.5	160	9	10	0.11	0.5	0.01	19	7	5
per lb as purchased (refuse 13%)	342	2.4	0.4	–	47	2.4	182	3470	0.08	0.20	0.08	3.9	0.5	28	biotin 0.008; FA 0.016	1460	1460	trace	B	2	631	36	39	0.43	2.0	0.04	74	28	20
canned, sweetened	79.1	0.4	0.1	–	20.1	0.4	78	430	0.01	0.02	0.02	0.6	0.05	4	biotin 0.0002; FA 0.0005	–	–	1.2	B	5	107	4	6.3	0.04	0.3	0.06	12	1	4
dried	25.0	3.0	0.7	–	68.3	3.1	262	3900	0.01	0.2	0.15	5.3	–	18	–	–	–	–	B	12	1100	48	54	0.67	6.0	0.3	117	240	11
Pears (Pirus communis)	83.2	0.5	0.4	–	15.5	1.5	61	20	0.02	0.04	0.02	0.1	0.05	4	biotin 0.0001; FA 0.002	120	240	3	B	2	129	8	9	0.06	0.3	0.13	11	7	4
per lb as purchased (refuse 9%)	343	2.1	1.7	–	64	6.2	252	83	0.08	0.17	0.08	0.4	0.21	17	biotin 0.0004; FA 0.008	495	991	12	B	8	532	33	37	0.25	1.2	0.54	46	29	17
canned, sweetened	79.8	0.2	0.2	–	19.6	0.6	76	trace	0.01	0.02	–	0.1	0.02	1	–	160	420	1.7	B	2	52	5	6	–	0.2	0.04	7	1.3	3
Persimmons, Japanese (kaki) (Diospyros kaki)	78.8	0.7	0.4	–	19.7	1.6	77	2710	0.03	0.02	–	0.1	–	11	–	–	–	–	B	6	174	6	8	–	0.3	–	26	–	–
Pineapples (Ananas sativus)	86.7	0.4	0.2	–	12.2	0.5	47	70	0.08	0.03	0.08	0.2	0.17	17	FA 0.004	120	770	–	B	0.3	210	17	17	1.07	0.5	0.07	8	2.5	46
canned, sweetened	79.9	0.3	0.1	–	19.4	0.3	74	50	0.08	0.02	0.07	0.2	–	7	FA 0.001	–	–	6.3	B	1	120	11	8	–	0.3	0.05	5	2.7	4.2
Pineapple juice, canned	85.6	0.4	0.1	–	13.5	0.1	55	50	0.05	0.02	0.1	0.2	0.1	7	FA 0.001	–	–	–	B	1	140	15	15	–	0.5	–	9	–	–
Plums (Prunus domestica)	85.7	0.7	0.1	–	12.3	0.7	50	250	0.07	0.04	0.05	0.5	0.13	6	biotin trace; FA 0.002	360–2390	30	10	B	2	167	13	13	0.1	0.4	0.3	23	5	2
per lb as purchased (refuse 6%)	365	3.0	0.4	–	52	3.0	213	1070	0.30	0.17	0.21	2.1	0.55	26	biotin trace; FA 0.008	1540–10200	128	43	B	8	712	55	55	0.4	1.7	1.3	98	21	8
canned, sweetened	77.4	0.4	0.1	–	21.6	0.3	83	230	0.02	0.02	0.027	0.4	0.08	2	FA 0.001	–	–	–	B	1	142	9	5	0.07	0.9	0.16	10	–	–
Prunes dried, uncooked	28.0	2.1	0.6	–	67.4	1.6	255	1600	0.1	0.17	0.5	1.6	0.35	3	FA 0.005	–	–	–	B	6	700	51	32	0.18	3.9	0.16	79	28	9
Quinces (Cydonia oblonga, Cydonia vulgaris)	84	0.3	0.3	–	14.9	2.4	57	30	0.03	0.02	–	0.2	–	15	–	680–1590	–	–	B	3	203	14	6	0.04	0.3	0.13	19	5	2

* To convert to kJ (kilojoule) multiply the values given by 4.1855.
** Vitamin A activity due to vitamin A = 0.0006 mg β-carotene.
*** FA = folic acid; E = α-tocopherol unless otherwise stated.
† In canned juice 12.

Content per 100 grammes edible portion (unless otherwise stated)	Water g	Proteins g	Fats Total g	Fats Poly-unsaturated g	Carbo-hydrates Total g	Carbo-hydrates Fibre g	Calories* kcal	Vitamins A** IU	B1 mg	B2 mg	B6 mg	Nicotinic acid mg	Pantothenic acid mg	C mg	Other vitamins*** mg	Malic acid mg	Citric acid mg	Oxalic acid mg	Excess acid A / Excess base B	Sodium Na mg	Potassium K mg	Calcium Ca mg	Magnesium Mg mg	Manganese Mn mg	Iron Fe mg	Copper Cu mg	Phosphorus P mg	Sulphur S mg	Chlorine Cl mg
Raisins (*Vitis vinifera*) dried	18.0	2.5	0.2	–	77.4	0.9	289	20	0.11	0.08	0.3	0.5	0.09	1	biotin 0.005; FA 0.01	–	–	–	B	31	725	62	42	0.32	3.5	0.2	101	42	9
Raspberries (*Rubus idaeus*)	84.2	1.2	0.5	–	13.6	3.0	57	150	0.03	0.09	0.09	0.9	0.2	25	biotin 0.0019; FA 0.005	40	1300	15	B	3	190	49	23	0.51	1.0	0.13	22	18	22
frozen, sweetened	74.3	0.7	0.2	–	24.6	2.2	98	70	0.02	0.02	–	0.6	–	21	–	–	–	–	B	1	100	13	11	–	0.6	–	17	–	–
Raspberry juice, fresh	88	0.2	0	–	11	trace	40	120	0.02	–	–	0.6	–	20	–	–	–	–	B	7	141	29	18	–	1.0	–	14	7	10
Strawberries (*Fragaria sp.*)	89.9	0.7	0.5	–	8.4	1.3	37	60	0.03	0.07	0.04	0.6	0.26	60	biotin 0.0011; FA 0.005	160	1080	19	B	1	145	21	12	0.06	1.0	0.13	21	12	11
frozen, sweetened	75.7	0.4	0.2	–	23.5	0.6	92	30	0.02	0.06	–	0.5	–	53	–	–	–	–	B	1	104	13	–	–	0.6	–	16	–	–
Tangerines (*Citrus nobilis*)	87	0.8	0.2	–	11.6	0.5	46	420	0.07	0.02	0.07	0.2	–	31	–	–	–	–	B	2	110	40	11	0.04	0.4	0.1	18	10	2
per lb as purchased (refuse 26%)	*292*	*2.7*	*0.7*	–	*39*	*1.7*	*154*	*1410*	*0.23*	*0.07*	*0.23*	*0.7*	–	*104*	–	–	–	–	*B*	*7*	*369*	*134*	*37*	*0.13*	*1.3*	*0.3*	*60*	*34*	*7*
Whortleberries (*Vaccinium myrtillus*)	83.2	0.7	0.5	–	15.3	1.5	62	100	0.03	0.06	0.091	0.5	0.12	14	FA 0.008	100	1560	15	B	1	89	15	10	2.3	1.0	0.11	13	11	8
frozen, sweetened	72.3	0.6	0.3	–	26.5	0.9	105	30	0.04	0.05	–	0.4	–	8	–	–	–	–	B	1	66	6	4	–	0.4	–	11	–	–
Vegetables																													
Artichokes (*Cynara scolymus*)	85.5	2.7	0.2	–	10.6	2.4	49	160	0.08	0.05	–	1.0	0.4	9	–	170	100	–	B	43	430	51	–	0.36	1.3	0.2	94	20	22
Asparagus (*Asparagus officinalis*)	92.9	2.1	0.2	–	4.1	0.8	21	900	0.18	0.20	0.14	1.5	0.62	33	E 2.5; FA 0.11	100	110	5.2	B	2	240	22	20	0.19	1.0	0.14	62	46	53
per lb as purchased (refuse 44%)	*236*	*5.3*	*0.5*	–	*10.4*	*2.0*	*53*	*2290*	*0.46*	*0.51*	*0.36*	*3.8*	*1.57*	*84*	*E 6.4; FA 0.28*	*254*	*279*	*13.2*	*B*	*5*	*610*	*56*	*51*	*0.48*	*2.5*	*0.36*	*157*	*117*	*135*
canned, drained solids	92.5	2.4	0.4	–	3.4	0.8	21	800	0.06	0.10	0.03	0.8	0.15	15	biotin 0.002; FA 0.03	–	–	–	B	236[1]	166	19	–	–	1.9	–	53	–	–
Beans																													
kidney (*Phaseolus vulgaris*)	11.6	21.3	1.6	–	61.6	4.0	338	0	0.6	0.22	0.28	2.1	0.98	2	E 4	–	–	–	B	2	1310	106	132	2.0	6.1	–	429	–	25
lima (*Phaseolus lunatus*)	67.5	8.4	0.5	–	22.1	1.8	123	290	0.24	0.12	0.55	1.4	1.3	29	biotin 0.01; FA 0.13	–	–	–	B	1	680	52	66	–	2.8	0.86	142	60	9
canned, drained solids	74.7	5.4	0.3	–	18.3	1.8	96	190	0.03	0.05	0.08	0.5	0.11	6	FA 0.013	–	–	–	B	236[2]	210	28	–	–	2.4	–	70	–	–
string (*Phaseolus vulgaris*)	90.1	1.9	0.2	–	7.1	1.0	32	600	0.07	0.11	0.14	0.5	0.2	19	E<0.1: K 0.29; FA 0.028	130	30	30	B	1.7	256	56	26	0.45	0.8	0.07	44	30	33
per lb as purchased (refuse 12%)	*360*	*7.6*	*0.8*	–	*28.3*	*4.0*	*128*	*2400*	*0.28*	*0.44*	*0.56*	*2.0*	*0.8*	*76*	*E<0.4; K 12; FA 0.12*	*519*	*120*	*120*	*B*	*7*	*1020*	*223*	*104*	*1.8*	*3.2*	*0.28*	*176*	*120*	*132*
canned, drained solids	91.9	1.4	0.1	–	5.2	1.0	24	470	0.03	0.05	0.043	0.3	0.07	4	biotin 0.001; FA 0.012	–	–	–	B	236[3]	95	45	13	–	1.5	–	25	–	–

* To convert to kJ (kilojoule) multiply the values given by 4.1855.
** Vitamin A activity due to vitamin A + carotenes; 1 IU vitamin A = 0.0006 mg β-carotene.
*** FA = folic acid; E = α-tocopherol unless otherwise stated.

[1] Unsalted 3. [2] Unsalted 4. [3] Unsalted 2.

Content per 100 grammes edible portion (unless otherwise stated)	Water g	Proteins g	Fats Total g	Fats Poly-unsaturated g	Carbohydrates Total g	Carbohydrates Fibre g	Calories* kcal	A** IU	B_1 mg	B_2 mg	B_6 mg	Nicotinic acid mg	Pantothenic acid mg	C mg	Other vitamins*** mg	Malic acid mg	Citric acid mg	Oxalic acid mg	Excess acid A / Excess base B	Na Sodium mg	K Potassium mg	Ca Calcium mg	Mg Magnesium mg	Mn Manganese mg	Fe Iron mg	Cu Copper mg	P Phosphorus mg	S Sulphur mg	D Chlorine mg
Bets (beetroots) (Beta vulgaris) peeled	87.3	1.6	0.1	–	9.9	0.8	43	20	0.03	0.04	0.05	0.4	0.12	10	FA 0.02	0	110	338	B	84	303	25	23	0.94	0.7	0.19	33	15	61
tops	90.9	2.2	0.3	–	4.6	1.3	24	6100	0.05	0.17	–	0.3	0.26	30	biotin 0.003; FA 0.06	–	–	916	B	130	570	119	71	1.3	3.3	0.09	40	35	40
Broadbeans, mature, dry (Vicia faba)	12.6	24.0	2.2	–	58.2	5.9	339	30	0.53	0.30	–	2.5	–	6	–	–	–	–	B	–	–	77	–	–	6.3	–	374	–	–
Broccoli (Brassica oleracea var. botrytis)	89.1	3.6	0.3	–	5.9	1.5	32	2500	0.1	0.23	0.17	0.9	1.3	113	FA 0.05	120	210	–	B	15	400	103	24	0.15	1.1	1.4	78	137	76
frozen	90.7	3.3	0.2	–	5.1	1.1	28	1900	0.07	0.13	–	0.6	–	78	–	–	–	–	B	13	244	43	21	–	0.7	–	60	–	–
Brussels sprouts (Brassica oleracea var. gemmifera)	84.8	4.7	0.4	–	8.7	1.2	47	550	0.1	0.16	0.16	0.9	0.72	100	E 1; K 0.8–3; biotin 0.0004; FA 0.05	200	240	–	B	12	450	29	20	0.27	1.5	0.1	80	78	40
per lb as purchased (refuse 8%)	*354*	*19.6*	*1.7*	–	*36.3*	*5.0*	*196*	*2300*	*0.4*	*0.67*	*0.67*	*3.8*	*3.0*	*417*	*E 4; K 3.3–13; biotin 0.0016; FA 0.21*	*835*	*1000*	–	*B*	*50*	*1880*	*121*	*83*	*1.13*	*6.3*	*0.4*	*334*	*326*	*167*
Cabbage red (Brassica oleracea var. capitata rubra)	91.8	1.5	0.2	–	5.9	1.1	26	50	0.07	0.05	0.15	0.4	0.32	50	E 0.2; biotin 0.002	–	–	–	B	4	266	35	18	0.1	0.5	0.06	30	–	100
white (var. capitata alba)	92.1	1.4	0.2	–	5.7	1.5	25	70	0.05	0.04	0.11	0.32	0.26	46	E 0.7; FA 0.08	–	–	–	B	13	227	46	23	0.1	0.5	0.06	27.5	–	37
Carrots (Daucus carota)	88.6	1.1	0.2	–	9.1	1.0	40	11000[1]	0.06	0.06	0.12	0.6	0.27	2–10	E 0.45; biotin 0.003; FA 0.008	24	90	33	B	50	311	37	21	0.06–0.25	0.7	0.08	36	21	40
per lb as purchased (refuse 18%)	*330*	*4.1*	*0.7*	–	*33.8*	*3.7*	*149*	*41000*	*0.22*	*0.22*	*0.45*	*2.2*	*1.0*	*7–37*	*E 1.67; biotin 0.011; FA 0.03*	*89*	*335*	*123*	*B*	*186*	*1160*	*138*	*78*	*0.22–0.93*	*2.6*	*0.30*	*134*	*78*	*149*
canned, drained solids	91.2	0.8	0.3	–	6.7	0.8	30	15000	0.03	0.02	0.04	0.3	0.11	3	biotin 0.002; FA 0.03	–	–	–	B	236[2]	110	26	5	–	0.7	0.04	22	–	445
Cauliflower (Brassica oleracea var. botrytis)	91.0	2.7	0.2	–	5.2	1.0	27	60	0.11	0.10	0.2	0.6	1.0	78	E 0.15; K 3.6; biotin 0.017; FA 0.022	390	210	0	B	16	400	25	7	0.17	1.1	0.14	56	29	30
Celery (Apium graveolens) leaves and stalks	94.1	0.9	0.1	–	3.9	0.6	17	240	0.05	0.03	–	0.4	–	9	E 0.7	170	10	50	B	96	291	39	25	0.16	0.5	0.01	40	22	137
root	88.4	1.8	0.3	–	8.5	1.3	40	16	0.03	0.03	–	0.7	–	8	E 0.2	–	–	34	B	100	300	60	12	0.16	0.9	0.15	60	–	50
Chard, Swiss (Beta vulgaris var. cicla)	90.8	1.6	0.4	–	5.6	1.0	27	6500	0.03	0.09	–	0.4	0.17	34	E 1.5; FA 0.03	–	–	690	B	147	550	110	65	0.3	2.7	0.11	29	–	–
Chicory (Cichorium intybus)	96.2	0.8	0.1	–	3.7	0.6	16	–	0.07	0.12	–	0.40	–	10	–	–	–	–	B	10	182	18	13	0.30	0.69	0.14	21	18	25

* To convert to kJ (kilojoule) multiply the values given by 4.1855.
** Vitamin A activity due to vitamin A + carotenes; 1 IU vitamin A = 0.0006 mg β-carotene.
*** FA = folic acid; E = α-tocopherol unless otherwise stated.
[1] Vitamin A content in dark varieties; in light varieties 2000.
[2] Unsalted 39.

Content per 100 grammes edible portion (unless otherwise stated)	Water g	Proteins g	Fats Total g	Fats Poly-unsaturated g	Carbohydrates Total g	Carbohydrates Fibre g	Calories* kcal	A** IU	Vitamins B1 mg	B2 mg	B6 mg	Nicotinic acid mg	Pantothenic acid mg	C mg	Other vitamins*** mg	Malic acid mg	Citric acid mg	Oxalic acid mg	Excess acid A / Excess base B	Sodium Na mg	Potassium K mg	Calcium Ca mg	Magnesium Mg mg	Manganese Mn mg	Iron Fe mg	Copper Cu mg	Phosphorus P mg	Sulphur S mg	Chlorine Cl mg
Chives (*Allium schoenoprasum*)	91.3	1.8	0.3	–	5.8	1.1	28	5800	0.04	0.11	–	0.3	–	22	–	–	–	1.1	B	3	250	76	32	–	0.9	0.11	26	–	–
Corn (sweet). See Maize																													
Cress, garden (*Lepidium sativum* ssp. *sativum*)	89.4	2.6	0.7	–	5.5	1.1	32	9300	0.08	0.26	–	1.0	–	69	biotin 0.001; FA 0.001	240	–	–	B	14	606	81	–	–	1.3	–	76	–	–
Cucumbers (*Cucumis sativus*)	95.6	0.8	0.1	–	3.0	0.6	13	300	0.04	0.05	0.04	0.2	0.3	8	–	–	10	25	B	5	140	25	9	0.15	1.1	0.06	27	12	30
per lb as purchased (refuse 5%)	*412*	*3.4*	*0.4*	–	*13*	*2.6*	*56*	*1290*	*0.17*	*0.22*	*0.17*	*0.9*	*1.3*	*34*	*biotin 0.004; FA 0.004*	*1030*	*43*	*108*	*B*	*22*	*603*	*108*	*39*	*0.65*	*4.7*	*0.26*	*116*	*52*	*129*
Dandelion greens (*Taraxacum officinale*)	85.6	2.7	0.7	–	9.2	1.6	45	14000	0.19	0.26	–	–	–	36	–	170	–	25	B	76	430	187	–	0.3	3.1	0.15	66	17	99
Eggplants (*Solanum melongena*)	92.4	1.2	0.2	–	5.6	0.9	25	10	0.05	0.05	–	0.6	0.23	5	–	–	0	6.9	B	0.9	190	17	10	0.11	0.4	0.08	26	9	24
Endives (*Cichorium endivia*)	93.1	1.7	0.1	–	4.1	0.9	20	3300	0.10	0.20	–	0.72	–	10	–	–	–	27.3	B	18	400	104	13	0.22	1.7	0.09	38	26	71
Fennel (*Foeniculum vulgare*)	90	1.5	0.1	–	6.4	0.5	27	3500	0.23	0.11	0.10	0.2	0.25	31	FA 0.1; biotin 0.003	–	–	–	B	331	784	100	–	–	2.7	–	51	–	–
Garlic (*Allium sativum*) bulbs	63.8	5.3	0.2	–	29.3	1.1	129	trace	0.21	0.08	–	0.6	–	9	–	–	–	–	B	32	515	38	36	–	1.4	–	134	–	–
Horse-radishes (*Armoracia lapathifolia*)	76.6	2.8	0.3	–	18.1	2.8	80	30	0.06	0.11	0.18	0.6	0.1–1.4	120	–	–	350	–	B	9	554	105	33	–	2.0	0.44	70	212	18
Kale (*Brassica oleracea* var. *acephala*)	87.5	4.2	0.8	–	6.0	1.3	38	8900	0.16	0.26	0.19	2.0	–	115	E 8; biotin 0.0005; FA 0.05	50	–	13	B	75	410	179	37	0.5	2.2	0.09	73	115	122
Kohlrabi (*Brassica oleracea* var. *gongylodes*), tubers	90.3	2.0	0.1	–	6.6	1.0	29	20	0.06	0.04	0.12	0.3	0.1	53	FA 0.01	170	20	7.1	B	10	392	41	48	0.11	0.5	0.14	51	–	57
Leeks, leaves (*Allium porrum*)	87.8	2.0	0.3	–	9.4	1.2	44	50	0.06	0.04	–	0.5	–	18	E 1.0	–	–	–	B	5	300	60	18	0.07	1.0	0.3	50	72	40
Lentils, dried (*Lens esculenta*)	11.1	24.7	1.1	–	60.1	3.9	340	60	0.50	0.25	0.49	2.0	1.5	–	biotin 0.013; FA 0.1	–	–	–	A	36	810	79	77	–	8.6	0.7	377	122	64
Lettuce (*Lactuca sativa*), headed	95.1	1.3	0.2	–	2.5	0.5	14	970	0.06	0.07	0.07	0.3	0.1	8	E 0.6; biotin 0.003; FA 0.02	170	20	7.1	B	12	140	35	10	0.80	2.0	0.07	26	12	39–74
Maize (*Zea mays*)	72.7	3.5	1.0	–	22.1	0.7	96	400[f]	0.15[f]	0.12	0.22	1.7	0.89	12	E 0.6; biotin 0.006; FA 0.03	0	0	5.2	B	0.4	300	3	38	0.15	0.7	0.06	111	32	14
canned, drained solids	75.9	2.6	0.8	–	19.8	0.8	84	350[f]	0.03	0.05	0.27	0.9	0.28	4	biotin 0.003; FA 0.008	–	–	–	B	236[2]	97	5	19	–	0.5	–	49	–	–
Mushrooms (champignons) (*Psalliota campestris*)	90.8	2.8[3]	0.24	–	3.7	0.9	22	0	0.1	0.44	0.05	6.2	2.1	5	E 0.834; biotin 0.016; FA 0.03; D 150 IU	–	–	–	B	5	520	9	13	0.08	0.8	1.8	116	34	25

* To convert to kJ (kilojoule) multiply the values given by 4.1855.
** Vitamin A activity due to vitamin A + carotenes; 1 IU vitamin A = 0.0006 mg β-carotene.
*** FA = folic acid; E = α-tocopherol unless otherwise stated.
[f] Based on yellow maize; white contains only a trace.
[2] Unsalted 2.
[3] 3 ⅔ N × 6.25.
[4] Total tocopherol.

Content per 100 grammes edible portion (unless otherwise stated)	Water g	Proteins g	Fats Total g	Fats Poly-unsaturated g	Carbohydrates Total g	Carbohydrates Fibre g	Calories* kcal	A** IU	B₁ mg	B₂ mg	B₆ mg	Nicotinic acid mg	Pantothenic acid mg	C mg	Other vitamins*** mg	Malic acid mg	Citric acid mg	Oxalic acid mg	Excess acid B / Excess base B	Sodium Na mg	Potassium K mg	Calcium Ca mg	Magnesium Mg mg	Manganese Mn mg	Iron Fe mg	Copper Cu mg	Phosphorus P mg	Sulphur S mg	Chlorine Cl mg
Onions (*Allium cepa*) ripe	89.1	1.5	0.1	—	8.7	0.6	38	40	0.03	0.04	0.1	0.2	0.17	10	E 0.26; biotin 0.004; FA 0.01	170	20	23	A	10	130	27	8	0.36	0.5	0.13	36	51	24
dried	4	8.7	1.3	—	82.1	4.4	350	200	0.25	0.18	—	1.4	—	35	—	—	—	—	A	88	1383	158	—	—	3.1	—	256	—	156
Parsley (*Petroselinum crispum*)	85.1	3.6	0.6	—	8.5	1.5	44	8500	0.12	0.26	0.2	1.2	0.03	172	biotin 0.0004; FA 0.04	—	—	190	B	28	880	203	52	0.94	6.2	0.21	63	190	30
Parsnips (*Pastinaca sativa*)	79.1	1.7	0.5	—	17.5	2.0	76	30	0.08	0.09	0.1	0.2	0.5	16	biotin 0.0001; FA 0.02	350	130	10	B	17	342	50	22	0.03–0.34	0.7	0.10	77	26	30
Peas (*Pisum sativum*) green, unripe	75.0	6.3	0.4	—	17.0	2.0	84	640	0.32	0.15	0.18	2.5	0.82	27	E 0.6; K 0.3; biotin 0.009; FA 0.025	80	110	1.3	B	2	370	26	30	0.41	2.0	0.23	116	50	33
green, frozen	80.7	5.4	0.3	—	12.8	1.9	73	680	0.32	0.10	—	2.0	—	19	E 0.02; FA 0.025	—	—	—	B	129	150	20	24	—	2.0	—	90	—	—
canned	82.3	3.4	0.4	—	12.7	1.3	67	450	0.11	0.06	0.05	0.9	0.17	9	—	—	—	0.8	B	2607†	201	25	25	—	1.8	0.21	67	44	318
dried, split	9.3	24.2	1.0	—	62.7	1.2	348	120	0.87	0.29	0.05	3.0	2.1	—	E 0.02; biotin 0.002; FA 0.01	—	—	16	B	42	880	73	116	2.0	6.0	0.8	303	129	60
edible, podded	86.2	2.6	0.1	—	10.5	1.5	53	55	0.06	0.10	—	0.8	—	30	biotin 0.02; FA 0.03	—	—	—	B	—	—	44	—	—	1.4	—	54	—	—
Peppers (*Capsicum* spp.) green chillies	92.8	1.2	0.2	—	5.3	1.4	24	420	0.08	0.08	—	0.4	—	128	—	—	—	—	B	4.2	186	9	12	0.13	0.4	0.11	25	19	13
Potato chips	1.8	5.3	39.8	—	50.0	1.6	568	trace	0.21	0.07	—	4.8	—	16	—	—	—	—	B	340	880	40	48	—	1.8	0.36	139	—	—
Potatoes (*Solanum tuberosum*) raw	79.8	2.1	0.1	—	17.7	0.5	76	trace	0.11	0.04	0.2	1.2	0.3	20	E 0.06; K 0.08; biotin 0.0001; FA 0.006	0	510	5.7	B	3	410	14	27	0.17	0.8	0.16	53	29	35
per lb as purchased (refuse 19%)	*293*	*7.7*	*0.4*	—	*65*	*1.8*	*279*	*trace*	*0.40*	*0.15*	*0.7*	*4.4*	*1.1*	*73*	*E 0.22; K 0.29; biotin 0.0004; FA 0.022*	*0*	*1874*	*21*	*B*	*11*	*1510*	*51*	*99*	*0.62*	*2.9*	*0.59*	*195*	*107*	*129*
dried	7.1	8.3	0.6	—	80.4	1.4	352	trace	0.25	0.10	—	4.8	—	26	—	150	—	—	B	84	1600	44	—	—	2.4	—	203	—	—
Pumpkins (*Cucurbita* spp.)	95.0	0.8	0.1	—	3.5	0.6	15	1600	0.05	0.11	—	0.6	—	9	—	—	0	—	B	1	457	21	12	0.04	0.8	0.08	44	10	37
Purslane (*Portulaca oleracea* var. *sativa*)	92.5	1.7	0.4	—	3.8	0.9	21	2500	0.03	0.10	—	0.5	—	25	—	—	—	—	B	2	754	103	151	—	3.5	—	39	—	—
Radishes (*Raphanus sativus*)	93.7	1.1	0.1	—	3.6	0.7	18	10	0.04	0.04	0.1	0.3	0.18	26	FA 0.01	—	—	—	B	15	260	30	15	0.05	1.0	0.13	31	37	37
Rhubarb (*Rheum undulatum*)	94.9	0.5	0.1	—	3.8	0.7	16	100	0.01	0.03	0.03	0.1	0.08	9	E 0.2; FA 0.003	1770	410	230–500	B	3.5	286	96	14	0.15	0.8	0.05	18	8	53

* To convert to kJ (kilojoule) multiply the values given by 4.1855.
** Vitamin A activity due to vitamin A + carotene; 1 IU vitamin A = 0.0006 mg β-carotene.
*** FA = folic acid; E = α-tocopherol unless otherwise stated.
† Unsalted 3.

Content per 100 grammes edible portion (unless otherwise stated)	Water g	Proteins g	Fats Total g	Fats Poly-unsaturated g	Carbohydrates Total g	Carbohydrates Fibre g	Calories* kcal	A** IU	B₁ mg	B₂ mg	B₆ mg	Nicotinic acid mg	Pantothenic acid mg	C mg	Other vitamins*** mg	Malic acid mg	Citric acid mg	Oxalic acid mg	Excess acid A / Excess base B	Sodium Na mg	Potassium K mg	Calcium Ca mg	Magnesium Mg mg	Manganese Mn mg	Iron Fe mg	Copper Cu mg	Phosphorus P mg	Sulphur S mg	Chlorine Cl mg
Rutabagas (*Brassica napus* var. *napobrassica*)	87.0	1.1	0.1	–	11.0	1.1	46	580	0.07	0.07	–	1.1	–	43	–	–	–	–	B	5	239	66	15	0.04	0.4	0.08	39	–	–
Salsify (*Scorzonera hispanica*)	79	3.2	0.6	–	16.4	1.8	77	10	0.04	0.04	–	0.2	–	12	–	–	–	–	B	5	320	40	–	–	1.5	–	76	–	–
Sauerkraut	92.8	1.0	0.2	–	4.0	0.7	18	50	0.03	0.04	–	0.2	0.08	14	–	–¹	–¹	–¹	B	650	140	36	–	–	0.5	0.1	18	–	–
Soybeans (*Glycine hispida*), dried	10.0	34.1	17.7	10.7	33.5	4.9	403	80	1.14	0.31	0.64	2.1	1.68	trace	E 6-11; biotin 0.06; FA 0.22	–	–	–	B	4	1900	226	235	–	8.4	0.11	554	–	–
Spinach (*Spinacia oleracea*)	90.7	3.2	0.3	–	4.3	0.6	26	8100	0.10	0.20	0.20	0.6	0.3	51	E 2.5; K 0.04-3; biotin 0.007; FA 0.075	90	80	460	B	62	662	106	62	0.82	3.1	0.20	51	27	65
per lb as purchased (refuse 8%)	*379*	*13.4*	*1.3*	–	*17.9*	*2.5*	*109*	*33800*	*0.42*	*0.83*	*0.83*	*2.5*	*1.3*	*213*	*E 10.4; K 0.17-12.5; biotin 0.029; FA 0.031*	*376*	*334*	*1920*	*B*	*259*	*2760*	*442*	*259*	*3.42*	*12.9*	*0.83*	*213*	*113*	*271*
canned	93.0	2.0	0.4	–	3.0	0.7	19	5500	0.02	0.06	0.095	0.3	0.06	14	biotin 0.002; FA 0.05	–	–	364	B	320²	260	85	–	–	2.1	–	26	–	–
frozen	91.3	3.0	0.3	–	4.2	0.8	25	8100	0.1	0.16	–	0.5	–	35	–	–	–	–	B	53	385	105	–	–	2.5	–	45	–	–
Squash, summer (zucchini) (*Cucurbita pepo* var. *medullosa*)	94.6	1.2	0.1	–	3.6	0.6	17	320³	0.05	0.09	–	1.0	–	19	–	–	–	–	B	1	202	28	–	0.14	0.4	–	29	–	–
Sweet potatoes (*Ipomoea batatas*)	70.6	1.7	0.4	–	26.3	0.7	114	8800	0.10	0.06	0.32	0.6	0.93	21	E 4.0; biotin 0.004; FA 0.012	0	70	56	B	5	530	32	31	0.15-0.52	0.7	0.15	47	15	85
canned	70.7	1.0	0.2	–	27.5	0.6	114	5000	0.03	0.03	–	0.6	–	8	–	–	–	–	B	48	120	13	–	–	0.7	–	29	–	–
Tomatoes (*Lycopersicon esculentum*)	93.5	1.1	0.2	–	4.7	0.5	22	900	0.06	0.04	0.1	0.6	0.31	23	E 0.27; biotin 0.004; FA 0.008	150	390	7.5	B	3	268	13	11	0.19	0.6	0.10	27	11	51
canned	93.7	1.0	0.2	–	4.3	0.4	21	900	0.06	0.03	0.07	0.7	0.2	17	biotin 0.0018; FA 0.003	–	–	–	B	130⁴	217	6	–	0.04	0.5	0.09	19	–	–
Tomato juice, canned	93.6	0.9	0.1	–	4.3	0.2	19	800	0.05	0.03	0.19	0.7	0.30	16	FA 0.007	23	336	–	B	230⁴	230	7	7	–	0.9	–	18	–	–
Tomato ketchup	68.6	2.0	0.4	–	25.4	0.5	106	1400	0.09	0.07	–	1.6	–	15	–	–	–	–	B	1042	363	22	21	–	0.8	–	50	15	–
Tomato puree	86.0	2.3	0.5	–	9.5	0.5	44	1200	0.09	0.06	0.18	1.5	–	9	–	–	–	–	B	590	1160	60	–	–	1.0	–	34	–	–
Turnips (*Brassica rapa*)	91.5	1.0	0.5	–	6.6	0.9	30	trace	0.04	0.07	0.11	0.6	0.02	36	E 0.02; biotin 0.0001; FA 0.004	230	0	0	B	37	230	39	7	0.04	0.5	0.07	30	22	41
greens	90.3	3.0	0.3	–	5.0	0.8	28	7600	0.21	0.39	0.98	0.8	0.38	139	E 2.3; FA 0.04	–	–	15	B	10	440	260	19	1.4	1.8	0.09	58	54	168
Watercress (*Nasturtium officinale*)	93.3	2.2	0.3	–	3.0	0.7	19	4000	0.1	0.27	–	0.9	0.1	75	–	–	–	–	B	60	301	151	17	0.54	2.0	0.04	46	147	109

* To convert to kJ (kilojoule) multiply the values given by 4.1855.
** Vitamin A activity due to vitamin A + carotenes; 1 IU vitamin A = 0.0006 mg β-carotene.
*** FA = folic acid; E = α-tocopherol unless otherwise stated.
¹ Lactic acid 1.6 g.
² Unsalted 34.
³ Including skin.
⁴ Unsalted 3.

Content per 100 grammes edible portion (unless otherwise stated)	Water g	Proteins g	Fats Total g	Fats Poly-unsaturated g	Carbohydrates Total g	Carbohydrates Fibre g	Calories kcal	A** IU	B1 mg	B2 mg	B6 mg	Nicotinic acid mg	Pantothenic acid mg	C mg	Other vitamins*** mg	Malic acid mg	Citric acid mg	Oxalic acid mg	Excess acid A / base B	Sodium Na mg	Potassium K mg	Calcium Ca mg	Magnesium Mg mg	Manganese Mn mg	Iron Fe mg	Copper Cu mg	Phosphorus P mg	Sulphur S mg	Chlorine Cl mg
Yeast (Saccharomyces cerevisiae) baker's, compressed	71.0	12.1	0.4	–	11.0	–	86	trace	0.71	1.65	1.2	11.2	5.3	trace	biotin 0.4; FA 0.5	–	–	–	A	16	610	13	59	–	4.9	–	394	–	–
brewer's, dried	5.0	38.8	1.0	–	38.4	1.7	283	trace	15.6	4.28	4.2	37.9	9.5	trace	biotin 0.08; FA 2.4	–	–	–	A	121	1700	210	231	0.53	17.3	3.32	1753	–	–
Yeast, torula (Torulopsis utilis)	6.0	38.6	1.0	–	37.0	3.3	277	trace	15.0	5.0	3.5	50.0	10.0	–	biotin 0.1; FA 3.0	–	–	–	A	15	2046	424	165	–	20	–	1713	–	–
Nuts																													
Almonds (Amygdalus communis) dried	4.7	18.6	54.2	10.8	19.5	2.6	598	75	0.25	0.92	0.10	3.5	0.58	trace	E 15; biotin 0.02; FA 0.045	–	–	–	B	3	690	234	252	1.9	4.7	0.14	504	150	2
Brazil nuts (Bertholletia excelsa)	4.6	14.3	66.9	18.4	10.9	3.1	654	10	1.0	0.07	0.11	7.7	0.23	2	–	–	–	–	A	2	670	127	225	2.8	2.8	1.1	600	198	61
Cashew nuts (Anacardium occidentale)	5.2	17.2	45.7	3	29.3	1.4	561	100	0.43	0.25	–	1.8	–	–	E 6.5; FA 0.005	–	–	–	–	15	464	38	267	–	3.8	–	373	–	–
Chestnuts (Castanea sativa)	48	3.4	1.9	–	45.6	1.3	213	0	0.23	0.22	0.29	0.5	0.3	6	–	–	–	–	B	2	410	46	42	3.7	1.4	0.06	74	29	11
dried	9.0	6.7	4.1	–	78.8	2.5	378	0	0.34	0.39	–	0.8	–	0	E 0.5; biotin 0.0013	–	–	–	B	4	875	57	–	–	3.3	–	170	–	–
Coconuts (Cocos nucifera)	48	4.2	34	0.6	12.8	3.3	351	0	0.06	0.03	0.06	0.6	0.33	2	E 1; FA 0.028	–	–	–	B	17	363	13	39	1.31	1.7	0.32	95	44	114
dried	3.5	7.2	64.9	0.6	23.0	3.9	662	0	0.06	0.04	–	0.6	–	0	–	–	–	–	B	29	588	26	90	–	3.3	0.55	187	76	196
Coconut water	94.2	0.3	0.2	–	4.7	trace	22	0	trace	trace	–	0.1	–	–	–	–	–	–	–	25	147	20	28	–	0.3	–	13	–	–
Hazelnuts (Corylus avellana)	6.0	12.7	60.9	23	18	3.5	627	100	0.47	0.55	0.54	1.6	1.15	7.5	E 21; FA 0.067	–	–	–	B	3	618	250	150	4.2	4.5	1.35	320	198	10
Peanuts (Arachis hypogaea) roasted	1.8	26.2	48.7	14.0	20.6	2.7	582	0	0.32	0.13	0.3	17.1	2.14	0	biotin 0.034; FA 0.057; E 6.5	–	–	–	A	3	740	74	181	1.51	2.2	0.27	407	377	7
Pecans (Carya illinoiensis)	3.4	9.2	71.2	14	14.6	2.3	687	130	0.86	0.13	0.19	0.9	–	2	E 1.5	–	–	–	–	trace	603	73	142	3.5	2.4	–	289	–	–
Pine nuts (Pinus pinea)	3.1	13.0	60.5	–	20.5	1.1	635	30	1.28	0.23	–	4.5	–	•	–	–	–	–	–	–	–	12	–	–	5.2	–	604	–	–
Pistachio nuts (Pistacia vera)	5.3	19.3	53.7	10	19.0	1.9	594	230	0.67	–	–	1.4	–	–	–	–	–	–	–	–	972	131	158	–	7.3	–	500	–	–
Walnuts (Juglans regia)	3.5	14.8	64.0	47.5	15.8	2.1	651	30	0.3	0.13	1.0	1.0	0.7	2	biotin 0.077; FA 1.5	–	–	–	A	4	450	99	134	1.8	3.1	0.31	380	146	–
Cereals, Cereal products																													
Barley (Hordeum spp.), pearled	12.0	9.0	1.4	0.8	76.5	0.8	346	0	0.12	0.05	–	3.1	0.5	0	–	–	70	–	A	3	160	16	37	1.68	2.0	0.4	189	116	105
Bread white, enriched, 4% nonfat milk solids	35.6	8.7	3.2	–	50.8	0.2	270	trace	0.25	0.21	–	2.4	–	trace	–	–	–	–	A	507	105	84	22	0.31	2.5	0.2	97	–	–
toasted	25.1	10.1	3.7	–	58.8	0.2	314	trace	0.23	0.24	–	2.8	–	trace	–	–	–	–	A	590	122	98	28	–	2.9	–	113	–	1040
wholemeal	36.4	9.1	2.6	–	49.3	1.5	241	trace	0.30	0.10	–	2.8	–	trace	–	–	–	–	A	530	256	84	78	–	9.1	–	254	–	–
French, unenriched	30.6	9.1	3.0	–	55.4	0.2	290	trace	0.08	0.08	–	0.8	–	trace	–	–	–	–	A	580	90	43	22	–	0.7	–	85	–	–

* To convert to kJ (kilojoule) multiply the values given by 4.1855.   ** Vitamin A activity due to vitamin A + carotenes; 1 IU vitamin A = 0.0006 mg β-carotene.   *** FA = folic acid; E = α-tocopherol unless otherwise stated.

Content per 100 grammes edible portion (unless otherwise stated)	Water g	Proteins g	Fats Total g	Fats Poly-unsaturated g	Carbohydrates Total g	Carbohydrates Fibre g	Calories* kcal	A** IU	B1 mg	B2 mg	B6 mg	Nicotinic acid mg	Pantothenic acid mg	C mg	Other vitamins*** mg	Malic acid mg	Citric acid mg	Oxalic acid mg	Excess acid A / Excess base B	Sodium Na mg	Potassium K mg	Calcium Ca mg	Magnesium Mg mg	Manganese Mn mg	Iron Fe mg	Copper Cu mg	Phosphorus P mg	Sulphur S mg	Chlorine Cl mg
Bread (continued)																													
pumpernickel	34.0	9.1	1.2	–	53.1	1.1	246	–	0.23	0.14	–	1.2	–	–	–	–	–	–	A	569	454	84	71	1.3	2.4	–	229	–	–
rye, American	35.5	9.1	1.1	–	52.1	0.4	243	–	0.18	0.07	–	1.4	–	–	–	–	–	–	A	557	145	75	42	1.3	1.6	0.28	147	–	1025
zwieback	5.0	10.7	8.8	–	74.3	0.3	423	40	0.05	0.07	–	1.3	–	0	–	–	–	–	A	250	150	13	–	–	0.6	–	69	–	–
Cornflakes	3.8	7.9	0.4	–	85.3	0.7	385	0	0.43[1]	0.1	0.005	2.1	0.19	0	–	–	–	5.6	A	660	160	10	17	0.05	1.4	0.17	45	93	6
Cornstarch	12.0	0.3	trace	–	87.9	0.1	362	0	trace	0.08	–	0.03	–	–	FA 0.006	–	–	–	A	4	4	trace	2	–	0.5	–	30	–	–
Flour																													
buckwheat	14.1	11.7	2.7	–	70	2.6	327	0	0.58	0.15	–	2.9	1.5	0	–	–	–	–	A	1	680	33	–	2.09	2.2	0.7	263	–	–
farina, unenriched	10.3	11.4	0.9	–	77.0	0.4	371	0	0.06	0.10	–	0.7	–	–	–	–	–	–	A	2	83	25	25	–	1.5	–	107	–	–
maize (corn)	12.0	7.8	2.6	–	76.8	0.7	368	340[2]	0.20	0.06	0.06	1.4	0.55	0	–	–	–	–	A	1	120	6	–	–	1.8	–	164	–	–
rye, light	11	9.4	1.0	–	77.9	0.4	357	0	0.15	0.07	–	0.6	–	–	biotin 0.006; FA 0.01; E 0.3	–	–	–	A	1	156	22	73	–	1.1	–	185	–	–
rye, medium	11	11.4	1.7	–	74.8	1.0	325	0	0.30	0.12	–	0.9	–	–	–	–	–	–	A	1	203	27	83	–	2.6	–	262	–	–
soybean, full fat	8.0	36.7	20.3	–	30.2	2.4	347	110	0.85	0.31	0.66	2.1	1.68	0	biotin 0.07; FA 0.43	–	–	–	B	1	1660	199	235	–	8.4	–	558	–	–
medium fat	8.0	43.4	6.7	–	36.6	2.5	264	80	0.83	0.36	–	2.6	–	0	–	–	–	–	B	3	2025	244	286	–	9.1	–	634	–	–
wheat, whole	12	13.3	2.0	–	71.0	2.3	333	0	0.55	0.12	–	4.3	–	0	–	–	–	–	A	3	370	41	113	–	3.3	0.2	372	–	–
white, unenriched	12	10.5	1.0	–	76.1	0.3	363	0	0.06	0.05	–	0.9	–	0	–	–	–	–	A	2	95	16	25	–	0.8	–	87	–	–
white, enriched	12	10.5	1.0	–	76.1	0.3	364	0	0.44	0.26	–	3.5	–	0	–	–	–	–	A	2	95	16	25	–	2.9	–	87	–	–
white, self-raising, enriched	11.5	9.3	1.0	–	74.2	0.4	352	0	0.44	0.26	–	3.5	–	0	–	–	–	–	A	1079	90	16	–	–	2.9	–	466	–	–
Muffins (enriched flour)	38.0	7.8	10.1	–	42.3	0.1	294	100	0.17	0.23	–	1.4	–	trace	–	–	–	–	A	441	125	104	–	–	1.6	–	151	–	–
Noodles, unenriched, dry	10.1	13.0	2.9	–	73	0.4	376	100	0.2	0.08	–	2.1	–	–	–	–	–	–	A	7	157	20	–	–	2.1	–	196	–	–
Oatflakes	10.3	13.8	6.6	2.7	67.6	1.4	387	–	0.55	0.14	0.75	1.1	0.92	0	E 0.25	–	–	–	A	2	340	53	145	4.9	3.6	0.74	407	199	49
Pancakes (enriched flour)	50.1	7.1	7.0	–	34.1	0.1	231	120	0.17	0.22	–	1.3	–	trace	–	–	–	–	A	425	123	101	–	–	1.3	–	139	–	–
Piecrust,plain,unenriched,unbaked	20.9	5.7	31.0	–	40.7	0.1	464	0	0.03	0.03	–	0.5	–	0	–	–	–	r.	–	568	46	13	–	–	0.4	–	47	–	–
Popcorn, popped	4.0	12.7	5.0	2.0	76.7	2.2	386	–	0.39	0.12	–	2.2	–	0	E 1.2	–	–	–	A	3	240	11	–	–	2.7	0.31	281	–	–
Pretzels	4.5	9.8	4.5	–	75.9	0.3	390	0	0.02	0.03	–	0.7	–	0	E 0.35	–	–	–	A	1680	130	22	–	–	1.5	–	131	–	–
Rice																													
whole	12.0	7.5	1.9	–	77.4	0.9	360	0	0.29	0.05	0.15	4.7	0.63	0	–	–	–	–	A	9	150	32	119	1.7	1.6	0.36	221	121	–
polished	12.0	6.7	0.4	–	80.4	0.4	362	0	0.07	0.03	–	1.6	–	0	–	–	–	–	A	6	113	24	28	1.08	0.8	0.06	94	79	27
polished, cooked	72.6	2.0	0.1	–	24.2	0.1	109	0	0.02	0.01	–	0.4	–	0	–	–	–	–	A	2[3]	38	10	8	–	0.2	0.19	28	–	9[3]
Semolina																													
maize	11.0	8.8	1.1	–	78.0	–	365	440[2]	0.15	0.05	0.05	0.5	–	0	–	–	–	–	A	–	80	4	20	–	1.0	–	73	–	–
wheat	13.1	10.3	0.8	–	76	–	362	0	0.12	0.04	0.085	1.3	–	–	E 1.8[4]	–	–	–	A	–	112	17	–	–	1	–	87	27	–
Spaghetti, unenriched, dry	10.4	12.5	1.2	–	75.2	0.3	369	0	0.09	0.06	–	2.0	–	–	–	–	–	–	A	5	–	22	–	–	1.5	–	165	4	16
Tapioca, dry	12.6	0.6	0.2	–	86.4	0.1	360	0	0	0.1	–	0	–	0	–	–	–	–	0	4	20	12	2	0.69	1.0	0.07	12	–	–
Wheat germ	11.5	26.6	10.9	2.9	46.7	2.5	363	650	2.0	0.68	0.92	4.2	2.2	0	FA 0.31; E 15[4]	–	340	–	A	2	780	72	336	–	9.4	1.3	1118	–	70

* To convert to kJ (kilojoule) multiply the values given by 4.1855.

** Vitamin A activity due to vitamin A + carotenes; 1 IU vitamin A = 0.0006 mg β-carotene.

*** FA = folic acid; E = α-tocopherol unless otherwise stated.

[1] Enriched.   [2] Based on yellow maize.   [3] Unsalted.   [4] Total tocopherol.

Content per 100 grammes edible portion (unless otherwise stated)	Water g	Proteins g	Fats Total g	Fats Poly-unsaturated g	Carbohydrates Total g	Carbohydrates Fibre g	Calories* kcal	A** IU	B1 mg	B2 mg	B6 mg	Nicotinic acid mg	Pantothenic acid mg	C mg	Other vitamins*** mg	Malic acid mg	Citric acid mg	Oxalic acid mg	Excess acid A / Excess base B	Sodium Na mg	Potassium K mg	Calcium Ca mg	Magnesium Mg mg	Manganese Mn mg	Iron Fe mg	Copper Cu mg	Phosphorus P mg	Sulphur S mg	Chlorine Cl mg
Confectionery, Sugar																													
Caramel[1]	7.6	4.0	10.2	–	76.6	0.2	399	10	0.03	0.17	–	0.2	–	trace	–	–	–	–	B	226	192	148	–	–	1.4	–	122	–	–
Chocolate																													
milk, sweetened	0.9	7.7	32.3	–	56.9	0.4	520	270	0.06	0.34	–	0.3	–	0	E 1.1	–	–	–	A	86	420	228	58	–	1.1	1.1	251	67	151
plain, sweetened	0.9	4.4	35.1	1.2	57.9	0.5	528	10	0.02	0.14	–	0.3	–	trace	–	–	–	–	A	19	397	63	107	–	1.4	1.1	142	32	–
Cocoa, dry powder	5.6	19.8	24.5	0.4	43.6	5.7	299	60	0.09	0.11	–	1.9	–	0	E 3.1[2]	–	–	450	A–B	–	900–3200	114	420	3.53	12.5	3.4	709	203	51
Dextrose, anhydrous	trace	0	0	–	99.5	0	385	0	0	–	–	0	–	0	–	–	–	–	0	1	0.4	5	3	–	–	–	–	–	–
Honey	17.2	0.3	0	–	82.3	–	304	0	trace	0.04	0.01	0.3	0.06	1	FA 0.003	–	–	–	0	7	51	5	10	0.03	0.5	0.2	6	–	29
Jams	29	0.6	0.1	–	70.0	1.0	272	10	0.01	0.03	–	0.2	–	2	–	–	–	–	B	16	112	12	–	–	1.0	0.23	9	–	9
Maple syrup	33	–	–	–	65	–	252	–	–	–	–	–	–	0	–	–	–	–	B	10	176	104	–	–	1.2	–	8	–	–
Molasses	24.0	–	–	–	60.0	–	232	–	0–0.08	0–0.16	0.27	2.8	0.5	0	biotin 0.009; FA 0.01	–	–	–	B	40	1500	273	–	0.04	6.7	1.9	69	6.5	317
Sugar																													
brown	2.1	0	0	–	96.4	–	373	0	0.01	0.03	–	0.2	–	0	–	–	–	–	B	24	230	85	–	–	3.4	–	19	–	–
cane or beet, white	trace	0	0	–	99.5	0	385	0	0	0	–	0	–	0	–	–	–	–	0	0.3	0.5	0	–	–	0.04	–	0	–	–
Beverages, nonalcoholic																													
Carbonated soft drinks[3]	88	–	–	–	12	0	46	–	–	–	–	–	–	–	–	–	–	–	0	1–15	1	5	–	–	–	–	–	–	–
Cola drinks	90	–	–	–	10	–	39	–	–	–	–	–	–	–	–	–	–	–	B	0	–	–	–	–	–	–	–	–	–
Coffee[4] (unsweetened)	98.5	0.3	0.1	–	0.8	–	5	0	0.01	0.01	–	0.9	–	0	–	18	29	1	B	1–6	80	0.3–5	1–13	0.09	0.2	–	5	–	0.6
Tea[5] (unsweetened)	99	0.1	0	–	0.4	–	2	0	0	0.04	–	0.1	–	1	–	–	–	10	0	0–2	16	5	–	0.69	0.2	–	1–4	–	0.4
Beverages, alcoholic			Alcohol[6]																										
Beer	90.6	0.5	3.6	–	4.8[7]	–	47	0	0.004	0.03	0.05	0.88	0.08	–	biotin 0.0005	–	–	–	B	5	38	4	9	–	–	–	–	–	8
Brandy	–	–	35–40	–	–	–	245–280	–	–	–	–	–	–	–	–	–	–	–	0	3	4	–	–	–	–	–	–	–	–
Fruit wine	–	–	5.2	–	1.0	–	40	–	–	–	–	–	–	–	–	–	–	–	B	7	72	5	–	–	–	–	–	–	–
Port wine	–	0.2	15.0	–	14.0	–	161	–	–	–	–	–	–	–	–	–	–	–	B	4	75	–	10	–	0.3	0.09	11	–	–
Rum	–	–	35.1	–	–	–	246	–	–	–	–	–	–	–	–	–	–	–	0	2	3	–	–	–	–	–	–	–	–
Whisky	–	–	35	–	–	–	245	–	–	–	–	–	–	–	–	–	–	–	0	0.3	1	–	–	–	–	–	–	–	–
Wine[8]	–	–	8.8–12.5	–	0.2–8.0	–	60–120	–	0.001 to 0.005	0.01	0.09	0.05	0.04	–	FA 0.001	0–280	6–58	–	B	4–7	20–120	7	7–16	0.3	0.3–3.5	0.05–0.25	10	15	2

* To convert to kJ (kilojoule) multiply the values given by 4.1855. ** Vitamin A activity due to vitamin A + carotenes; 1 IU vitamin A = 0.0006 mg β-carotene. *** FA = folic acid; E = α-tocopherol unless otherwise stated.

[1] Full-cream products. [2] Total tocopherol. [3] Not true mineral waters. [4] Caffeine 75–100 mg, trigonelline 100 mg, acetic acid 20 mg, formic acid 12 mg, chlorogenic acid and other phenolic acids 200 mg. [5] Caffeine 40–60 mg. [6] Alcohol has a calorific value of 7 kcal/g. [7] Extract. [8] Tartaric acid 163–234 mg, lactic acid 90–130 mg, aromatic acids 56–136 mg, glycerol 0.8–2.6 g.

Content per 100 grammes edible portion (unless otherwise stated)	Water g	Proteins g	Fats Total g	Fats Poly-unsaturated g	Fats Cholesterol g	Carbohydrates g	Calories* kcal	A** IU	B1 mg	B2 mg	B6 mg	Nicotinic acid mg	Pantothenic acid mg	C mg	Other vitamins*** mg	Purine nitrogen mg	Excess acid A / base B	Na mg	K mg	Ca mg	Mg mg	Mn mg	Fe mg	Cu mg	P mg	S mg	Cl mg
Fats, Oils																											
Butter	17.4	0.6	81.0	4	0.28	0.7	716	3300	trace	0.01	trace	0.1	trace	trace	E 2.4; D 40 IU	–	A	10[7]	23	16	1	0.04	0.2	0.03	16	9	–
Cod-liver oil	0	0	99.9	–	0.85	0	901	85000	–	0	–	0	–	–	E 26[2]; D 8500 IU	–	0	0.1	–	–	–	–	–	–	–	–	–
Corn oil	trace	0	99.9	56	0	0	883	–	–	–	–	–	–	–	E ~ 19	–	0	–	–	–	–	–	–	–	–	–	–
Cottonseed oil	trace	0	99.9	50	0.1	0	883	–	–	–	–	–	–	–	E ~ 30	–	0	–	–	–	–	–	–	–	–	–	–
Lard	1.0	trace	99.9	10	0.1	0	901	0	0	0	–	0	–	0	E 2	–	A	0.3	0.2	1	–	–	0.1	0.02	3	25	4
Margarine, salted	15.5	0.6	81.0	14	–	0.4	720	3300	0.02	0.04	–	trace	–	0	–	–	A	987	23	20	2	–	0	–	16	–	–
Mayonnaise	15.1	1.1	78.9	32[3]	–	3.0	718	280	–	–	–	–	–	0	–	–	A	702	53	18	48	–	0.5	–	28	–	–
Mustard, brown	78.1	5.9	6.3	–	–	5.3	91	–	–	0	–	–	–	–	–	–	0	1307	130	124	–	–	1.8	0.07	134	–	–
Olive oil	trace	0	99.9	8	0	0	883	0	0	–	–	0	–	0	E ~ 3	–	A	0.1	trace	0.5	–	–	0.08	–	–	–	–
Palm oil	trace	0	99.9	9	0	0	883	–	–	–	–	–	–	–	E 30	–	0	–	–	–	–	–	–	–	–	–	–
Peanut butter	1.8	27.8	49.4	11.9	0	17.2	581	–	0.13	0.13	0.30	15.7	2.5	0	biotin 0.04; FA 0.06	–	A	607[4]	670	63	178	–	2.0	–	407	225	–
Peanut oil	trace	0	99.9	29	0	0	883	–	–	–	–	–	–	–	E 13	–	0	–	–	–	–	–	–	–	–	–	–
Safflower oil	trace	0	99.9	72	0	0	883	–	–	–	–	–	–	–	E 31	–	0	–	–	–	–	–	–	–	–	–	–
Soybean oil	trace	0	99.9	60	0	0	883	–	–	–	–	–	–	–	E 18	–	0	–	–	–	–	–	–	–	–	–	–
Sunflower oil	trace	0	99.9	63	0	0	883	–	–	–	–	–	–	–	E 22	–	0	–	–	–	–	–	–	–	–	–	–
Vegetable fat	0	0	100	7	–	0	884	–	0	0	0	0	0	0	–	0	0	0	0	0	0	0	0	0	0	0	0
Dairy products, Eggs																											
Butter. See under 'Fats', above																											
Cheese																											
Camembert	51.3	18.7	22.8	–	–	1.8	287	1010	0.05	0.45	0.25	1.45	0.1–0.9	0	biotin 0.005	–	A	1150[4]	109	382	18	–	0.5	0.08	184	–	–
Cheddar	37	25.0	32.2	1	0.10	2.1	398	1310	0.03	0.46	0.07	0.1	0.40	0	biotin 0.004; FA 0.016	–	A	700	82	750	43	–	1.0	–	478	–	–
cottage, creamed	78.3	13.6	4.2	trace	0.015	2.9	106	170	0.03	0.25	–	0.1	–	0	–	–	A	229	85	94	–	–	0.3	–	152	–	–
cottage, uncreamed	79.0	17.0	0.3	–	–	2.7	86	10	0.03	0.82	0.01	0.1	–	1	–	–	A	290	72	90	–	–	0.4	–	175	–	–
cream	51	8.0	37.7	1	0.12	2.1	374	1540	0.02	0.24	–	0.1	–	0	–	–	A	250	74	62	–	–	0.2	–	95	–	–
Parmesan	30.0	36.0	26.0	–	–	2.9	393	1060	0.02	0.73	–	0.2	–	0	biotin 0.002; FA 0.03	–	A	755[4]	153	1140	50	–	0.4	0.36	781	251	1110[4]
Roquefort	40.0	21.0	32.0	–	–	1.8	378	800	0.06	0.3–0.7	–	0.4–0.9	0.5–0.7	0	biotin 0.003	–	A	–	–	700	–	–	1	–	–	–	–
Swiss (Emmentaler)	34.9	27.4	30.5	–	–	3.4	398	1140	0.05	0.33	0.09	0.1	–	0.5	E 0.35[2]; D 100 IU	–	A	620[4]	100	1180	55	–	0.9	0.13	860	–	1210[4]
Cream																											
heavy 30%	64.1	2.2	30.4	0.8	–	2.9	288	1100[5]	0.025	0.17	0.035	0.07	–	1	D 40 IU	–	B	38	78	75	–	–	0–0.1	–	63	–	–
Eggs																											
whole, raw	74.0	12.8	11.5	2.3	0.46	0.7	162	1180	0.12	0.34	0.25	0.1	1.6	0	D 200 IU; B₁₂ 0.002; E 1; K 0.002; biotin 0.02; FA 0.005	–	A	135	138	54	13	0.05	2.3	0.03	205	197	159

* To convert to kJ (kilojoule) multiply the values by 4.1855.
** Vitamin A activity due to vitamin A + carotenes; 1 IU vitamin A = 0.0006 mg β-carotene.
*** FA = folic acid; E = α-tocopherol unless otherwise stated.

[1] Unsalted.
[2] Total tocopherol.
[3] Prepared with corn oil.
[4] Variable, depends on salt content.
[5] In summer; in winter 500.

Content per 100 grammes edible portion (unless otherwise stated)	Water g	Proteins g	Fats Total g	Fats Poly-unsaturated g	Fats Cholesterol g	Carbohydrates g	Calories* kcal	A** IU	B1 mg	B2 mg	B6 mg	Nicotinic acid mg	Pantothenic acid mg	C mg	Other vitamins*** mg	Purine nitrogen mg	Excess acid A / base B	Na mg	K mg	Ca mg	Mg mg	Mn mg	Fe mg	Cu mg	P mg	S mg	Cl mg
Egg white, raw	87.6	10.9	0.2	–	0	0.8	51	0	0.02	0.23	0.22	0.1	0.14	0	biotin 0.007; FA 0.001	–	A	192	148	9	11	0.04	0.2	0.03	17	208	161
Egg yolk, raw	50.0	16.1	31.9	6.7	1.6	0.6	360	3400	0.32	0.52	0.30	0.02	4.2	0	B12 0.002; E 3; D 350 IU; biotin 0.05; FA 0.013	–	A	50	123	141	16	0.09	7.2	0.02	569	194	142
1 egg, medium (48 grammes)	35.5	6.1	5.5	1.1	0.22	0.4	77	580	0.06	0.16	0.12	0.04	0.8	0	–[3]	–	A	66	67	26	6	0.02	1.3	0.02	98	95	69
1 egg white, medium (31 grammes)	27.0	3.3	0.1	–	0	0.3	16	0	0.01	0.07	0.07	0.03	0.04	0	–[3]	–	A	57	46	3	3	0.01	0.06	0.01	5	64	48
1 egg yolk, medium (17 grammes)	8.5	2.8	5.4	1.1	0.22	0.1	61	580	0.05	0.09	0.05	trace	0.7	0	–[3]	–	A	57	21	23	–	0.01	1.2	0.01	93	32	21
Egg powder	4.1	47.0	41.2	–	2.14	4.1	592	4460	0.35	1.23	0.08	0.2	7.4	0	D 240 IU	–	A	519	483	190	41	–	8.7	0.18	800	630	592
Milk (cow's)[1,2] pasteurized, whole	88.5	3.2	3.7	0.1	0.01	4.6	64	140	0.04	0.15	0.05	0.07	0.33	1	E 0.06[4]; B12 0.0006; biotin 0.002; D 0.5–4 IU; FA 0.0001	–	B	75	139	133	13	0.002	0.04	0.01	88	29	105
buttermilk, cultured	91.2	3.5	0.5	–	–	4.0	35	35	0.04	0.18	0.04	0.1	0.36	1	B12 0.0003; E 0.054	–	B	57	147	109	16	–	0.1	0.02	95	30	100
condensed (sweetened)	27.1	8.1	8.7	0.2	–	54.3	321	350	0.1	0.38	0.06	0.2	0.85	1	biotin 0.002; FA 0.0003	–	B	140	340	262	25	–	0.1	–	206	–	–
canned, evaporated (unsweetened)	73.8	7.0	7.9	0.2	–	9.7	138	350	0.06	0.36	0.03	0.2	0.85	1	biotin 0.003; FA 0.0007; D 3.5 IU	–	B	100	270	252	25	–	0.2	–	205	–	–
dried, whole	2.0	26.4	27.5	0.7	–	38.2	502	1200	0.28	1.2	0.3	0.7	2.7	10	B12 0.002; biotin 0.013	–	B	410	1330	909	112	–	0.6	0.16	708	234	784
nonfat	3.0	35.9	1.0	–	–	52.0	362	30	0.35	1.80	0.4	0.9	3.5	10	B12 0.002; biotin 0.016; + FA 0.0024	–	B	525	1335	1300	111	–	0.6	–	1016	300	1130
skimmed	90.9	3.5	0.07	–	0.003	4.8	34	7	0.038	0.17	0.05	0.1	0.28	2	E 0.034; biotin 0.002	–	B	53	150	123	14	–	0.1	0.003	97	–	100
Breast milk[1,2]	87.7	1.03	4.4	0.3	0.01–0.02	6.9	70	330	0.01	0.04	0.02	0.18	0.24	5	E 0.234; B12 trace; biotin 0.0001; FA 0.0001; D 0.4–9.7 IU	–	B	17	50	33	3	trace	0.05	0.05	14	14	36
Camel's milk[1]	87.1	3.7	4.2	–	–	4.1	69	120	0.05	0.12	–	–	–	6	B12 0.0001; biotin 0.002; FA 0.0002; D 2 IU	–	B	34	180	129	13	0.008	0.1	0.04	103	16	150
Goat's milk[1]	86.6	3.6	4.2	–	–	4.8	71	120	0.05	0.12	0.027	0.2	0.35	2	B12 0.0003; FA 0.0001	–	B	–	70	100	10	–	0.1	–	60	–	–
Mare's milk[1]	91.1	2.1	1.25	–	–	6.3	44	45	0.03	0.02	–	0.05	0.30	10	B12 0.0003; biotin 0.009; FA 0.0002	–	B	30	190	190	–	–	0.1	–	150	–	20
Sheep's milk[1]	81.6	5.6	7.5	–	–	4.4	107	200	0.07	0.50	–	0.50	0.35	3	biotin 0.002; E 0.024	–	B	45	190	50	1	–	–	–	53	–	140
Whey	93.3	0.9	0.3	–	–	4.7	25	8–16	0.04	0.08	0.02	0.07	0.35	1.5	–	–	B	62	129	150	–	–	0.2	–	135	–	–
Yoghurt[5]	86.1	4.8	3.8	–	–	4.5	71	145	0.045	0.024	0.05	0.18	–	2	–	–	B	–	190	150	–	–	–	–	–	–	–
Meat, Poultry (raw unless otherwise stated)																											
Bacon medium fat	20.0	9.1	65.0	6.5	0.22	trace	625	0	0.36	0.11	0.35	1.8	–	0	E 0.4	28	A	1770	225	13	15	–	1.2	–	108	–	–

[1] Values per 100 g	pH	Spec. grav.	Total protein	Casein	Albumin	Nonprotein-N	Ash
Breast milk	6.97	1.031	1.0–6.0 g	0.40 g	0.30 g	32.4 mg	0.21 g
Camel's milk		1.031	3.4–3.7 g	2.90 g	0.90 g	–	0.68 g
Cow's milk	6.60	1.031	2.0–6.0 g	2.80 g	0.40 g	13–14 mg	0.72 g
Goat's milk		1.031	3.6–3.8 g	2.87 g	0.89 g	40 mg	0.85 g
Mare's milk	7.20	1.034	2.13 g	1.40 g	–	–	0.36 g
Sheep's milk	6.54	1.036	4.5–5 g	4.17 g	0.98 g	42.5 mg	0.93 g

* To convert to kJ (kilojoule) multiply the values given by 4.1855.
** Vitamin A activity due to vitamin A + carotene; 1 IU vitamin A = 0.0006 mg β-carotene
*** FA = folic acid; E = α-tocopherol unless otherwise stated.
[2] See also pages 687–689.
[3] Can be calculated from the 100 g values.
[4] Total tocopherol.
[5] Citric acid 232 mg, lactic acid 487 mg, acetic acid 44 mg.

Content per 100 grammes edible portion (unless otherwise stated)	Water g	Proteins g	Total g	Poly-unsaturated g	Cholesterol g	Carbohydrates g	Calories* kcal	A** IU	B₁ mg	B₂ mg	B₆ mg	Nicotinic acid mg	Pantothenic acid mg	C mg	Other vitamins*** mg	Purine nitrogen mg	Excess acid A / Excess base B	Sodium Na mg	Potassium K mg	Calcium Ca mg	Magnesium Mg mg	Manganese Mn mg	Iron Fe mg	Copper Cu mg	Phosphorus P mg	Sulphur S mg	Chlorine Cl mg
Beef																											
loin, lean	69.7	21.1	8.2	–	–	0	164	20	0.09	0.19	–	5.1	–	–	–	–	A	65	355	12	24	–	3.2	–	196	–	–
rib, lean	66.8	20.7	11.6	–	–	0	193	20	0.09	0.18	–	5.0	–	–	–	–	A	65	355	12	24	–	3.1	–	208	–	–
round, total edible	69.0	19.5	12.5	0.3	0.12	–	196	–	0.08	0.17	0.50	4.7	0.52	–	biotin 0.003; FA 0.01; B₁₂ 0.002	50	A	68	400	11	22	–	2.9	0.08	180	–	–
rump, total edible	56.5	17.4	25.3	–	–	0	303	50	0.08	0.16	–	4.2	–	–	–	–	A	–	–	10	21	–	2.6	–	160	–	–
sirloin, lean	71.8	21.5	5.7	–	–	0	143	10	0.09	0.19	–	5.2	–	0	–	–	A	–	355	12	24	–	3.2	–	200	–	–
canned, corned	59.3	25.3	12.0	–	0.23	0	216	0	0.02	0.2	–	3.4	–	0	–	36	A	1300	60	20	–	–	4.3	–	106	–	–
dried, salted	47.7	34.3	6.3	–	–	0	203	0	0.11	0.32	–	3.7	–	0	–	–	A	4300	200	20	–	–	5.1	–	404	161	256
hamburger, cooked	54.2	24.2	20.3	–	2.36	0.8	364	40	0.09	0.21	–	5.4	1.8	–	–	–	A	47	450	11	21	–	3.2	0.2	194	–	–
brain	79.4	10.4	8.0	0.2	0.15	0.6	120	580	0.15	0.23	0.16	4.0	2.0	14	biotin 0.007; B₁₂ 0.005; FA 0.012	–	A	104	191	11	12	–	1.6	0.3	265	–	–
heart	75.5	16.8	6.0	–	0.41	0.9	128	20	0.53	0.88	0.29	6.8	4	6	biotin 0.007; B₁₂ 0.01; FA 0.003	94	A	85	286	5	17	0.08	4.0	0.35	195	–	–
kidneys	75.9	15.4	6.7	–	0.32	0.9	130	1000	0.25	2.1	0.39	6.4	4	11	B₁₂ 0.04; FA 0.06; biotin 0.009	94	A	245	231	11	11	0.27	5.5	2.1	219	–	–
liver	69.9	19.7	3.8	0.7	0.35	5.9	136	20000	0.30	2.9	0.7	13.6	7.3	31	E 1; B₁₂ 0.065; biotin 0.10; FA 0.29	110	A	116	292	7	15	–	6.5	–	352	–	–
lungs	80.1	16.9	2.0	–	–	trace	90	–	0.09	0.32	–	4.0	1.4	0	B₁₂ 0.003; biotin 0.006	–	A	–	–	12	–	–	6.6	0.06	196	–	–
pancreas	70.6	14.6	12.3	–	trace	trace	173	17	0.10	0.40	–	4.2	3.5	58	B₁₂ 0.014; FA 0.02; biotin 0.014	–	A	62	249	9	15	–	1.0	–	335	–	–
spleen	77	18.1	3.4	–	–	trace	108	0	0.13	0.28	0.12	4.2	1.2	6	B₁₂ 0.005; biotin 0.006	55	A	99	379	7	–	–	8.9	0.07	236	–	–
tongue	68.0	16.4	15.0	–	0.15	0.4	207	0	0.14	0.27	0.13	5.0	2.0	0	biotin 0.003; B₁₂ 0.003	22	A	80	260	8	10	–	3.0	–	182	–	–
tripe	78	19.0	2.0	–	–	0.2	99	0	0.01	0.09	–	3.0	–	0	–	–	A	46	19	69	–	–	0.9	–	132	–	13
Calf (see also 'Veal')																											
brain	79.4	10.2	8.3	–	–	0.8	122	–	0.20	0.20	0.16	3.7	2.5	18	biotin 0.06; B₁₂ 0.002	–	A	172	265	9	14	–	2.6	0.14	353	–	–
heart	78.3	12.2	7.6	0.5	–	0.8	124	30	0.6	1.05	–	6.3	2.8	5	biotin 0.015; B₁₂ 0.01; FA 0.01	–	A	120	230	16	18	–	2.2	0.34	350	–	–
kidneys	75.0	16.7	6.4	0.18	0.36	0.8	132	70	0.37	2.5	0.5	6.5	4.0	13	B₁₂ 0.025; FA 0.04	–	A	200	290	10	–	0.34	3.4	0.51	171	–	–
liver	70.7	19.2	4.7	–	–	4.1	140	22500	0.28	2.72	1.2	17	9.7	32	E 0.9–1.6†; K 0.15; FA 0.05; B₁₂ 0.06; biotin 0.075; D 50 IU	120	A	84	295	8	15	–	5.4	4.4	311	–	–
sweetbreads	75	19.6	3.0	–	0.28	0	111	–	0.08	0.17	–	2.6	–	–	–	400	A	73	519	–	–	–	0.9	0.08	–	–	–
tongue	74.3	18.5	5.3	–	–	0.9	130	–	–	–	–	–	–	–	–	A	84	200	9	–	–	3.0	–	190	–	–
Chicken, flesh and skin																											
fryers	72.7	20.6	5.6	1.2	0.09	–	138	170	0.1	0.2	0.50	6.8	0.80	2.5	E 0.21; biotin 0.011; FA 0.003	60	A	83	359	12	37	0.02	1.8	0.3	200	–	85
roasters	66.9	19.5	12.6	–	–	0	197	410	0.08	0.12	–	7.4	–	–	–	–	A	–	–	11	–	–	1.5	–	191	–	–
hens and cocks	61.3	19.0	18.8	–	–	0	251	610	0.06	0.13	–	9.2	–	–	–	–	A	–	–	11	–	–	1.3	–	182	–	–
liver	72.2	19.7	3.7	1.0	0.20	2.9	141	12100	0.4	2.5	0.80	10.8	4.1	35	FA 0.38; B₁₂ 0.004; D 50 IU	–	A	85	179	12	13	0.18	7.9	0.32	236	–	–
Duck, medium fat	54.0	16	28.6	6.9	0.07	0	326	–	0.10	0.24	–	5.6	–	8	–	60	A	85	285	15	–	0.03	1.8	0.4	188	–	85

* To convert to kJ (kilojoule) multiply the values given by 4.1855.
** Vitamin A activity due to vitamin A + carotenes; 1 IU vitamin A = 0.0006 mg β-carotene.
*** FA = folic acid; E = α-tocopherol unless otherwise stated. † Total tocopherol.

Content per 100 grammes edible portion (unless otherwise stated)	Water g	Proteins g	Fats Total g	Fats Poly-unsaturated g	Fats Cholesterol g	Carbohydrates g	Calories* kcal	A** IU	B_1 mg	B_2 mg	B_6 mg	Nicotinic acid mg	Pantothenic acid mg	C mg	Other vitamins*** mg	Purine nitrogen mg	Excess base A / Excess acid B	Sodium Na mg	Potassium K mg	Calcium Ca mg	Magnesium Mg mg	Manganese Mn mg	Iron Fe mg	Copper Cu mg	Phosphorus P mg	Sulphur S mg	Chlorine Cl mg
Gelatin, dry	13.0	85.6	0.1	—	—	0	335	0	0	0	—	0	—	0	—	0	0	—	—	—	—	—	—	—	—	—	—
Goat	70	18.7	9.4	0.4	—	0	165	0	0.17	0.32	—	5.6	—	0	—	—	A	—	—	11	—	—	2.2	—	—	—	—
Goose, medium fat	51.0	16.4	31.5	2.5	—	0	354	—	0.10	0.24	0.6	5.6	—	—	—	100	A	85	420	15	—	0.05	1.8	0.3	188	—	—
liver	66	17	10	—	0.49	5.5	184	—	0.02	—	0.9	4	—	—	—	—	A	140	230	10	—	—	—	—	180	—	—
Ham raw	53.0	15.2	31.0	—	0.07–0.1	0	345	0	0.74	0.18	0.44	4.0	0.64	0	FA 0.01; B_{12} 0.001; biotin 0.005	49	A	76	339	9	18	0.06	2.3	0.31	168	—	—
boiled	57.0	19.5	20.6	2.0	0.07	0	269	0	0.54	0.26	—	4.2	0.53	0	—	45	A	876	348	10	—	—	2.5	—	150	—	—
smoked, raw	42.0	16.9	35.0	—	0.11	0.3	389	0	0.7	0.19	0.40	4.0	—	0	—	—	A	2530	248	10	20	—	2.5	—	207	—	2060
canned, spiced	65.0	18.3	12.3	—	0.07	1.5	193	0	0.53	0.19	—	3.8	—	0	—	—	A	1150	293	11	20	—	2.7	0.09	156	—	—
Hare	73	22.3	0.9'	—	0.08	0.2	103	0	0.09	0.19	—	5.0	0.50	1	—	—	A	50	400	12	—	—	3.2	—	157	—	—
Horse flesh, lean	74.3	21.7	2.6	—	—	0.9	120	—	0.07	0.12	—	4.3	—	—	E 0.5'; B_{12} 0.003	—	A	44	332	10	—	—	2.7	—	150	—	—
Lamb (medium fat) chop	52.0	14.9	32.0	0.7	0.07	0	352	—	0.13	0.18	0.33	4.3	0.59	—	E 0.6	65	A	90	345	9	14	—	2.2	—	138	—	—
leg	64.0	18.0	18.0	0.5	0.07	0	239	—	0.16	0.22	0.32	5.2	0.62	—	biotin 0.006; B_{12} 0.003; FA 0.003	81	A	78	380	10	16	—	2.7	—	213	—	—
Pork (see also 'Bacon' and 'Ham') cutlets	53.9	15.2	30.6	2.8	0.07	0	341	0	0.8	0.19	0.48	4.3	0.40	0	E 0.6; B_{12} 0.001; biotin 0.005; FA 0.002	—	A	62	326	10	19	0.06	2.6	—	193	—	—
loin, lean	71.2	18.6	9.9	—	—	—	168	—	1.1	0.31	—	6.5	—	—	—	—	A	74	348	12	22	—	3.0	—	234	—	—
loin or chops, cooked	50	23.0	26.0	1.6	—	0	333	0	0.83	0.24	—	5.0	0.65	0	B_{12} 0.001	—	A	—	—	11	—	—	3.0	—	235	—	—
ribs	52.6	14.6	32.0	2.8	0.10	0	351	—	0.92	0.18	—	3.9	0.40	2	—	—	A	76	252	5	—	0.06	2.2	—	157	—	—
spare ribs, total edible	53.9	15.2	30.6	—	0.07	0	341	0	0.8	0.19	0.48	4.3	—	0	E 0.6; B_{12} 0.001; biotin 0.005; FA 0.002	—	A	62	326	9	19	—	2.3	—	170	—	—
canned, strained	75.7	17.1	6.0	—	—	0	127	0	0.35	0.28	—	4.7	—	18	—	—	A	153	312	14	11	—	1.7	0.3	180	—	—
brain	78.0	10.6	9.0	—	—	trace	126	—	0.16	0.28	—	4.3	2.8	3	B_{12} 0.003	—	A	80	257	10	15	—	3.6	0.3	300	—	—
heart	76.8	16.9	4.8	0.27	—	0.4	117	30	0.43	1.24	0.43	6.6	2.5	12	E 1.4'; B_{12} 0.003; biotin 0.02	—	A	—	—	6	—	—	3.3	0.3	132	198	113
kidneys	77.8	16.3	5.2	0.29	—	0.8	120	130	0.34	1.8	0.55	9.8	3.1	27	B_{12} 0.015; biotin 0.13	—	A	173	242	11	16	0.10	6.7	0.38	218	—	190
liver	71.6	20.6	4.8	—	—	2.6	131	10900	0.43	2.7	0.85	16.4	7.0	—	E 1.0'; B_{12} 0.010; FA 0.22; biotin 0.10	—	A	77	350	10	18	0.3	19	0.85	316	228	102
tongue	66.1	16.8	15.6	—	—	0.5	215	—	0.17	0.29	0.35	5.0	—	0	—	—	A	93	234	9	—	—	1.4	—	186	—	—
Rabbit	70.4	20.4	8.0	1.5	0.12	0	159	30	0.04	0.18	0.6	12.8	0.8	0	E 1.0'	38	A	40	385	18	—	—	2.4	0.17	210	199	51
Sausages beef	49.2	13.8	18.4	—	—	15.7	286	—	—	—	—	—	—	—	—	—	A	1130	255	21.2	16.6	—	4.1	—	168	163	1770
bologna	56.2	12.1	27.5	—	—	1.1	304	—	0.16	0.22	—	2.6	—	—	—	—	A	1300	230	7	—	—	1.8	—	128	—	—
frankfurter, cervelat	55.6	12.5	27.6	—	—	1.8	256	0	0.16	0.20	—	2.7	—	0	—	—	A	1100	230	7	—	—	1.9	—	133	—	—
canned	65.7	13.0	19.6	—	—	—	232	—	0.03	0.08	—	3.1	—	—	—	—	A	711	—	10	—	—	2.7	—	185	—	1100

* To convert to kJ (kilojoule) multiply the values given by 4.1855.
** Vitamin A activity due to vitamin A + carotenes; 1 IU vitamin A = 0.0006 mg β-carotene.
*** FA = folic acid; E = α-tocopherol unless otherwise stated.
' Total tocopherol.

Content per 100 grammes edible portion (unless otherwise stated)	Water g	Proteins g	Fats Total g	Fats Poly-unsaturated g	Cholesterol g	Carbohydrates g	Calories* kcal	A** IU	B1 mg	B2 mg	B6 mg	Nicotinic acid mg	Pantothenic acid mg	C mg	Other vitamins*** mg	Purine nitrogen mg	Excess acid A / base B	Sodium Na mg	Potassium K mg	Calcium Ca mg	Magnesium Mg mg	Manganese Mn mg	Iron Fe mg	Copper Cu mg	Phosphorus P mg	Sulphur S mg	Chlorine Cl mg
Sausages (continued)																											
mortadella	52.3	12.4	32.8	–	–	–	349	–	0.10	0.15	–	3.1	–	0	–	–	A	668	207	12	–	–	3.1	–	238	–	920
pork English	50.7	8.8	28.8	–	–	9.8	335	–	–	–	–	–	–	–	–	–	A	770	158	15	11.5	–	2.5	0.12	108	73	1070
American	38.1	9.4	50.8	–	–	trace	498	0	0.43	0.17	–	2.3	–	–	–	–	A	740	140	5	–	–	1.4	–	92	–	–
salami	27.7	17.8	49.7	–	–	–	524	–	0.18	0.20	–	2.6	–	–	E 0.11	–	A	1260	302	35	–	–	–	–	–	–	2390
Sheep (see also 'Lamb' and 'Mutton')																											
kidneys	77.7	16.8	3.3	–	0.12	0.9	105	690	0.51	2.42	0.37	7.4	4.3	15	B12 0.063	103	A	151	205	13	13	0.09	7.5	0.3	218	–	–
liver	70.8	21.0	3.9	–	–	2.9	136	50500	0.4	3.28	–	16.9	7.1	33	biotin 0.13; FA 0.28	–	A	51	170	13	14	0.23	10.9	6.3	349	–	–
Turkey	64.2	20.1	14.7[f]	3.0	0.015	0.4	218	trace	0.13	0.14	–	7.9	0.75	0	FA 0.01	79	A	66	315	8	–	0.03	1.5	0.2	212	–	123
Veal																											
rib	70.0	19.5	9.0	0.6	–	0	164	–	0.14	0.26	0.43	6.5	0.50	0	biotin 0.002; FA 0.005; B12 0.0007	50	A	90	301	11	16	0.03	2.9	0.25	200	–	–
round with rump	73.0	19.9	6	–	–	0	139	–	0.15	0.26	–	6.7	–	–	–	–	A	90	320	12	–	–	3.0	–	206	–	–
Venison	73.0	21.4	3.6	0.3	–	0	124	0	0.37	0.28	–	7.4	–	0	–	–	A	70	336	19	29	–	5.0	–	183	211	41
Whale meat	71	20.6	4.0	–	–	1	125	1860	0.03	0.1	–	4.4	–	8	–	–	A	78	–	12	–	–	2.4	–	144	–	–
Fish, Sea foods (raw unless otherwise stated)																											
Carp (*Cyprinus carpio*)	72.4	18.9	7.1	–	–	0	145	300	0.08	0.04	–	1.5	–	1	–	54	A	51	285	34	15	–	1	–	220	–	62
Caviar, pressed	46.0	26.9	15.0	–	0.3	3.3	262	–	0	–	–	–	–	–	–	40	A	2200	180	276	–	–	11.8	–	355	–	1800
Clams, long and round (*Mya arenaria, Ensis americana*)	83.1	10.5	1.3	–	0.12	3.1	70	–	0.1	0.19	0.08	1.5	0.6	–	biotin 0.002; FA 0.003	–	A	121	235	12	63	–	0.6	–	208	–	–
Cod (*Gadus callarias*)	81.2	17.6	0.3	–	0.05	–	78	–	0.06	0.07	0.20	2.2	0.12	2	biotin 0.0002; FA 0.001; B12 0.0005	62	A	86	339	11	28	0.01	0.5	0.5	190	–	97
Crab (*Cancer pagarus*), canned or cooked, meat only	77.2	17.4	2.5	–	0.15	1.1	101	–	0.08	0.08	0.35	2.5	0.5	trace	biotin 0.005; B12 0.0005; FA 0.0003	61	A	1000	110	45	48	–	0.8	1.3	182	–	–
Eel (*Anguilla anguilla*)	60.7	12.7	25.6	–	0.05	0	285	2000	0.15	0.31	0.28	2.2	–	1.8	D 5000 IU	–	A	78	247	18	18	0.03	0.7	0.03	166	130	35
smoked	50.3	18.6	27.8	–	–	0.8	333	2500	0.14	0.35	0.15	3.8	–	–	B12 0.006; D 6400 IU	–	A	798	239	95	50	0.03	0.7	–	211	–	–
Flounder (*Platichthys flesus, Pleuronectes flesus*)	81.3	16.7	0.8	–	0.06	0	79	30	0.22	0.21	0.25	3.8	–	–	–	86	A	68	332	12	31	0.02	0.8	0.18	195	–	151
Frog legs (*Rana spp.*)	81.9	16.4	0.3	–	0.04	0	73	–	0.14	0.25	–	1.2	–	5	–	–	A	55	308	18	–	–	1.5	–	147	163	40
Haddock (*Melanogrammus aeglefinus*)	80.5	18.3	0.1	–	0.06	0	79	60	0.06	0.17	0.2	3.0	0.14	0	E 0.6; biotin 0.0003; B12 0.001; FA 0.001	67	A	99	301	18	24	0.02	0.7	0.23	197	238	241
smoked	72.6	23.2	0.4	–	–	0	103	–	0.06	0.05	–	2.1	–	–	–	–	–	–	–	–	–	–	–	–	–	–	–
Halibut (*Hippoglossus hippoglossus*)	75.2	18.6	5.2	–	0.06	0	126	440	0.09	0.18	0.42	6	0.30	0	biotin 0.002; B12 0.001; FA 0.002	68	A	56	340	13	–	0.01	0.7	0.23	211	–	–

* To convert to kJ (kilojoule) multiply the values given by 4.1855.

*** FA = folic acid; E = α-tocopherol unless otherwise stated.

f Lean meat 6.6.

† Lean meat.

Content per 100 grammes edible portion (unless otherwise stated)	Water g	Proteins g	Fats: Total g	Fats: Poly-unsaturated g	Fats: Cholesterol g	Carbohydrates g	Calories* kcal	A** IU	B₁ mg	B₂ mg	B₆ mg	Nicotinic acid mg	Pantothenic acid mg	C mg	Other vitamins*** mg	Purine nitrogen mg	Excess acid A / Excess base B	Sodium mg	Potassium K mg	Calcium Ca mg	Magnesium Mg mg	Manganese Mn mg	Iron Fe mg	Copper Cu mg	Phosphorus P mg	Sulphur S mg	Chlorine Cl mg
Herring (*Clupea harengus*)	62.8	17.3	18.8	–	–	0	243	130	0.06	0.24	0.45	4.3	1.0	0.5	E 2¹·²; B₁₂ 0.01; D 900 IU⁷	119	A	118	317	57	26	0.02	1.1	0.3	240	202	122
pickled	60.2	18.3	14	–	–	0	204	150	–	0.08	0.15	3.3	–	–	–	–	A	1000	–	30	9	–	–	–	150	–	1600
smoked	61.0	22.2	12.9	–	–	1	211	40	0.04	0.28	0.35	3.3	–	–	B₁₂ 0.01	73	A	720	285	66	–	0.04	1.4	–	254	–	230
Lobster (*Homarus vulgaris*)	78.5	16.9	1.9	–	0.20	0.5	91	–	0.15	0.13	–	1.5	1.3	5	biotin 0.005; FA 0.0005	–	A	300	260	29	22	–	0.6	2.2	200	170	500
canned	77.2	18.4	1.3	–	–	0.4	92	–	0.16	0.14	–	2.2	–	4	–	–	A	–	–	65	–	–	0.8	–	192	–	–
Mackerel (*Scomber scombrus*)	67.2	19.0	12.2	–	0.08	0	191	450	0.15	0.35	0.70	7.7	0.46	0	E 1.6²; B₁₂ 0.01; biotin 0.002; FA 0.001; D 50 IU	–	A	144	358	5	33	0.02	1.0	0.16	239	197	170
smoked	59.4	23.8	13.0	–	–	0	219	–	–	–	–	–	–	–	–	–	A	–	–	–	–	–	–	–	–	–	–
Mussels (*Mytilus edulis*)	84.1	11.7	1.9	–	0.15	2.2	76	180	0.16	0.22	–	1.6	–	–	–	199	A	290	315	88	23	0.25	5.8	3.2	250	367	460
Ocean perch, Atlantic (*Sebastes marinus*)	77.9	18.9	3.0	–	–	0	108	30	0.09	0.08	–	2.5	–	–	–	–	A	94	345	46	–	–	1.0	–	212	–	–
Octopus (*Octopus bimaculatus*)	82.2	15.3	0.8	–	0.17	0	73	–	0.02	0.06	–	1.8	–	3	–	–	A	–	–	29	–	–	0.19	0.44	173	–	–
Oysters (*Ostrea spp.*)	83.0	9.0	1.2	–	0.11–0.33	4.8	68	310	0.18	0.23	0.11	2.5	0.5	trace	biotin 0.001; FA 0.004; B₁₂ 0.015; D 5 IU	29	A	73	110	94	42	0.2	5.5	1.2–3.7	143	–	–
Perch (*Perca fluviatilis*)	79.5	18.4	0.8	–	0.07	0	86	30	0.075	0.12	–	1.7	–	–	–	–	A	67	238	20	–	0.02	1	–	198	–	–
Pike (*Esox lucius*)	80.2	18.2	1.2	–	–	0	89	–	0.15	0.07	–	1.7	–	–	E 0.2	45	A	70	300	20	30	0.02	0.7	0.25	210	–	100
Salmon Atlantic (*Salmo salar*)	65.5	19.9	13.6	5.3	0.06	0	208	220	0.17	0.17	0.98	7.5	0.8	1	biotin 0.001; D 650 IU; FA 0.002; B₁₂ 0.003	47	A	48	391	29	29	–	0.8	0.2	266	190	64
canned, solids and liquid	64.2	21.7	12.2	–	–	0	203	60	0.03	0.18	0.45	6.5	0.5	trace	biotin 0.015; FA 0.0005; D 500 IU	101	A	540	330	67	30	–	1.3	0.05	285	–	865
chinook	64.2	19.1	15.6	–	–	0	222	310	0.10	0.23	0.13	–	–	–	–	–	A	45	399	–	–	–	–	–	301	–	–
canned, solids and liquid	64.4	19.6	14.0	–	–	0	210	230	0.03	0.14	0.11	7.3	–	–	–	–	A	–	366	154	27	–	0.9	–	289	–	–
sockeye canned, solids and liquid	67.2	20.3	9.3	–	–	0	171	–	0.04	0.16	–	7.3	–	–	–	–	A	522	344	259	29	–	1.2	–	344	–	–
smoked	58.9	21.6	9.3	–	–	0	176	–	–	–	–	–	–	–	–	–	A	–	–	14	–	–	–	–	245	–	–
Sardines, canned in oil solids and liquid	50.6	20.6	24.4	–	–	0.6	311	180	0.02	0.16	0.16	4.4	0.5	0	biotin 0.005; D 300 IU; B₁₂ 0.01; FA 0.002	234	A	510	560	354	29	–	3.5	0.04	434	–	–
drained solids	61.8	24.0	11.1	3	0.07	1.2	214	290	0.03	0.20	0.28	5.4	0.6	0	biotin 0.02; D 250 IU	–	A	823	590	437	–	–	2.9	0.04	499	–	–
Scallops (*Pecten spp.*)	79.8	15.3	0.2	–	–	3.3	79	0	0.04	0.06	–	1.3	0.14	–	biotin 0.0003; FA 0.0005	117	A	150	420	26	–	–	1.8	–	208	342	–
Shrimps (*Crangon spp.*)	78.2	18.7	2.2	–	0.14	–	97	10	0.07	0.05	0.13	1.25	0.21	2	B₁₂ 0.001	–	A	140	258	63	42	–	2.0	0.43	300	–	–
canned, drained solids	70.4	24.2	1.1	–	0.15	0.7	116	60	0.01	0.03	0.11	1.5	–	0	FA 0.002; D 105 IU	–	A	–	122	115	–	–	3.1	–	263	–	–
Snails (*Helix*)	82	15	0.8	–	–	2	75	–	–	–	–	–	–	0	–	–	A	–	–	170	250	1.6	3.5	0.4	–	140	–
Trout (*Salmo trutta*)	77.6	19.2	2.1	–	–	0	101	150	0.09	0.25	–	3.5	–	–	–	92	A	39	470	19	–	0.03	1.0	0.33	220	–	–
Tunny (*Thunnus thynnus*) canned, solids and liquid	52.5	23.8	20.9	–	–	0	290	90	0.05	0.06	0.25	10.8	0.2	0	biotin 0.0005; B₁₂ 0.001; FA 0.003	–	A	361	343	7	–	–	1.2	–	294	–	–

* To convert to kJ (kilojoule) multiply the values given by 4.1855.
** Vitamin A activity due to vitamin A + carotenes; 1 IU vitamin A = 0.0006 mg β-carotene.
*** FA = folic acid; E = α-tocopherol unless otherwise stated.
⁷ Without gonads.
² Total tocopherol.

INDEX